CHILDREN'S LITERATURE AND
IMAGINATIVE GEOGRAPHY

CHILDREN'S LITERATURE AND IMAGINATIVE GEOGRAPHY

Aïda Hudson, *editor*

Wilfrid Laurier University Press acknowledges the support of the Canada Council for the Arts for our publishing program. We acknowledge the financial support of the Government of Canada through the Canada Book Fund for our publishing activities. This work was supported by the Research Support Fund.

  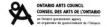

Library and Archives Canada Cataloguing in Publication

    Children's literature and imaginative geography / Aïda Hudson, editor.

Includes bibliographical references and index.
Issued in print and electronic formats.
ISBN 978-1-77112-325-9 (hardcover).—ISBN 978-1-77112-326-6 (EPUB).—
ISBN 978-1-77112-327-3 (PDF)

    1. Children's literature—History and criticism. 2. Geography in literature.
3. Imaginary places in literature. I. Hudson, Aïda, [date], editor

PN1009.5.G48C55 2018   809'.89282   C2018-902148-9
                                                C2018-902149-7

Front-cover photo by Kim LaFave. Cover and text design by Lime Design Inc.

This book is printed on FSC® certified paper and is certified Ecologo. It contains post-consumer fibre, is processed chlorine free, and is manufactured using biogas energy.

Printed in Canada

Every reasonable effort has been made to acquire permission for copyright material used in this text, and to acknowledge all such indebtedness accurately. Any errors and omissions called to the publisher's attention will be corrected in future printings.

Acknowledgements ix
Introduction ❧ AÏDA HUDSON *1*

═══════════

One / GEOGRAPHICAL IMAGINARIES
The Old World and the New

1) Pullman and Imperialism: Navigating the Geographic
Imagination in *The Golden Compass* ❧ CORY SAMPSON *25*

2) Nineteenth-Century British Children's Literature and the North
❧ COLLEEN M. FRANKLIN *45*

3) Envisioning Ireland: Landscape and Longing in Children's
Literature ❧ MARGOT HILLEL *67*

4) From Vanity to World's Fair: The Landscape of John Bunyan's
Allegory in Frances Hodgson Burnett's *Two Little Pilgrims' Progress*
❧ SHANNON MURRAY *85*

5) Old World, New World, Other World: Overcoming Prosaic
Landscape with *The Golden Pine Cone* ❧ LINDA KNOWLES *101*

6) Healing Relationships with the Natural Environment
by Reclaiming Indigenous Space in Aaron Paquette's
Lightfinder ❧ PETRA FACHINGER *121*

Interlude

7) History, Hills, and Lowlands: In Conversation
with Janet Lunn ❧ AÏDA HUDSON *139*

Two / GARDENS AND GREEN PLACES

8) How Does Your Garden Grow? The Eco-Imaginative Space
of the Garden in Contemporary Children's Picture Books
❧ MELISSA LI SHEUNG YING *153*

9) Into the (Not So) Wild: Nature Without and Within in
Kenneth Grahame's *The Wind in the Willows* ❧ ALAN WEST *169*

Interlude

10) Earth, Sea, and Sky Writing in *Becca at Sea* ❧ DEIRDRE F. BAKER *185*

Three / FANTASY WORLDS AND RE-ENCHANTMENT

11) The Imaginary North in Eileen Kernaghan's *The Snow Queen*
❧ JOANNE FINDON *197*

12) Camping Out on the Quest: The Landscape of Boredom
in *Harry Potter and the Deathly Hallows*
❧ SARAH FIONA WINTERS *215*

13) Sky Sailing: Steampunk's Re-enchantment of Flight
❧ CHRISTINE BOLUS-REICHERT *229*

14) Mythic Re-enchantment: The Imaginative Geography of
Madeleine L'Engle's Time Quintet ❧ MONIKA HILDER *243*

Four / SPACE AND GENDER

15) Female Places in *Earthsea* ❧ PETER HYNES 263

16) Dancing and Hinting at Worlds in Theatre for Young Audiences
❧ HEATHER FITZSIMMONS FREY 281

Postlude

17) Following the Path of the Unconscious in the Owen Skye Books,
and Others ❧ ALAN CUMYN 301

———

Works Cited 317
About the Contributors 335
Index 339

Acknowledgements

I wish to thank the University of Ottawa and the APTPUO (Association of Part-time Professors of the University of Ottawa) for awarding an Academic Development Fund Publication Grant for this volume. My very able research assistant Amy Einarsson, who completed an MA in English from the University of Ottawa, helped with the manuscript from the very beginning. Special thanks to her for the list of contributors. I also want to thank Michael Hudson for his fine technical assistance. Many thanks to the Department of English. Tom Allen advised me during the various stages of this book and, previously as chair, gave support and financial aid for the symposium from which many of these chapters grew.

Together with all the scholars and writers who contributed to this volume, I would like to thank Wilfrid Laurier University Press. Thanks to Lisa Quinn, now director, who accepted the manuscript; to Siobhan McMenemy, who, though new to the press, hit the ground running and guided me with such conscientiousness through the editing and review process; to Clare Hitchens for her Web and marketing work, to Mike Bechthold, who oversaw the fine cover design for the illustration by the celebrated Kim LaFave, and to Robert Kohlmeier and his team who transformed the manuscript into a book. We are grateful to the reviewers who gave all of us in this volume such excellent advice. Finally, we wish to thank our families and friends for their support.

Introduction ❧ AÏDA HUDSON

T he roots of the words "imaginative geography" together are refreshingly graphic. "Imaginative" is derived by way of Old French from
the Latin *imaginatio(n-)*; from the verb *imaginary*, "picture to oneself"; and
from *imago, imagin-*, "image." "Geography" comes from the Greek *geōgraphia*,
gē for "earth" and -*graphia* for "writing." "Imaginative geography" means
"image or picture to oneself of earth writing" or "imaged earth writing."
On a literal level, "imaginative geography" can bring Earth to mind as a
whirling globe of continents and oceans, fiery at its core, enveloped in air,
cloud, and wind, and circling our galaxy's sun in a universe of galaxies.

"Image or picture to oneself" means that imaginative geography relies
on one's perspective, that is, on what is seen in the mind's eye. As this passage from John Gardner's *Art of Fiction* makes clear, this is also true about
the reader of fiction. "If we carefully inspect our experience as we read, we
discover that the importance of physical detail is that it creates for us a

kind of dream, a rich and vivid play in the mind. We read a few words at the beginning of the book or the particular story, and suddenly we find ourselves seeing not words on a page but a train moving through Russia, an old Italian crying, or a farm-house battered by rain" (30-31). A reader "sees," whether it is a moving place like a train, a character like the old Italian, or a stationary place like the farmhouse created within the imaginative geography of the author.

Place is essential to fiction, for every story begins with a "where." "The truth is," wrote Eudora Welty, "fiction depends for its life on place. Location is the crossroads of circumstance, the proving ground of 'What happened? Who's here? Who's coming?'—and that is the heart's field" ("Place in Fiction"). This is true whether authors write for adults, teenagers, or children.

The geography of an imagined world may be realistically representative of our own cosmos or not. The imaginary geographies in children's literature can be fantastic, even mythic, like Middle-earth or Earthsea, or realistic to varying degrees like Green Gables or the Ontario North or the "bush" in Tim Wynne-Jones's *The Maestro*. In her chapter in this present collection, Monika Hilder quotes from Stephen Prickett's "Fictions and Metafictions": "realism and fantasy are two sides of the same coin: ... realism is as much an arbitrary and literary convention as fantasy, and ... fantasy is as dependent on mundane experience as realism" (225). Works of the imagination, in whatever genre, begin with what happens in our world; they rely on the temporal to be believable. John Gardner insists on the similarities between a "fable" and a realistic story:

> The fabulist—the writer of non-realistic yarns and tales, or fables—may seem at first glance to be doing something quite different; but he is not. Dragons, like bankers and candy-store owners, must have firm and predictable characters. A talking tree, a talking refrigerator, a talking clock must speak in a way we learn to recognize, must influence events in ways we can identify as flowing from some definitive motivation: and since characters can come only from one of two places, books or life, the writer's aunt is as likely to show up in a fable as in a realistic story. The process by which once writes a fable, on one hand, or realistic story, on the other, is not much different." (21-22)

Fantasy is rooted in reality in order to be believable. Alice's fall down the rabbit hole into Lewis Carroll's Wonderland is made real because she situates herself within Earth. "I must be getting somewhere near the centre of the earth ... I wonder what Longitude and Latitude I've got to? ... I wonder if I shall fall right through the earth! How funny it will seem to come out among the people that walk with their heads downward! The antipathies I think" (Haughton, ed., 10–11). In an aside, the author makes clear that Alice does not know what Longitude or Latitude are, but she knows enough to use them when "placing" herself as she falls, and her mispronunciation of Antipodes completes her imaging of the globe. She lands unhurt on a "heap of sticks and dry leaves" and finds herself in "a long low hall, which was lit up by a row of lamps hanging from the roof" (12). This hall in her dreamscape is made "real" because of her previous temporal placing. Carroll's heroine has imaged a geography for the reader based on reality and consequently makes believable a fantastic place, that long low hall.

∾ Edward Said and Imaginative Geography

JUST AS IN OTHER LITERATURES, in children's and young adult literature, whether realistic or fantastic or a blending of the two, place varies in scale from geographies as large or larger than our Earth, like the parallel worlds of *His Dark Materials*, or the universe of Madeleine L'Engle's Time Quintet, or epic geographies like Narnia, or the magically overlapping worlds of the Muggles and Wizards in the Harry Potter books, to the Toronto-scapes to Canada-scapes of Dennis Lee's rhymes in *Alligator Pie*, or the Ojibway canoe route in *Last Leaf First Snowflake to Fall*, or Tom Sawyer's town St. Petersburg, or even smaller places like Mr. McGregor's vegetable garden in *The Tale of Peter Rabbit*. However, whatever the scale, "imaged earth writing" can also be dualistic; it can involve the perception of both familiar and unfamiliar geography, which is the basis of Edward Said's seminal book, *Orientalism*. In order to examine imaginative geography in children's literature from the early nineteenth century to the present, we will first look at how the term was introduced by Said in postcolonial discourse, how it influenced and was also reshaped in other geographical approaches to literature, such as environmentalism, and how it was preceded by Tolkien's approach to place and myth.

Edward Said defines imaginative geography in the following passage:

> It is perfectly possible to argue that some distinctive objects are made by the mind, and that these objects, while appearing to exist objectively, have only a fictional reality. A group of people living on a few acres of land will set up boundaries between their land and its immediate surroundings and the territory beyond, which they call "the land of the barbarians." In other words, this universal practice of designating in one's mind a familiar space which is "ours" and an unfamiliar space beyond ours which is "theirs" is a way of making geographical distinctions that *can be* entirely arbitrary. I use the word "arbitrary" here because imaginative geography of the "our land–barbarian land" variety does not require that the barbarians acknowledge the distinction. It is enough for "us" to set up these boundaries in our own minds; "they" become "they" accordingly, and both their territory and their mentality are designated as different from "ours." (54)

What is "made by the mind" can define an "object," yet be a "fictional reality." Using the image of a domicile, "a group of people living on a few acres of land," Said presents an "arbitrary" boundary created between "a familiar space" and "an unfamiliar space" to mark what is "ours" and what is "theirs." "Imaginative geography" here is a dichotomy—"of the 'our land–barbarian land' variety." For Said, his dualistic vision of imaginative geography helped explain Western views of the "Orient" and the establishment of Western imperial dominance. He makes it clear that the barbarians' understanding of themselves has no part in this "fictional reality." His full sentence reads, "I use the word 'arbitrary' here because imaginative geography of the 'our land–barbarian land' variety does not require that the barbarians acknowledge the distinction." We have here what makes "our" view of "them," and what makes "them," people in the Orient, "Other." "Our" perception is key in the relationship between familiar and unfamiliar people and their cultures. Imaginative geography of this kind also speaks to the human condition; it is "a universal practice." Every individual whether young or old has been exposed to familiar places and unfamiliar places, the *topoi* or physical places themselves (*topos* comes from the Greek word meaning "place") or *topoi* associated with other individuals and cultures, or both, whether in life or in literature.

The prominent environmentalist Lawrence Buell, who has recently turned his attention to children's literature, acknowledged the profound impact of *Orientalism* in his study published in 2005, *The Future of Environmentalism*, but with an important caveat. "Said's *Orientalism* would never have had the impact it did without being a first-rate work of scholarship and intellection, despite its allegoristic limitations" (133). On the contrary, it is Said's "allegoristic" expression of imaginative geography, his imaging "a group of people living on a few acres of land" setting up boundaries between their land and the "'the land of the barbarians'" that brings home the full meaning of West/non-West dualism, of "ours" and "theirs." He pictures this dualism in a way that everyone can relate to whether in real life or while reading. Even when a place is far, strange, or unknown in our minds, it can have a preconceived geography that can stress difference from "our" geography whether "true" or not. Said insists "that imaginative geography and history help the mind to intensify its own sense of itself by dramatizing the distance and difference between what is close to it and what is far away" (55).

In *Orientalism* (published in 1978) and his later work *Culture and Imperialism* (published in 1993), which he considered a sequel, Said argued that the Western canon cannot be fully understood without taking into account the imperialistic dominating perception that can grow out of the "our land–barbarian land" duality. Even Jane Austen's *Mansfield Park*, which explores the family life of English country gentry in the "pre-imperialist age" cannot be fully appreciated, according to Said, without the geographical context of the leisured class she depicts relying on the riches that came from a plantation in the colony of Antigua worked by slaves (*Culture and Imperialism* 59). Yet Austen herself protested that the focus of her work was narrow—indeed, so limited that she compared it to "the little bit of ivory (two inches wide)," one on which she worked on "with so fine a brush as produces little effect after much labour" (*Jane Austen's Letters*, 337). In each of her novels, she writes about a small circle of families, landed British gentry, whose wealth and power are based on their nation's colonial domination. Nevertheless, the geographical map of her work does indeed include life abroad, plantations, trade, travel, and war, albeit offstage.

Said's study of *Mansfield Park* in *Culture and Imperialism* "worlds" this novel. The geographer Karen Morin notes that for Said, "all texts, literary or otherwise, are political, and must be 'worlded'—located in the world and exposed for the geographical imaginations from which they arise" (238). In his article "Imaginative Geographies"—a title he borrows from Said—the geographer Derek Gregory writes that Said was "one of those rare critics for whom a geographical imagination is indispensible" (447). According to Gregory, his geography is "also derived from the spatialization of cultural and social theory" (453). Morin observes how difficult it is to categorize Said's work "in disciplinary terms" (238) and that his work "initiated debates across the humanities and social sciences, from literature to music to anthropology to political science to geography" (238). Recent articles inspired by Said's imaginative geography, from studies of Irish and South American literature to geopolitical views of the Balkans after the First World War to climate justice between the Northern and Southern Hemispheres, show the continuing global reach of his paradigm.[1]

ᔇ Imaginative Geography Travels

SAID'S CONCEPT OF IMAGINATIVE GEOGRAPHY has crossed into the study of children's literature in Margot Stafford's chapter "*Journeys Through Bookland's* Imaginative Geography: Pleasure, Pedagogy, and the Child Reader," in which she explores that early-twentieth-century home library series, which included fiction, history, mythology, poetry, travelogues, nature writing, and biography for young readers from preschool age to adulthood. According to Stafford, *Journeys Through Bookland* came into being "when geography and cartography were being used as part of the imperialist project in the belief that to map and represent the globe offered the ability to know and master it" (149). She compares books to the continents of Africa and Antarctica; for, though they are an imaginary territory, they can be "measured, charted, and represented as a knowable, quantifiable, conquerable landscape" (149). This series of books was intended to encourage and to steer readers through a course of reading throughout their youth. Stafford also states that "Bookland must also be understood as a geography of the imagination, an attempt to understand and depict interiority through spatial representations" (149). The world of books is "imaged earth-writing" and as a concept relies as much on the root meaning of imaginative geography as it does on Said's dualistic use of it. Although Stafford does not explore Said's "Other" in relationship to *Journeys Through Bookland,* she does compare children's reading to an imperialist mastery that captures the postcolonial spirit of Said's paradigm.

ENCOURAGED BY STAFFORD'S STUDY and keeping in mind what Said has done in *Culture and Imperialism to* "world" *Mansfield Park*, I will attempt to "world" a children's classic written in 1911 when Britain still ruled its colony India.

To "world" *The Secret Garden* by Frances Hodgson Burnett is to unroll its geographical map, one that includes colonial India of the early twentieth century where the "Contrary" Mary Lennox first lived and where her English parents and so much of her household die of cholera, as well as her final destination in Yorkshire, Misselthwaite Manor, and its secret garden, remote and neglected by its absentee master, her uncle Archibald Craven. Antigua may be offstage in *Mansfield Park*, but India in *The Secret Garden* is not. The novel opens with Mary's lonely life there, shut up in a big bungalow and cared for by a parade of *ayahs* who allow her, the plain daughter of the pretty Mem Sahib mother who neglects her, to dominate them and even abuse them, conditions that are the basis of her initial imperious behaviour at Misselthwaite towards the household. There the Pan-like Dickon, though from a simple rustic Yorkshire family, helps her revitalize the walled garden where Mr. Craven's wife had her fatal fall, the reason why the garden is shut up and why Mr. Craven shuns Misselthwaite. As Mary improves in health and spirits while tending the garden, she helps her invalid cousin Colin back to health. "Mary Quite Contrary" grows into a likeable happy child. However, physical dichotomies are also imaged for the reader. Back in India, pale sour Mary spent most of her time hidden in the house, only to pretend garden with blossoms in the dirt now and again. The heat, the strange culture of the servants, and Mary's sallow face and thin, sickly body back in India are recurring images later as she explores the other gardens of Misselthwaite and runs and skips in the wind from the moors, moors that fill so many of the conversations in the book. They also geographically frame the secret neglected garden that Mary, Dickon, and Colin grow to love.

The imaged earth-writing of the secret garden is some of Burnett's best. The change from the tangled unkempt climbing roses and the choked green shoots of crocuses and daffodils in the early spring to the flowering splendour of summer and the golden glory of autumn, the robin's visits, and Dickon's animal menagerie are inextricable from Mary's growth and later Colin's growth into good health and belongingness. The sour sickly

Mary of India is little by little transformed into the more genial, healthy, romping Yorkshire Mary. Her playfulness with the Yorkshire dialect is a measure of this metamorphosis. She exults about the happy afternoon when she first brings Colin, pushed in a wheelchair by Dickon, into the garden: "'Aye, it s a graidley [grand] one,' said Mary, and she sighed for sheer joy. 'I'll warrant it's th' graidliest one as ever was in this world'" (93). The secret garden becomes famil(y)-iar; the garden at Misselthwaite is the natural catalyst for the bonding between cousins Mary and Colin and their bonding later with their uncle/father Archibald Craven.

The imaginative geography of the *The Secret Garden*, which relies on the double-worlding of India and Yorkshire, highlights English physical beauty and vigour and the "natural" superiority of its class system, because it is more equitable and genial. There is a real kinship between Mary and Dickon and between Mary and, to a lesser degree, the "Rajah" Colin, heir of Misselthwaite, and the old gardener, Ben Weatherstaff. Yorkshire becomes Mary's "home land," one that she can claim together with her uncle and cousins as a Saidian "our land." Paradoxically, there is also a twist; it is in Yorkshire that Mary loses her imperious ways.

ᔓ *Place-Attachment and Environmentalism*

WE HAVE SEEN HOW IMAGINATIVE GEOGRAPHY has been used to explore a children's home library and is reflected in a novel set in colonial India and in imperial Britain. However, there are tracings of Said's use of imaginative geography in Buell's concept of "place-attachment" in *The Future of Environmental Criticism*. Said's paradigm is territorial, marking the boundary between our land and that of others, whereas Buell's "place-attachment" concentrates on places that one cares for. Buell considers how "one also becomes attached to place by the power of imaging alone—a power as ancient as folk stories told at bedtime and the bardic performances that produced the tales synthesized by Homeric epic and *Beowulf*" (72). He continues that the "places that haunt one's dreams and to some extent define one's character can range from versions of actual places to the utterly fictitious—Alaska's wild north slope, Robin [*sic*] Crusoe's 'desert island,' the 'little house on the prairie' of Laura Ingalls Wilder's children's books, the promised land of the ancient Israelites imagined from Egyptian or Babylonian captivity, the Hopi *Túwanasavi* or origin-place to which the people are called to return" (73). Discussing "place-attachment" led

Buell to look back at ancient epics and then to a variety of places, both real and fictitious, including ones from books commonly read by children. It is a bookland of his own, made up of imaginary places and real places that collectively make an imaginative geography because of "imaging." This is imaged earth-writing. As readers we can "see" these places in Buell's text. His visual sketches range from Nature in the raw, "Alaska's wild north slope," and Crusoe's "'desert island,'" to "'the little house on the prairie,'" to those well-known American settler narratives about the trials and tribulations of conquering the land, to the "promised land" of the enslaved Israelites in the Old Testament, to the origin-place of a First Nation tribe, the Hopi. The very selection of these images has a postcolonial dimension to them, tracings of Said's land of the "Other" as well as "our land." Buell also includes the importance of place in his childhood: "My memory of the place where I grew up has affected my response to all the places where I lived since, and so too I find for those who led a more wandering existence when young" (73). This too is reminiscent of Said's "familiar space."

Cheryll Glotfelty's often quoted definition of environmentalism appears both in Buell's book and in Melissa Li Sheung Ying's account of modern environmentalism in her chapter in this present volume; it is "'the study of the relationship between literature and the physical environment.'" Environmentalism has led to activism. Man-made disasters like the indiscriminate spraying of DDT in the United Sates became the subject of Rachel's Carson's 1962 classic environmental study *Silent Spring,* which in turn led to the founding of Greenpeace and Friends of the Earth. The nuclear power accidents at Three Mile Island in 1978 and at Chernobyl in 1986 and the *Exxon Valdez* oil spill in 1989 have led to protests over nuclear energy and over pollution more generally. In the present day, there are concerns about oil pipelines, deep-sea oil rigs, and the ocean transport of oil and gas along coastlines. Natural disasters like the 2004 Indian Ocean earthquake and tsunami and the alarming evidence of climate change such as the increasingly rapid melting of Arctic sea ice in recent decades provide a backdrop to a proliferation of environmental literature for the young, as well as book awards like the Green Book Award for Children's Literature and the Environment Award of Children's Literature. If imaginative geography can be thought of as a way of seeing the world, then environmentalism can be perceived as a way of doing something to save it.

Environmental saving is often based on a deep sense of catastrophe. In his recent article "Environmental Writing for Children: A Selected

Reconnaissance of Heritages, Emphases, Horizons," Buell observes that Dr. Seuss's *The Lorax* has been criticized for "gratuitous doom-crying" (9). In a dark but outrageously whimsical fashion, Dr. Seuss pictures Once-ler's relentless deforestation of Truffala Trees to make Thneeds in biggering factories, which leads to the exodus of Brown Bar-ba-loots, Swomee-Swans, and Humming-Fish. The one last hope is Once-ler's urging the child listener to act by planting the very last Truffala seed. The message is made clear for the reader: "UNLESS someone like you / cares a whole awful lot, / Nothing is going to get better, / it's not" (*The Lorax*). The imaginative geography in this groundbreaking picture book published in 1971 encapsulates the need for environmental activism that resonates to this day. A fantastic but allegoristically familiar green landscape is defamiliarized by the dominating Once-ler, and destroyed through deforestation and the consequent displacing of the creatures that once lived there.

Yet the emphases Buell makes in his article are twofold—the alarmist aspect of environmentalism, but also its greening hopefulness. He examines two "constellations": books about the interaction between animals and humans, both "real" animals and anthropomorphized creatures, in selected texts from the Golden Age of children's literature, including *The Wind and the Willows* and *Black Beauty,* and more recent works such as *Where the Wild Things Are,* and *The Lorax* discussed above; and books about the "catalytic" effect of often hidden outdoor places on children that bond them "to the natural environment and beyond that, by implication at least, in identity-formation over the long run" (2), such as *The Secret Garden* and *The Earth Is My Mother.* In his discussion of Burnett's novel, place-attachment grows to be a kind of "biophilia" shaped by "the power of active interaction with the living earth (birds, flowers, trees and animals) to reshape human being" (7).

Buell refers to an earlier study, Sidney I. Dobrin and Kenneth B. Kidd's *Wild Things: Children's Culture and Ecocriticism,* which also investigates the changing relationship between children and nature in children's literature and popular culture. This book does a remarkable job of raising awareness of environmental concerns and of the need for environmental activism: "we believe both that children are naturally close to nature and that nature education, even intervention, is in order" (7). Buell goes one step further by emphasizing two aspects of children's relationship to Nature—"biophilia" or "human responsiveness to non-human beings" and, secondly, "place imprints generally," reminiscent of his "place-attachment" in *The Future of Environmentalism*, a tracing of Said's imaginative geography.

∽ Imaginative Geography and the Mythical

THERE IS ONE MORE ASPECT OF IMAGINATIVE GEOGRAPHY that needs to be dealt with, the mythical. In fantasy and folklore, place is the locus of myth. Zeus resides on Mount Olympus, leprechauns are found at the end of rainbows, and Sauron rules in Mordor and reaches out to the rest of Middle-earth. Tolkien's canonical study of fantasy, "On Fairy-Stories," explores the concept of a Secondary World that is a fantastic imaged earth-writing. First presented as a lecture in 1939, decades before the publication of *Orientalism*, his essay redefined the importance of place in modern fantasy and fairy tales. "The definition of a fairy-story—what it is, or what is should be—does not, then, depend on any definition or historical account of elf or fairy, but upon the nature of *Faërie*: the Perilous Realm itself, and the air that blows in that country" (10). For Tolkien, it "contains many things besides elves and fays and besides dwarfs, witches, trolls, giants or dragons: it holds the seas, the sun, the moon, the sky; and the earth, and all things that are in it: tree and bird, water and stone, wine and bread, and ourselves, mortal men, when we are enchanted" (9). Tolkien also refers to *Faërie*, as a Secondary World, a Realm that relies on the geographical reality of the Primary World, our Earth. The Secondary World of Middle-earth in *The Hobbit* and *The Lord of the Rings* is a testament to this inspired imaged earth-writing. His Aragorn, Gandalf, and Galadriel, his hobbits, orcs, and Ents—all are alive in an epic geography that includes the gently rolling countryside of the Shire, the wonders of Rivendell, the plains of Gondor, the depths of Moria, and the desolation of Mordor; all possess "reality" because they are based on our world. Nevertheless, there is a pre-Saidian dualism in the sweeping imaginative geography of Tolkien's Middle-earth, where peoples and beings fight for domination.

In *Tolkien: A Critical Assessment*, Rosebury insists that a "further impetus to the creation of Middle-earth and its myths was given by the experience of war" (125). Tolkien had fought in the First World War at the Somme:

> The ineradicable memory of a land pulverized by "total war" is evident (though combined with images suggestive of industrial pollution, and fully absorbed into the wider imaginative geography) in the most

hauntingly repellent landscapes of his work—less, perhaps, in the desert plains of Mordor than in the Dead Marshes, with their preserved corpses, half-real, half-hallucinatory, staring out of deep water, and the "obscene graveyard" before the gates of Mordor, with its "gasping pits and poisonous mounds," and "pools … choked with ash and crawling muds, sickly white and grey." (126)

This passage encapsulates the influence of Tolkien's experiences in the First World War on his Secondary World of Middle-earth and refers to it as "imaginative geography." The horror and pollution of war in Tolkien's *Two Towers* is presented here. A dark Saidian imperialism comes alive in *Lord of the Rings* as Sauron's evil empire fails to defeat a confederacy of good men, elves, dwarves, ents, and hobbits inspired by the good wizard Gandalf as they deflect his attention from Mount Doom, where Frodo, Sam, and Gollum carry the ring and destroy it.

Imaginative geography has been explored both in its root meaning and in the postcolonial writings of Said. Also explored have been its effects on place-attachment and how it relates to Tolkien's thoughts on Fantasy as a Secondary World. Imaginative geography also plays a role in the works of Indigenous writers. In "Environmental Writing," Buell recognizes the emerging force of "cultural survival literature for children around the theme of intimacy between humans and the natural world by first peoples and other postcolonial authors around the world" (10). He believes that such literature will produce "some tempering of nonnormative eco-dissidence in children's literature" (10).

Indigenous literature for the young does at times image earth-writing in a holistic way, reconnecting mankind with the Earth. However, according to Doris Wolff and Paul De Pasquale (91–92), the children's novel *Little Voice* by the Anishinaabe writer Ruby Slipperjack, worlded in the traditional Anishinaabe territory of northwestern Ontario, and the young adult novel *Will's Garden* by Lee Maracle, worlded in Sto:lo (also known as Salish) territory on British Columbia's Fraser River, have some vestiges of "protest literature" that characterize Indigenous adult literature. They write that these two novels "do engage with political issues, past and present, and sometimes even with anger" (91). These novels may reflect the anger of peoples who have been dispossessed and displaced, but they also evoke rootedness to the land and an affirmation of Indigenous culture. In her essay "The Other Side of Me," Maracle writes that "no people ever totally deserts its ancestry" (385). Wolff and De Pasquale also observe that Indigenous picture books are more "idyllic"; they leave behind "the anger and siege mentality" that characterize Native adult literature (91), a result that can be perceived as a Saidian othering.

Paradoxically, the idyllic nature of many Indigenous picture books can be viewed as the best protest of all. As an example, in *Last Leaf First Snowflakes to Fall* the late Ojibway author/illustrator Leo Yerxa images both in word and illustration Ojibway traditions, ways of life, myths, and love of the land. "We don't have to 'go back to the land.' We never left it," writes Maracle of First Nations peoples (384). In his text, Yerxa opens with an impressionistic tribute to the Anishinaabe Creation story of Turtle Island (North America), that heralds the journey of a nishnawbe father and son at the end of autumn and the beginning of winter. Like the blanket that later wraps the young nishnawbe when night comes, Yerxa's story encloses the beingness of the North—trees, leaves, water, an island, a beaver pond, an overnight camp, a waking to see a snow-covered world. We travel through early morning to night and then to day again. The imaginative geography of this iconic picture book includes myth, place-attachment, and a total freedom from othering.

ᔥ Mapping the Chapters

IF WE WERE TO MAP the places examined in *Children's Literature and Imaginative Geography*, we would find them in the Northern Hemisphere. Of the seventeen chapters, fourteen are geographically based critical considerations of literature written in English from the early nineteenth century to the present, for the young in Canada, Britain, the United States, and Ireland. They explore realistic places like Owen's ramshackle Canadian farmhouse in Alan Cumyn's *The Secret Life of Owen Skye,* semi-fantastic but decidedly English places like the riverbank and the Wild Wood of Kenneth Grahame's *The Wind in the Willows,* or the fantastic but palpably real tropical island in the Pacificus in Kenneth Oppel's steampunk novel, *Airborn.* Other chapters examine mythical places like Earthsea created by the American fantasy writer Ursula Le Guin in her Earthsea series and the Alberta landscape transformed in a "Dreamtime" inspired by Indigenous myth in *Lightfinder.* There is the fantastic North, based on Scotland and Svalbard, in Pullman's *The Golden Compass*; the Canadian North portrayed in a selection of British nineteenth-century works for children inspired by

British travel writing; and a re-enchanted North in the Canadian Eileen Kernaghan's *Snow Queen* based on the North created by the Danish writer Hans Christian Andersen in his classic fairy tale. Three chapters, two that serve as interludes and one as a postlude, are reflections by authors on their own Canadian imaginative geographies and on how they came to create them, geographies that are inspired by the Maritimes, Prince Edward County in Ontario, and the Salish Sea off the coast of in British Columbia. Some of this volume's chapters consider Said's paradigm, or provide a nuanced approach to it, and some challenge it. Some scholars touch upon place-attachment whether in an environmental context or in a postcolonial one. Myth and its re-enchantment are explored in others. Also considered is imaginative geography in relation to particular *topoi* such as gardens, as well as to gender and to bodily space.

~ *The Chapters*

One / Geographical Imaginaries: The Old World and the New

CHAPTER 1, "Pullman and Imperialism: Navigating the Geographic Imagination in *The Golden Compass*," by Cory Sampson, examines the meeting of imperialism with geographic themes—exploration, adventure, and the culturally and ethnically "Other" peoples and creatures of Pullman's fantastic world that nevertheless "feels" like the British Empire of the late Victorian and early Edwardian period. This chapter underscores the imperialistic fervour found in Lyra Belacqua's push North, as well as the cruel colonization of children by the General Oblation Board, which Sampson compares to the experiences of Canadian First Nations children at Christian residential schools from the turn of the nineteenth century until the end of the twentieth.

In Chapter 2, "Nineteenth-Century British Children's Literature and the North," Colleen M. Franklin investigates the explosion of writing for children about the North after 1818, following the Napoleonic Wars, when the British Admiralty renewed the search for the Northwest Passage and launched a new search for the North Pole. Franklin surveys and discusses a number of nineteenth-century British children's works that embody the "sublime quest" into the Canadian North. Their imaginative geography is shaped by a curious combination of myth, British imperialistic ambition, and a lack of empirical knowledge. In this literature, the sublime quest became a trope for self-conquest and had educative value. She concludes that this fascination with the North has continued in modern British fantasy for children, as in *The Golden Compass* examined in Chapter 1.

In Chapter 3, "Envisioning Ireland: Landscape and Longing in Children's Literature," Margot Hillel investigates the attitudes towards Irish landscape in selected nineteenth-century Irish texts. The imaginative geography of Ireland in these works for the young is shaped by Irish emigrants' longing for their homeland and by their loyalty to their country and those they left behind. Too often Old World peoples are thought to be conquering imperialists (sometimes associated with a Saidian world view) when in reality ethnic groups like the Irish included hapless crofters forced to leave their home country because they were starving. Their yearning can result in a "geographic and sometimes nostalgic patriotism" (68) in their literature. This chapter shows how leaving a familiar place, an "our land"—in this case nineteenth-century Ireland—for an unfamiliar one in the New World creates a romanticized imaginative geography that can be read as characteristic not only of Irish emigrant literature, but also of the emigrant literature of other ethnic groups.

The first three chapters examined children's literature encapsulating viewpoints originating in the Old World, even though Chapter 2 concerned British imperialist perceptions of the Canadian North. Like these first three chapters, the next three critique children's literature from different time periods, but unlike them, they reflect viewpoints that are "landed" in the New World.

In Chapter 4, "From Vanity to World's Fair: The Landscape of John Bunyan's Allegory in Frances Hodgson Burnett's *Two Little Pilgrims' Progress*," Shannon Murray looks at how the imaginative geography of Bunyan's allegory, a nursery mainstay even in America, is superimposed by two young protagonists on the Chicago World's Fair with a surprising imaginative effect, a conflation that leads to a third imaginary landscape.

In Chapter 5, "Old World, New World, Other World: Overcoming Prosaic Landscape with *The Golden Pine Cone*," Linda Knowles considers the dearth of fantasy for children in late-nineteenth- and early-twentieth-century Canada, a consequence, she argues, of the harsh reality of the Canadian wilderness. She looks at how the British-born Canadian writer Catherine Anthony Clark solved this imaginative problem (one associated with a non-Indigenous view of the land in the 1950s) in her fantasy novel *The Golden Pine Cone* and shows that Indigenous myth was the solution.

In Chapter 6, "Healing Relationships with the Natural Environment by Reclaiming Indigenous Space in Aaron Paquette's *Lightfinder*," Petra Fachinger examines aspects of speculative fiction, the dystopian novel, and the captivity narrative to show how Paquette plays with "generic expectations" (125) to challenge the idea that Indigenous peoples are circumscribed by the past. She considers how Paquette turns to his Cree and Cherokee heritage to use a number of different decolonizing strategies to underscore the importance of "the imagined geographies of environmental health and a (trans)national Indigenous space." In his novel, two Cree teenage siblings are caught up in the survival of Indigenous knowledge, cultural identity, and the future of life on Earth. Fachinger shows how Paquette portrays the shortcomings of both the middle generation, who were traumatized by their experiences at Canadian residential schools, and the Elders, who were unable to keep up with a changing world. Nevertheless, these weaknesses do not stand in the way of transformation and empowerment for the young generation. Fachinger also deals with the novel's Alberta landscape, the specific legends associated with it, and the myth of its Aboriginal characters. This chapter reveals that *Lightfinder*'s Alberta landscape was never empty of myth for its Indigenous people.

First Interlude

CHAPTER 7 is the first of three chapters, two interludes and a postlude, written by three Canadian children's authors who provide an inside look on how imaginative geography in a work of fiction is created. Their focus is on how they imagine. How do images of place come to mind, and how are they put into words on the page? Each author has not only a unique sense of place, but also a distinct relationship with the geography of his or her imagination.

In Chapter 7, "History, Hills and Low Lands: In Conversation with Janet Lunn," by Aïda Hudson, the queen of young adult historical fiction in Canada, Janet Lunn, focuses on how she imagines mythology and geography in *Shadow in Hawthorn Bay*, the first novel in her trilogy. She discusses the importance of the difference in geography between the misty hills of Scotland and the forested lowlands of Upper Canada in the 1820s. This difference shapes the course of her heroine, Mairi Urquhart, who leaves her Scottish myths behind and, at the end, states that she and Luke Andersen will become the "old ones" of myth in this new land. There is nothing imperialistic or dominating about Mairi, who cared for sheep in Scotland, is terrified of the dark forests of Upper Canada, and accepts the help of Indigenous people. Her early settler experience challenges Said's paradigm. Janet Lunn also talks about her own love of the hills of Vermont, where she grew up, and of Prince Edward County (part of historic Upper Canada), where she lived for much of her adult life, giving us a glimpse into the relationship between well-loved places in her own life and the places in her fiction. These geographies serve as background for the time travel story in *The Root Cellar,* and the journey of the United Empire Loyalists to Canada during the American Revolution in *The Hollow Tree,* the last two novels in her trilogy.

Two / Gardens and Green Places

IN THIS PART, select *topoi* and how they are imaged are examined—places like gardens, a riverbank, and a rural countryside.

In Chapter 8, "How Does Your Garden Grow? The Eco-Imaginative Space of the Garden in Contemporary Children's Picture Books," Melissa Li Sheung Ying considers the rise of the eco-critical (or environmental) study of literature and the need for its further growth in the study of children's literature. She discusses three picture books about gardens, by three celebrated author-illustrators, two Americans and one Canadian. Place-attachment and memory are important aspects of one, and the greening of urban landscapes whether real or imaginary shapes the interconnection between a child and his or her family in the other two. The imaginative geography of each garden picture book maps identity both for the child and for the adult(s) in the books.

In Chapter 9, "Into the (Not So) Wild: Nature Without and Within in Kenneth Grahame's *The Wind in the Willows,*" Alan West examines how Grahame celebrates the English rural landscape in this classic semi-fantastic novel. The natural world is seen as benign and often idyllic, overseen by a benevolent Pan (renamed the Friend and Helper). The four friends, Ratty, Mole, Badger, and Toad, are an idealized blend of the human and the animal, though West shows that the author suggests that "human evolution has not been a process of unqualified progress" (174). Pan manifests a "dichotomy" in the text; he is sensually depicted, but without some darker aspects of the original Greek demigod. West explores the not-so-wild wildness in the book, including the taming of the animals of the Wild Wood. In this chapter, the "our land–barbarian land" duality helps us understand the creatures within the imaginative geography of Grahame's novel, between his tamed and untamed anthropomorphic animals, between the riverbankers and those creatures who live in the Wild Wood.

Second Interlude

CHAPTER 10, "Earth, Sea, and Sky Writing in *Becca at Sea*," grows out of the author Deirdre Baker's place-attachment to the Salish Sea off the coast of British Columbia. That attachment is so intense that it transforms what "place" means. "How can I invite readers to perceive at once that place is not 'a beautiful backdrop' for the 'real' story, the one involving humans, but au contraire *is* the story, every bit as much as the human characters?" (191). Baker's love of the Salish Sea, its geology and its rich diversity of sea creatures and plant life that inspired her novel, is what makes her also aware of a human danger, so that her vision is intrinsically antiphonal. She writes of "that energetic dialectic between intimacy and exclusion, identification and alienation, satisfaction and longing," that is "crucial to the deepest kind of wonder—and to a sense of responsibility to and about environment" (193). She wants readers to feel that wonder, but in words that recall Said, she continues with a warning about this "dialectic": "It is the way also to avoid a kind of mental colonisation or appropriation of place—the sentimentalizing or romanticizing we can impose on place once we've left it" (193). Baker's *Becca at Sea* celebrates the "Other" that is non-human Nature.

Three / Fantasy Worlds and Re-enchantment

SECONDARY WORLDS, myth, and the re-enchantment evoked in place are the subjects of this Part. In Chapter 11, "The Imaginary North in Eileen Kernaghan's *The Snow Queen*," Joanne Findon argues that Andersen's North is as much an idea as a geographical space in his famous fairy tale and that Kernaghan's *The Snow Queen* is a profound reimagining of it. Her Kai and Gerda are teenagers, not children, and she enlarges the emotional canvas of the tale through her development of the Little Robber Girl as the Saami girl Ritva. Moreover, Kernaghan infuses Andersen's North with the power of the Finnish epic *Kalevala,* whose heroes become role models for Ritva, and whose Dark Enchantress is reflected in Kernaghan's Snow Queen. The author's extensive knowledge of Canadian exploration of the North informs her landscapes. The Canadian Kernaghan reshapes the northern landscape as a fantastic realm of ancient myth and shamanic power. Whereas Knowles points out that Clark turns to Indigenous myth in her fantasy, Findon shows that Kernaghan looks back to the Old World for wonder.

In Chapter 12, "Camping Out on the Quest: The Landscape of Boredom in *Harry Potter and the Deathly Hallows*," Sarah Winters explores the long "camping section" of the seventh novel in the Harry Potter series. She shows that this "geography of boredom" (216) is one of stagnation, of exile, and yet, paradoxically, of quest in the beautiful but unnamed and unmapped natural landscapes of Britain. This part of the novel also evokes the epic fantasy and quest-romance *The Lord of the Rings*. Winters argues that the wanderings of Harry and Frodo in their vast and varied geographies are essential in the construct of a Christian heroism that embraces doubt, stagnation, and boredom as part of the suffering and sacrifice necessary in the struggle against evil.

In Chapter 13, "Sky Sailing: Steampunk's Re-enchantment of Flight," Christine Bolus-Reichert confirms that all steampunk depends on the imaginative geography of another time—"'The past is a foreign country.'" But steampunk for children particularly involves the extension of the child's imaginative geographies. She examines three recent novels for young readers—Kenneth Oppel's *Airborn* (2004), Philip Reeve's *Larklight* (2006), and Scott Westerfeld's *Leviathan* (2009)—that take up what has become one of steampunk's most persistent fantasies: the re-enchantment of flight. She borrows the idea of re-enchantment from John McClure's important 1994 study, *Late Imperial Romance*. Oppel, Reeve, and Westerfeld re-enchant flight by taking it back to the age of sail. Bolus-Reichert discusses the wonder of imaginary airborne machines and flight that recalls the late Victorian obsession with flying technology. She discusses the views of many theorists, including those of Frye, Tolkien, and George McDonald, on fantasy and wonder, but a wonder that leaves out a religious dimension. Wonder for Matt Cruise, the protagonist in *Airborn*, begins on the airship he calls home and in what he discovers in its travels, like cloud cats.

In contrast, Chapter 14, "Mythic Re-enchantment: The Imaginative Geography of Madeleine L'Engle's Time Quintet," Monika Hilder includes the numinous: "L'Engle invites readers to explore mythic reality through her imaginative geography: geography of earth and fictional planets, biblical and invented characters and species, from contemporary and from other times, and, above all, the 'inscape' or inner spiritual geography of place, event, and the individual soul" (244). She argues that the world of the spirit is part of the temporal world. For Hilder, "L'Engle's imaginative geography is her creative response to de-sacralization: hers is a work of recovery in mythic imagination in which the material and the spiritual are shown as being one indivisible reality" (245).

Four / Space and Gender

THE NEXT TWO CHAPTERS explore how place and space can be transformed through their relations to gender and bodily movement. In Chapter 15, "Female Places in *Earthsea*," Peter Hynes examines the connection between female places and masculine places in the first three Earthsea novels, such as the boy's school for wizardry on the Isle of Roke and the underground tunnels and quarters of the priestesses of Atuan, as well as what he identifies as the "palinode" *Tehanu* (278). This chapter mines the richness of combining the study of imaginative geography with the study of gender in children's literature.

In the final critical essay in this volume, Chapter 16, "Dancing and Hinting at Worlds in Theatre for Young Audiences," Heather Fitzsimmons Frey looks at how dancers moving across a stage transform space. "Some directors I spoke to acknowledged the literal limitations of the physical theatre space, and explicitly created movement motifs to guide the audience into the imaginary world of the play and to define the space—temporarily imposing a kind of embodied map, but one that expands the literal space of the theatre" (288). She also suggests that embodied space helps a spectator see imaginary geographies in the mind's eye—indeed, that the "empathetic collaboration" between performer and audience makes the spectator *feel* those geographies. This radical way of exploring geography in contemporary plays for young adults opens a new area for scholarly study.

Postlude

IN CHAPTER 17, "Following the Path of the Unconscious in the Owen Skye Books, and Others," Alan Cumyn, the award-winning author of thirteen wide-ranging novels, including three for children and three for young adults, narrates his own story about the role his unconscious plays both in his life and in his writing, revisiting the places in which he has lived, dreamed, and written. Summoning the winding Kennebecasis River in

New Brunswick, next to which he lived as a child, and on whose banks the unnamed places of the Owen Skye novels originated, Cumyn shows how indivisible life and the writing life are for him. His narrative illustrates the truth of Neil Gaiman's words in the introduction to *Fragile Things: Short Fiction and Wonders*: "The tale is the map which is the territory" (xix).[2] For Cumyn, "territory" is his unconscious, the map for his story-making. Story, the geography of places in his life and work, memory, and the unconscious make up an inseparable whole.

∾ Conclusion

GEOGRAPHICAL APPROACHES to literature for the young promise a new critical frontier. Collections like Dobrin and Kidd's *Wild Things: Children's Culture and Ecocriticism*, Doughty and Thompson's *Knowing Their Place? Identity and Space in Children's' Literature*, and Cecire, Field, Finn, and Roy's *Space and Place in Children's Literature, 1789 to the Present*, as well as Clare Bradford's *Unsettling Narratives: Postcolonial Readings of Children's Literature*, have shown the way.

Without imaginative geography, readers have only the voice of the narrator. Stripped of "imaged earth-writing," a text is unimaginable. Eudora Welty argued that fiction's life is based on place, just as Tolkien's idea of a Secondary World, his Realm of Faërie, establishes "where" as the beginning place of fantasy. Said's dualistic view of imaginative geography, encapsulated in his term "our land–barbarian land," maps "familiar" and "unfamiliar" space, a model that can help in the study of any young protagonist's adventure from the known to the unknown. But that paradigm, too, has been challenged in this volume. Modern environmental crises have fostered environmental writing for children that is "doom-crying" but also hopeful, showing both the preciousness of non-human geography and how greening urban spaces brings renewal. Place-attachment brings energy, sometimes biophilic, to solving the identity problems of young protagonists. Imaged earth-writing can also be mythical and re-enchanting. In addition, spatial scale promises a better critical understanding of the sense of infinite possibility that is unlocked in the universes of children's fantasy, science fiction, and steampunk; it also provides a window on smaller places, such as bodies in motion on a stage in young adult plays, such as Drew Hayden Taylor's version of *Raven Stole the Sun* (see Frey's chapter). Imaginative geography allows a reader to see and feel Harry Potter stand in his beloved and besieged Hogwarts waiting for Voldemort to kill him, Ratty and Mole shiver outside Badger's door, the Skye brothers fall through the hole in the floor of the haunted house, the nishnawbe son remove his blanket to greet the first snow in the morning, and Peter pick up sheers in the topiary garden ready to snip away and sculpt his own bush in the image of his great-grandfather. Imaginative geography fosters a better appreciation of the mimetic truth of fiction. Characters have adventures that often include self-discovery, not because they are *in* the worlds the authors have created, but because they are *of* those worlds.

<hr />

Notes

1) See the following articles, which give a sense of the global impact of Said's work: Mary Kelly, "When Things Were 'Closing-In' and 'Rolling Up': The Imaginative Geography of Elizabeth Bowen's Anglo-Irish War Novel *The Last September*," *Journal of Historical Geography* 38, no. 3 (July 2012): 282–93; David Tavares and Pierre-Martineau LeBel, "Forcing the Boundaries of Genre: The Imaginative Geography of South America in Luis Sepulveda's *Patagonia Express*," *Area* 40, no. 1 (March 2008): 45–54; James Perkins, "Peasants and Politics: Re-Thinking the British Imaginative Geography of the Balkans at the Time of the First World War," *European History Quarterly* 4, no. 1 (2017): 55–77; and Ashley Dawson, "Edward Said's Imaginative Geographies and the Struggle for Climate Justice," *College Literature* 40, no. 4 (2013): 33–51.

2) I am grateful to Robert Tally for having introduced me to this wonderful quotation in his fine conclusion to *Spatiality* (149). It comes from the section titled "The Mapmaker" from Gaiman's Introduction to *Fragile Things: Short Fictions and Wonders*. This is in itself a cautionary tale about a king who, in mapping his story, uses his riches to reconstruct an island to make it fit that map, with tragic results for him, but not for the island, which, in the end, reverts to Nature.

One

GEOGRAPHICAL IMAGINARIES

The Old World and the New

1) Pullman and Imperialism

Navigating the Geographic Imagination in *The Golden Compass*

❧ CORY SAMPSON

There are many alternate universes in Phillip Pullman's *His Dark Materials*, yet the one that remains the most intriguing is the first one we see: Lyra's universe. There is much to arrest our imaginations in it, of course—a part of everyone's soul is visible and embodied in their daemon, magic and witches are real (and mysterious), and sinister, theocratic government agencies are ready, willing and able to do unspeakably horrible things to children. It is equally attractive as a space for adventure. *His Dark Materials*, and particularly *The Golden Compass* (published as *The Northern Lights* in the UK), hearken back to the heyday of the children's adventure story. This particular genre formed the bulk of children's fiction in the latter half of the nineteenth century. It's no mistake then that Lyra's Britain looks and feels so much like our universe's Late Victorian Britain. In borrowing from Late Victorian culture and literary representation, Pullman effectively signals the mode of his oeuvre; we can reasonably expect that a novel set in a time and place closely resembling a heavily stylized Late Victorian Britain will contain more than a bit of globetrotting adventure.

The manner of its stylization, however, should not go overlooked. In addition to more familiar modes of evoking the Victorian era, such as the architecture of Jordan College, or the patriarchal culture of the Master's Retiring Room, Phillip Pullman uses the concept of geographic inquiry itself—which Thomas Richards calls "the queen of all imperial sciences in the nineteenth century" (13)—to mark Late Victorian culture. The privileged science of geography not only establishes the backdrop for the action but also drives the plot forward. The Arctic expedition by Stanislaus Grumman, for example, serves as the catalyst for Lord Asriel's own expedition to the north—and, subsequently, for the discovery of how windows to other universes may be opened.

Pullman's alternate version of Earth furthermore constitutes an imaginative geography. While we can presume that its geology (in terms of the placement of land masses, oceans, rivers, and other geological features) is identical to our Earth's, its human geography—the patterns of human migration and settlement, the spread of culture and language, and the boundaries of nations—vastly differs. Pullman offers us a world where Tartary still exists as a geographic region, where Arctic exploration is new and exciting, and where nomadic Gyptians travel by boat from place to place. Echoes of our world, displaced in space and time, make once-familiar geography unfamiliar, strange, enticing, and novel. Pullman's imaginative geography is Earth as it might have been, given a complete reconfiguration of the European geopolitics that continue to inform our own concept of geography to this day.

One element is conspicuously absent from this world where geography plays such a (privileged) role. Pullman's alternate universe is missing a British Empire.[1] Instead, Pullman imagines a pan-European Catholic empire called "The Magisterium", which fulfills the role of primary antagonistic force throughout the first novel of the series. Geopolitically, Lyra's England controls no empire on which the sun never sets; yet in spite of the lack of a political British Empire, the novel contains, nevertheless, a sense of empire *manqué* latent in the ideological values held by its characters. Furthermore, Lyra's story of growth and development relies on tropes inherited from a Late Victorian literary tradition that not only draws on but is inextricable from the British Imperial project. This is not to say that Pullman is wholly uncritical of imperialism—in several instances, he does gesture towards an argument against it—but for all the text's explicit redressing of wrongs, there remain some relics of imperialism's literary legacy implicit in the values that drive the text forward. In creating his atmosphere, Pullman romanticizes the spirit of rational, post-enlightenment Western geographic expansionism that made British Imperialism possible. Such a celebration warrants critical attention. In 1993, in *Culture and Imperialism*, Edward Said declared that the entire field of literary criticism had reached "a point ... when we can no longer ignore empires and the imperial context in our studies" (6); Pullman's *The Golden Compass*, published a scant few years later, makes an excellent proof text for Said's claims. Said furthermore states:

> We must take stock of the nostalgia for empire, as well as the anger and resentment it provokes in those who were ruled, and we must try to look carefully and integrally at the culture that nurtured the sentiment, rationale, and above all the imagination of empire. And we must also try to grasp the hegemony of the imperial ideology, which by the end of the nineteenth century had become completely embedded in the affairs of cultures whose less regrettable features we still celebrate. (12)

For all that the novel (as well as the series at large) opposes absolutism, theocracy, totalitarianism, and fundamentalism, it nonetheless privileges, as the mode for Lyra's education in becoming an effective protagonist, the type of occidental peregrination through the margins of civilization that plays out in nineteenth-century texts, which, in large portion, are uncritically affirming of Britain's right to administer and rule nearly one-quarter of the Earth's landmass. Thus, Pullman partakes of the same nostalgia for empire that Said warns us about—even without explicitly authoring the British Empire into his work.

The Golden Compass plays out imperialist themes in three principal areas: the romance of geographic exploration; the nature of child adventurers and their education abroad; and the patterns of ethnic "othering" that leave Europeans in a position of ascendency over all other ethnicities. These themes, while never as overtly propagandistic as their Victorian counterparts, nevertheless remain unresolved loci of tension. By examining to what degree *The Golden Compass* replicates the literary mechanisms (if not the spirit) of British Imperialism, I hope to demonstrate that Pullman's novel is a cautionary tale in authorial responsibility to darker histories of literary production.

WHETHER past, present, or future, talked about or actually undertaken, travel takes up the bulk of the novel. The romance of geography is particularly important in Late Victorian Imperialist texts, literature for children being no exception. While its literary counterpart was steeped in imagination, real-life geographic expeditions served several practical purposes. Thomas Richards writes that Late Victorian Imperial hegemony was less a product of military or economic dominance than a product of knowledge control; by assembling and classifying "facts," the British military, civil service, and educational institutions created a "paper empire: an empire built on a series of flimsy pretexts that were always becoming texts" (4). Chief among the mechanisms for gathering "facts" was the geographic expedition. Not only did it serve the practical purpose of surveying the territory that ostensibly belonged to Britain, but it also provided fodder for outright propaganda or inspiration for any number of adventure narratives, real and fictional. Often the writers of these narratives were the very explorers themselves:

> In the 1870s and 1880s the Empire had produced solitary travellers who had risked everything to explore the unknown world: Richard Burton had gone in disguise to Mecca and Medina, Fred Burnaby had ridden to Khiva in Samarkand, and again, in another year, to furthest Turkey and on to Persia. They had both written up their trips, and had become famous, as much for their attitude of insouciant daring, their splendid presumption that the world was their—and therefore England's—as for the exoticism of the places they penetrated ... Their writing typified an attitude peculiar to their class and professions: a compound of confidence, superiority, and bravado, self-confidence, pride in their own strength, a delight in life, and an occasional sympathy with the indigenous peoples they so cheerfully patronised. (MacDonald 13)

This romance of exploration, born during the high period of British territorial conquest, persisted even past the latter's decline. "It is no coincidence," writes M. Daphne Kutzer, "that at the end of the era of colonial expansionism we find the beginnings of the so-called Heroic Age of Exploration"; concerned primarily with polar exploration and mountain climbing, these expeditions offered mere "bragging rights to being the noblest and strongest nation in the world. There was absolutely nothing to gain at the pole except national pride and reputation" (xx). While polar exploration may have been symptomatic of the British Empire's death throes, it was nevertheless carried out (and written about) with the same ideological gusto as the geographic surveys of a prior generation.

Pullman appropriates the romance, the vigour, and the self-actualizing potential of polar exploration, while leaving out the rearguard jingoism. Significantly though, the novel retains exploration's lure of fame and prestige—which makes it a highly desirable enterprise. Pullman builds the reader's interest in exploration, chiefly by focusing our attention on Lyra's interest in it. The romance of polar exploration—its surrounding narrative of glory and heroic accomplishment—enchants her and, by proxy, the reader. There is some indication, before she has even left Oxford, that she dreams of international travel. While eavesdropping on Lord Asriel's meeting in the Retiring Room (in the second chapter, aptly titled "The Idea of North")[3] she listens to the Scholars talk and decides that "mighty dull talk it was, too; almost all of it politics, and London politics at that, nothing exciting about Tartars. The smells of frying poppy and smoke-leaf drifted pleasantly in through the wardrobe door, and more than once Lyra found herself nodding" (18). Lyra wants nothing to do with boring domestic intrigues; her focus is outward, on the exotic (and dangerous). The reader too is drawn into the same kind of reverie; invoked by the frying poppy (broadly hinted to be opium poppy), which is itself a commodity linked to foreign and Oriental spaces, this passage invites the reader into Lyra's travel-dream, infecting the reader with the same inarticulate wanderlust that forms the chief narrative desire. This wanderlust leaves Lyra particularly susceptible to temptation by Mrs. Coulter; having already been infected by "The Idea of North" from Lord Asriel's presentation, Lyra only needs to hear Mrs. Coulter say that she has been to the North "several times" before the enchantment takes hold. Once it does, we are told that "nothing and no one else existed now for Lyra. She gazed at Mrs. Coulter with awe, and listened rapt and silent to her tales of igloo building, of seal hunting, of negotiating with the Lapland witches" (68). Once Lyra becomes enchanted, Mrs. Coulter tightens her grasp, promising that "when [Lyra] comes back to Jordan College, [she'll] be a famous traveler" (71). Her encounter with the other members of the Royal Arctic Society (and its relics—"the harpoon with which the great whale Grimssdur had been killed; the stone carved with an inscription in an unknown language which was found in the hand of the explorer Lord Rukh, frozen to death in his lonely tent; a fire-striker used by Captain Hudson on his famous voyage to Van Tieren's Land") leaves Lyra with "admiration for these great, brave, distant heroes" (77). Pullman's aligning the Royal Arctic Society with the evil Mrs. Coulter takes a tentative step towards redressing the idea of an Arctic expedition as a heroic venture; however, the refutation is not satisfactory—nor could it possibly be, given the narrative expectations set out by Pullman's seduction of the reader to this "Idea of North." While of course she never ends up going on Mrs. Coulter's expedition (nominally keeping her untainted by the pseudo-scientific pretensions of the Royal Arctic Society), Lyra nonetheless fulfills her wish (and the reader's) of seeing the North. Though divorced from its geopolitical function of enshrining national heroes as proof to all other European nations of Britain's superiority, the romance of polar exploration is nevertheless preserved. Lyra enacts nineteenth-century geographic fantasies of claiming and mapping the final frontier—and the reader is drawn into the continuum of desire as well. Eventually her desire is fulfilled as she steps across the boundary between worlds, performing the ultimate act of Western geographic expansion.

❧ Lyra's Coming of Age Abroad

IF ADVENTURE ABROAD was the highest goal and proof of British manhood in Imperialist propaganda, then growing up abroad was a uniquely effective crucible for the characters of British boys. The tale of a boy's adventures abroad became a literary genre in its own right. For writers like George Alfred Henty, who wrote some ninety books that "transported schoolboys to the far reaches of the British Empire" (Singh 4), these tales acted as surrogate character-builders for young Britons, who could emulate the fictional, daring young men and boys furthering the cause of Empire. As Rashna B. Singh writes, character would become a national concern and justification for ongoing Imperialist projects. It is important to note that the basis for the ascendency of British character over the national characters of all others was a combination of racist scientific theories and popular belief in the singular quality of British institutions (Singh 8–29). In her book, *Goodly Is Our Heritage*, Singh traces this thread of Imperialist propaganda through the development of British children's fiction:

The connection between qualities of character and national or racial identity is clearly delineated in the children's literature that was written in the era of the British Empire ... National character becomes imbricated with individual character, manners with morals. In character ethics, the interweaving strands of politics, culture, race, history, and pseudoscience are clearly discernible. Since character was emphasized so strongly and so coherently by parents, teachers, public schools, scouting groups, books, papers, and magazines, children would naturally have considered it one of the most important aspects of their development ... Resilience, raw courage, daring, endurance, hardiness, duty, loyalty, and sacrifice: these were the qualities that were considered the inheritance of a white, English especially, male child. (60–61)

In these adventure stories, the boy adventurers have a preternatural ability to adapt to any new environment; while they may make boyish blunders, they habitually end up getting the better of non-British adults in whatever Imperial setting they end up in. The corollary, however, is that the geographic "Other," lacking in British pluck, can serve only the narrative purpose of furthering the goals of the plucky young boy; vast pre-existing networks of relations and hierarchies are easily collapsed, as both friend and foe, whether in helping or attempting (unsuccessfully) to hinder the British protagonist, become mere accessories to the narrative of British character development, whatever status they might enjoy in their native culture. British pluck is endemic to the British; it is this combination of "pluck" and consummate character-building that gives Britain its right to Imperial dominance. Thus race, education, exploration, character, and dominion are all intertwined, particularly during the height of the British Empire. The literary phenomenon of young men raised on the margins of Empire brings together these notions; through their inherent British "pluck," young men and boys persevere in dangerous situations, learning all the while the skills they need to be competent Imperial administrators as well as the manners and proper deportment that ensure the propagation of British culture among the benighted races they are destined someday to rule. This ascendancy over the Other constitutes its own imaginative geography—author(iz)ing through an imagined world and an imagined people, easily bested by mere children (albeit British children), the moral imperative and justification for material conquest of the real-world analogues of those imaginary places.

Children's literature has inherited the pluck and adventurousness of Imperialist propaganda. *The Golden Compass*, however, with its conspicuous pseudo-Victorian setting, pays homage to the boy's adventure narrative even as it undermines its gender assumptions. One aspect that *The Golden Compass* retains, albeit in a roundabout fashion, is the ascendency of "pluck" as a uniquely British character trait—one that enables a particularly British child to gain the upper hand when dealing with ethnic others. In this regard, Lyra resembles no nineteenth-century character so much as Rudyard Kipling's eponymous hero in *Kim*. Edward Said identifies *Kim* as an Imperialist idyll and celebration of the limitless potential of living in India, at the centre of which is Kim, properly Kimball O'Hara, the Lahore-born-and-raised son of a deceased Irish soldier. What enables Kim to survive is not only his pluck, but also his protean quality—what Said calls his "chameleon-like progress dancing in and out of it all, like a great actor passing through many situations and at home in each" (158). Kim's quick wits and adaptability eventually grant him access to "the great game"— Kipling's metaphor for the British-administered network of espionage in India that facilitates British dominion. Kim's childhood on the streets of Lahore grants him his facility with native customs and languages; however, it is his British heritage that gives him a British education, and gives him a foot in the door of the Royal Geographic Survey under Colonel Creighton. His education combines the best of British boarding school tradition with street-smarts. He is left with an amalgam of social skills and practical non-academic knowledge meant to help the British subject settle into the various crevices of the unchartable human geographies of colonized nations, where strange environments and strange peoples create tensions that baffle occidental epistemology. What is notable, however, is that only Western European subjects can enact the kind of cultural appropriation that allows them to draw whatever is useful from the geographic fringes and synthesize this knowledge into determined and practical courses of action. Kim can do it; Colonel Creighton can do it; the unfortunate Hurree Babu, try as he might to emulate British customs, dress, and deportment, is an unconvincing figure of fun—the native who would be a sahib.

Lyra is Kim's inverse doppelgänger; she possesses the same protean quality, the same pluck, and the same wild childhood (battling the children of other colleges, of the town, of the brickburners). Like Kim, she is initially unaware of her true parentage—and like Kim, her parentage is what grants her access to elite "British" education (though Lyra's Oxford tutelage is certainly more desultory than Kim's boarding school education at St. Xavier's). Like Kim, who combines his formal education with his native wit to outsmart the Russian agents threatening the geographical survey, Lyra combines her academic knowledge and the folk wisdom of Farder Coram to learn how to read the alethiometer—which she does better than anyone else. Lyra's story is the inverse of Kim's, however, due to the fact that unlike Kim, who is raised as the ethnic other first and *then* "rescued" from obscurity by his enduring "whiteness," Lyra is reclaimed by her thoroughly English father almost from the start. It is all the more impressive then, when Lyra demonstrates the ease with which she appropriates the talents and traditions of other cultures, without even having, as Kim does, a lifetime of experience to guide her. When Lyra falls among the gyptians, we are told that she learns quickly how to perform their essential daily tasks:

> She cleaned and swept, she peeled potatoes and made tea, she greased the propeller shaft bearings, she kept the weed trap clear over the propeller, she washed dishes, she opened lock gates, she tied the boat up at mooring posts, and within a couple of days she was as much at home with this new life as if she'd been born gyptian. (110)

It later turns out that Lyra, in her infancy, was nursed by Ma Costa, and that she might have been raised gyptian. Her return to the gyptians carries some poetic justice—but the whole episode highlights how her Englishness, like Kim's, gives her the privilege to step in and out of ethnic communities—to simply "pass" as necessary, never having to wholly take part in order to employ what is useful. Ma Costa's advice to Lyra preserves the essentialism of the gyptian race: "You en't gyptian, Lyra. *You might pass* for gyptian with practice, but there's more to us than gyptian language. There's deeps in us and strong currents" (112; italics mine). This phrase contains some chilling implications, given the similarities between Lyra and Kim; just as it would have been unseemly to Victorian British sensibilities for Kim, a scion of Britain, to remain entrenched in Indian culture, so too do Ma Costa's words suggest some essential barrier to Lyra's adopting the gyptian culture—even when she herself might have been raised among them.

Other characters of Western European descent also have the ability to appropriate marginal knowledge—Lord Asriel, Mrs. Coulter, Lee Scoresby, and especially Major John Parry / Stanislaus Grumman, who, Kurtz-like, finds himself the shaman of an inscrutable native tribe. The ethnic "others," however, have no such faculty—the gyptians are gyptians, the Tartars are Tartars, even the witches are witches, and, as the novel makes abundantly clear, the bears are bears.

ᙈ *Putting the Noble Bear in His Place*

I ISSUE A CAVEAT before assigning the panserbjorne the status of ethnic "other": they are not human characters, but bears. The essentializing discourses of Victorian Imperialism that delineated racial properties as immutable, naturally determined characteristics almost make sense when comparing two different species—especially when one is a fantasy species having no intelligible real-world referent. However, the panserbjorne serve a structural purpose that is not entirely unlike that of the colonial native in Victorian Imperial texts—in fact, in many ways, it is an uncomfortably familiar one. The panserbjorne, particularly Iorek Byrnisson and Iofur Raknisson, play out the familiar drama of the noble savage versus the colonized native who would be a Westerner. The one is permitted to excel at certain things (provided they are aligned with essential, natural forces) in order to chastise and correct certain excesses of the Western world; the other is a parody of what happens when ethnic and racial hierarchies are inverted and the subaltern grotesquely puts on the mantle of the master. Both figures appear in Victorian literature, and in both cases, they exist to secure intellectual Anglo-European ascendency over some ethnic or racial "other". When we first meet Iorek Byrnisson, we learn that he is a "renegade"—the one armoured bear who is not in the employ of the Northern Progress Company (172). He is also, however, sunken into a state of degradation; his armour has been stolen, and he is doing odd jobs for "meat and spirits" (181). He essentially has to be rescued by Lyra, who discovers, using the alethiometer, where the armour is being kept. After this, they enter into a very complex power dynamic; Iorek is, of course, physically stronger. Lyra, however, knows how to lie. Iorek cannot be tricked during the fencing scene, but he makes an admission that turns out to be the key to overthrowing Iofur Raknisson's rule. After Lyra tries in vain to hit Iorek with her stick (which he constantly and expertly deflects), the following exchange occurs:

> "I bet you could catch bullets," [Lyra] said, and threw the stick away. "How did you *do* that?"
>
> "By not being human," [Iorek] said. "That's why you could never trick a bear. We see tricks and deceit as plain as arms and legs. We can see in a way humans have forgotten." (226)

The panserbjorne, due to essential characteristics, have a natural ability that humans have somehow lost; it's telling, however, that their ability undergoes a decline when they act too human. The witch Serafina Pekkala suggests that "when bears act like people, perhaps they can be tricked" (317)—that Iorek's drunkenness (a human characteristic) enables the Trollesund villagers to steal his armour and exploit him for his services. Iorek's first appearance as degraded subaltern is precisely due to this momentary lapse in his bear nature; indulging in drink leads to the loss of his armour, which functions as a symbol for his bear identity. He becomes much more noble—and more alien—once he puts on his armour. If we think of Iorek Byrnisson as the noble savage, whose moral pureness comes from his adherence to his essential nature, then Iofur Raknisson is the savage who apes the Western man. Kutzer identifies this particular trope—"native figures who aspire—comically—to be white" (84)—and suggests that they function to emphasize racial hierarchies; that though the savage must be civilized, he can never truly *be* a European. Though divorced from the context of colonial racism, the drama between Iorek Byrnisson and Iofur Raknisson follows very closely a set of narrative conventions that find their origin in white European ethnic superiority. Lyra's power over Iorek (and Iofur) is that she can lie. Iorek is indebted to Lyra already for restoring his armour to him; it is through her machinations that he is able to regain his throne. First, though, he must be wholly a bear, with no blemish to his pure nature. Iofur Raknisson, who wishes more than anything else to be human, who surrounds himself with the trappings of what he interprets as human culture (living in a palace, surrounding himself with ambassadors, wearing human-inspired jewellery and other finery), has sinned against his own nature and must be corrected. His own desire to be other than what he is proves his downfall. Lyra tricks Iofur because he is acting like a human and harbours illicit desires far beyond his savage nature. She convinces Iofur to accept Iorek's challenge of single combat by telling him she has become Iorek's daemon—something that is impossible—and that she will become his if he defeats Iorek. Iofur loses the fight primarily because his armour, vainly constructed of gold and steel and not iron, the traditional "bear-metal," is more ornamental than practical. Iorek's kingship sees a return to the true nature of the panserbjorne—all thanks to Lyra. Stripped down to its structural elements, this episode is the story of a privileged European who meets a degraded (yet noble) savage, reminds him of his identity, basks in awe at his savage and alien nature, and then again uses her European pluck to chastise and subjugate another savage who would seek to aspire to a position greater than what his nature would allow. It is a pastiche of imperialist narratives in which the reader is aligned with an anglocentric subject position, replete with "othering" discourses that, in the end, ratify the particular talents of good Britons, who alone have the mandate to order the world of other cultures and restore them to their proper function. In this way Pullman's imaginary Svalbard draws eerily on the heritage of the imaginary Africa, or India, or America, which so many European authors of the Imperial Era made the backdrop for their own dramas of the European trickster versus the hopeless native.

ᠵ Of the (Anti)Imperialists' Party

THOUGH I HAVE ARGUED SO FAR that Pullman's work, rather than critically engaging with imperialism, tends towards a nostalgic celebration of imperialist tropes, the mainstay of criticism, both academic and popular, for *His Dark Materials* has come from critics who find Pullman's theological credentials lacking. The trilogy has received a great deal of harsh criticism in particular from Christian theological commentators. This is perhaps unsurprising, given Pullman's stated intention to "undermine the basis of Christian belief" (Wartofsky), his animosity toward the works of C.S. Lewis, and the literal death of Yahweh in *The Amber Spyglass*. Negative criticism of *His Dark Materials* is not, however, solely the province of the theologian. One common thread that runs through the works of Pullman's detractors is the notion that, in certain spheres, Pullman's technique backfires—that in his deliberate attempt to say X (and X differs for each critic), Pullman uses the very discourses underpinning X, thus legitimizing them.

The harshest criticism of Pullman's *His Dark Materials* series tends to focus on the theological aspect of the series, the unifying argument being that Pullman's theological knowledge falls short in ways that undercut his message and throw the authority with which he questions theological concepts into doubt. Those critics with the clearest bias form the most blunt arguments; just so for Edward Higgins and Tom Johnson, writing for *Christian Century*, who accuse Pullman of reducing Christianity to "a fairly common straw man: the oppressiveness of organized religion" (29), stating that Pullman's Magisterium is more "akin to Tolkien's Mordor and the evil Sauron and to Lewis's White Witch of Narnia" (30). They identify the plot's thrust as sacrificial and redemptive, echoing the Christian story of redemption—ultimately reclaiming Pullman's narrative for the very Christianity Pullman sought to undermine. Jonathan and Kenneth Padley, an English professor and a clergyman, respectively, declare that Pullman's attempt to differentiate himself theologically from Christian doctrine ultimately fails in several areas, largely due to the simplistic theology of the series. They charge Pullman with ascribing to Christianity theological concepts that are in themselves contradictory to extant Christian theology—for example, the idea that Christianity is monolithically anti-materialist—and, just as Higgins and Johnson claim, of replaying the Christian story with the Authority cast as Satan, stating that "the conquering of the Authority in *His Dark Materials* has remarkable parallels to the defeat of the devil by Christ on the cross" (331). They identify Lyra's releasing the spirits of the world of the dead as a reformulation of the Harrowing of Hell (332), and that Will and Lyra's "restoration of the balance of creation through a new Adam and Eve has roots in Christian symbolism going back to its very earliest days" (332). Both of these critical excoriations contain a consistent, recurring motif. All of these critics, in their own way, accuse Pullman of the inverse of Blake's famous charge against Milton: that Pullman is of God's party without knowing it. They identify the ways in which the Authority and the Magisterium fulfill a Satanic role in the narrative, and point towards the redemptive actions of the protagonists as rehearsing any number of Christlike characteristics.

I shall, therefore, break with the tradition of Pullman's critics in suggesting that even though *The Golden Compass* does not engage critically with the borrowed imperialist tropes that power its narrative, this does not make Pullman an unwitting imperialist himself. In fact, if we read the Magisterium as an imperial power in itself—albeit a religious empire, rather than the empire of any one European nation-state—then Pullman's oeuvre astutely identifies one propensity in particular of nineteenth- and twentieth-century British Imperialism, one that was indeed justified by religious motives and given broad sanction. Canadians should be familiar with this shamefully recent part of our history: the residential school system, which from 1883 to 1998 preyed upon the most vulnerable members of Canada's already embattled First Nations people.[4] While Pullman's work is not allegorical of this real-world institution, the goals, methods, and justifications for the General Oblation Board offer some eerie parallels, which demonstrate that even as he borrows from the literary legacy of imperialism, Pullman's work denounces the tendency of imperial powers to use religion as a cloak to hide its treatment of already marginalized peoples.

Canada's residential schools were officially established by the federal government in 1883, though religious schools (some day schools and some boarding schools) existed prior to the federally established ones (TRC, *Canada's Residential Schools,* 4). The Truth and Reconciliation Commission of Canada describes the residential school system as "an education system in name only," whose purpose was "separating Aboriginal children from their families, in order to minimize and weaken family ties and cultural linkages, and to indoctrinate children into a new culture—the culture of the legally dominant Euro-Christian Canadian society" (TRC, *Honouring the Truth*, v). Children were taken from their homes, "torn from their parents, who often surrendered them only under threat of prosecution" (41), and taken to boarding schools where they were forbidden from speaking their native languages and wearing their native dress, subject to harsh discipline that, in some cases, manifested itself in physical and emotional abuse, and even sexual abuse. The last residential school was not closed until 1998 (TRC, *Honouring the Truth*, 106). The Truth and Reconciliation Commission of Canada was established in 2008 to "reveal to Canadians the complex truth about the history and the ongoing legacy of the church-run residential schools" and to "guide and inspire a process of truth and healing, leading toward reconciliation within Aboriginal families, and between Aboriginal peoples and non-Aboriginal communities, churches, governments, and Canadians generally" (TRC, *Honouring the Truth*, 23). Survivors' accounts describe countless tragedies; one such survivor, Wilbur Abrahams, describes his being separated, for the first time in his life, from his sisters:

Somebody guided us through the door, and going down the hallway, and I didn't realize it, but they were separating us, girls on this side, and boys on this side, and I was following my sisters. And all of a sudden this, I felt this little pain in my, my left ear, and this, I looked up, and I saw this guy with a collar, and he pulling me back with, by my ear, and telling me I was going the wrong way. You're going this way. Pull, still pulling my ear. I have always believed that, I think at that particular moment, my spirit left. (TRC, *Survivors Speak* 92, unedited from interview)

This feeling of loss of spirit, of separation from an essential part of one's being, is a recurring theme in survivor accounts. Yet for much of the life-span of the residential schools, the religious organizations running them, particularly the Roman Catholic, Anglican, United, Methodist, and Presbyterian churches (TRC, *Canada's Residential Schools*, 4), operated out of a profound belief that their mission was God-ordained, just, and useful. Milloy asserts that "in the church's estimation, their residential school mission, which purportedly transformed Aboriginal boys and girls into useful Christian Canadian men and women, was a most sublime Christian act" (xii), one that aligned with assimilationist policies established by the British government even before Canada's Confederation (Milloy 11–13). Indeed, the tenor of religious feeling towards Canada's Aboriginal population when the first residential schools were established is best summarized by politician Alexander Morris, negotiator of several treaties in Western Canada, who in 1880 prayed:

Let us have Christianity and civilization among the Indian tribes; let us have a wise and paternal government ... doing its utmost to help and elevate the Indian population ... and Canada will be enabled to feel, that in a truly patriotic spirit our country has done its duty by the red men. (qtd. in Milloy 6)

In establishing the residential schools, religious organizations and imperialism went hand in hand, the one supported by government policy and funding, the other providing moral justification and acting as a guard against censure.

The General Oblation Board's project is not a perfect analogy for the residential school system, for it is much kinder than the residential schools were. The Experimental Station at Bolvangar is described as hospital-like, "brilliantly lit, with the glint of shiny white surfaces and stainless steel" (237), and the children are well fed on familiar food like stew, mashed potatoes, tinned peaches, and ice cream (241); whereas children at residential schools were housed in "poorly constructed and dangerous buildings" (TRC, *Honouring the Truth*, 159) and given an insufficient diet of foods alien to them (85–90). Nevertheless, the General Oblation Board shares some of the residential schools' imperialist imperatives and methods—namely, it targets the children of marginalized people, destroys an essential aspect of the child's soul in order to produce ideal Christians, and uses its religious mandate as public justification for atrocities committed far from the public eye.

Pullman presents us with a model victim of the Gobblers: Tony Makarios. Throughout this section, Tony is constructed as a kind of subaltern; a not wholly British hybrid of several "other" ethnicities, living in abject poverty with an improvident, drunken mother. Gyptian children are targeted at a far greater rate than any other community, according to Tony Costa (Pullman 109). In either case, the General Oblation Board targets children who will not be missed by polite society. It should also be noted that aspects of Tony Makarios's home life—including his drunken and unaware mother—are similar to the frequent justifications used by residential school officers when removing Aboriginal children from their parents. The Truth and Reconciliation Commission likens this treatment to a self-reinforcing cycle, in that by "removing children from their communities and … subjecting them to strict discipline, religious indoctrination, and a regimented life more akin to life in a prison than a family, residential schools often harmed the subsequent ability of the students to be caring parents" (TRC, *Honouring the Truth*, 137–38); parents who themselves had been in residential schools a generation prior often turned to substance abuse to cope, leading to their own children's apprehension on child welfare grounds. That Pullman's model victim comes from a home rife with substance abuse is evocative of the myopic propensity of the Canadian government to treat the symptom rather than the cause of systemic poverty.[5]

Mrs. Coulter describes the intercision process clinically and euphemistically as an "operation" and "a little cut" (Pullman 283), yet its actual purpose is to sever a child's connection with their daemon—an essential piece of their soul—in order to prevent the accumulation of Dust, which the Magisterium has decided is the physical manifestation of original sin. As was the case with Aboriginal children in residential schools, the process is thought to produce demure, perfect Christians, exemplified by the nurses at Bolvangar. These ideal cases are not typical, however. Many children die from the process, as shown by the empty daemon cages that Lyra and Roger find (259–60) and by the pathetic death of Tony Makarios, clutching the dead fish he so desperately wishes was his daemon, Ratter (218). And the subjects who survive the intercision are shadows of their former selves: Sister Clara is described as having no imagination or curiosity, and Lyra feels a sense of abject unease at seeing her severed daemon following after her (238). The adults at Bolvangar, furthermore, are not afraid to break the great taboo that prevents the touching of another person's daemon, a process described as a violation almost akin to rape: "It was as if an alien hand had reached right inside where no hand had a right to be, and wrenched at something deep and precious" (275). Taken altogether, the Experimental Station at Bolvangar represents nothing less than an institution that runs roughshod over the human spirit, all in the name of a form of idealized and sanitized Christianity, while its day-to-day operations involve mistreatment, abuse, and even death.

Yet the General Oblation Board is not publicly excoriated for its activities—rather, those who know about it are approving of its philosophical and religious mission. The middle-aged man flirting with Adèle Starminster at Mrs. Coulter's party describes the work of the General Oblation Board as a continuation of the medieval practice of devoting sons and daughters to the church. Later, Lord Boreal defends the General Oblation Board, claiming that "What's done [to the children] is for their good as well as ours" (95), and among the chatter Lyra overhears at the party is the statement, "The children don't suffer, I'm sure of it" (96). Much later, Mrs. Coulter defends the General Oblation Board directly to Lyra; even, however, as she defends it, she reassures Lyra that she is safe from intercision, saying, "There, there ... You're safe, my dear. They won't ever do it to you" (282). What Mrs. Coulter is unabashedly willing to do to the poor children of the streets or the gyptians, she refrains from doing to her own daughter, Lyra. This religious hypocrisy echoes the hypocrisy behind the residential schools' ostensible religious mission; Milloy accuses the system of representing "the intolerance, presumption, and pride that lay at the heart of Victorian Christianity and democracy, that passed itself off as caring social policy and persisted, in the twentieth century, as thoughtless insensitivity" (xviii). Surely Pullman's representation of the General Oblation Board would be subject to the same criticism.

It is important to draw parallels between Pullman's description of the General Oblation Board and Canada's residential school system, not only because it raises further awareness of Canada's ongoing responsibility to redress the wrongs it created but also because it demonstrates the quality of Pullman's imperialist analysis. Pullman is often accused of gross religious intolerance and animosity towards organized religion; however, if we read *The Golden Compass* as in some part anti-imperialist, then critique of religion cannot be separated from critique of imperialism, given how intimately organized religion was tied into European imperialism of the nineteenth century. Residential schools were not the only means by which the British Empire attempted to assimilate people of other nations and ethnicities, and Pullman reminds us, through the Magisterium and the General Oblation Board, that there are valid criticisms of the excesses of organized religion when paired with self-serving imperialist policies.

∾ *Imperial Nostalgia and the Children's Book*

IN CLOSING, I issue the following caveat: while I have been critical of Pullman's imperial nostalgia, I do not mean to suggest that Pullman somehow made a mistake by engaging with the literary history of the British children's adventure narrative, inextricably linked as it is to imperial history. I believe that Pullman set out to intimate a Victorian *feel* to his text—and almost of necessity, given how interwoven imperialism and Victorian culture were, imperialist themes colonized Pullman's text (and I use the term "colonized" quite deliberately, to suggest that once a rhetorical foothold was gained, through Pullman's decision to make his setting pseudo-Victorian, further imperialist themes mushroomed throughout the text, in order to continue to support that pseudo-Victorian setting and narrative). For example, while Lyra's characteristic "pluck" is the inheritance of a children's literature tradition that finds its inception in the British imperial project, it does not disqualify that pluck from being a positive or even celebratory aspect of Lyra's character. Readers can appreciate this quality in Lyra even while being cognizant that such characterization has a darker past when its literary genealogy is scrutinized.

The Golden Compass, furthermore, offers a critique not of its imperialist heritage but of the grotesqueries of religious imperialism itself. Even in a world lacking in a British Empire, the text provides us with shadowy impe-rialist projects in some ways akin to those actually undertaken in our his-tory. Though this does not remedy the lack of critique of the text's other, less-examined imperialist tropes, it does demonstrate that Pullman, per-haps even knowing it, is of the anti-imperialists' party.

Said reminds us that the spectre of imperialism is by no means a black mark against a work of literature:

> The novels and other books I consider here [in *Culture and Imperialism*] I analyze because first of all I find them estimable and admirable works of art and learning, in which I and many other readers take pleasure and from which we derive profit. Second, the challenge is to connect them not only with that pleasure and profit but also with the imperial process of which they were manifestly and unconcealedly a part. (xiv)

Said refers here to texts from the European imperial era—I do not argue that Pullman's work is "unconcealedly" imperialist—but we can take to heart that *The Golden Compass* is a work from which we can take pleasure and derive profit even while being critical of its imperialist heritage. Our responsibility as literary critics, however, is not to let the shadows of impe-rialism go unchallenged, even in acclaimed works of literature.

Notes

1) The words "empire," "imperial," and "imperialism" are used in a few different ways in this chapter. Where the words refer specifically to the British Empire, they are capital-ized. Where they refer to the general idea of an empire, they remain lowercase. This is to preserve the necessary distinction between the actual entity that was and is the British Empire and the theoretical construct of "empire" as used, for example, by Said and other critics of imperialism.

2) For additional information on the general trend of polar exploration and its role in the British "boys' adventure" narrative, see Colleen Franklin's "Nineteenth-Century British Children's Literature and the North" (Chapter 2 in this volume), wherein she reproduces a brief history of how exploration transformed the British concept of the North from "a desolate region of marvels, mystery, magic, and even evil" (Franklin 2)

prior to the nineteenth century, to being the "ultimate testing-ground for the nation" (4) during the nineteenth century.

3)   The chapter title takes its name from the Glenn Gould radio documentary *The Idea of North*, which explores the role the North plays in the Canadian psyche.

4)   Attempts have been made at providing full histories of the Canadian residential school system, though its full effects in terms of human sorrow and suffering are immeasurable by academic discussion. My primary sources of information are John S. Milloy's *"A National Crime": The Canadian Government and the Residential School System, 1879 to 1986*, which provides a consummate history of relevant facts, figures, and dates; and the various reports from the Truth and Reconciliation Commission of Canada, which highlight first-hand accounts of the horrors of Canada's residential schools.

5)   I would also be remiss were I not to point out that although the federal government began to dismantle the residential school system from 1940 onward (TRC, *Canada's Residential Schools*, 9), the height of the residential schooling era was followed by the infamous "Sixties Scoop," which involved the broad apprehension of Aboriginal children into the foster care system, beginning in the 1960s. These children were to be sent not to residential schools but to non-Aboriginal families, where they were no less isolated from their families and native culture. Even today, rates of apprehension of Aboriginal children on child welfare grounds in Canada are still ten times higher than for non-Aboriginals; according to a 2011 Statistics Canada study, "14,225 or 3.6% of all First Nations children aged fourteen and under were in foster care, compared with 15,345 or 0.3% of non-Aboriginal children" (TRC, *Honouring the Truth*, 138).

2) Nineteenth-Century British Children's Literature and the North

❦ COLLEEN M. FRANKLIN

———

Britain's preoccupation with what is now known as the Canadian North has a long and vigorous history. Beginning in the fifteenth century, English merchants and explorers, barred by a Papal decree from the southerly routes to the east allotted to Spain and Portugal, sought a passage by the northwest to Cathay and its riches. The project was abandoned when several centuries of fruitless attempts persuaded the public that such a passage probably did not exist and that the cost of a continued search in terms of personnel and equipment outweighed the possible benefits of its discovery. However, natural philosophers, geographers, and Britons hungry for national glory continued to pine after the goal and finally, in 1818, the British Admiralty, swollen with pride and ships and crews from its victories in the Napoleonic wars, deemed the time ripe for a renewed search for the Passage and the initiation of a search for the North Pole. The dramatic events that ensued, most importantly the disappearance of Sir John Franklin's 1845 expedition for the Passage and the many subsequent searches for his lost ships and men, fuelled the century's explosion of writing about the North. The market was flooded with songs, poems, novels, and travelogues that celebrated the Arctic enterprise. The British reading public could not wait for new accounts of the region and its adventurers; as explorer after explorer tested himself against the Arctic and returned with his story, readers made northern narratives some of the bestselling books of the period (Loomis 95). The rapidly expanding market for children's literature capitalized on this hunger for tales of northern exploration and northern landscapes. The British reading public's fascination with the North was inscribed in the imaginative geography of "picture books, story books, and Tales for Boys and Girls," to employ F.J. Harvey Darton's criteria.

∾ The History of the British Representation of the North

THE RENEWAL OF THE SEARCH for the Passage ushered in a renewal of a trope particular to the history of writing about the North in Britain. The British representation of the North had always been marked by the lack of empirical knowledge of the area. In fact a land that teems with life and that supports its people and their rich culture, the North was imagined as a desolate region of marvels, mystery, magic, and even evil. In the Middle Ages, authorized by Aristotelian and Ptolemaic cosmology and the exegeses of the patristic writers of the Catholic Church, sermons, vision narratives, plays, and poetry forwarded the belief that the North was the location of hell or the entrance to hell, the home base of Satan and the rebel angels who fell with him from Heaven. Early modern writers such as William Shakespeare, Christopher Marlowe, and John Milton continued to engage with medieval representations of the North even as the drive to discover the Northwest Passage resulted in a trickle of empirical knowledge of the area; however, as the North and northern exploration became foci of early modern writers of the emerging genres of geography, history, biography, and the travel collection, the dearth of eyewitness accounts of the northern landscape meant that those writers were mainly dependent upon a single early modern text that was itself dependent upon the earlier representations of the North, *The Strange and Dangerous Voyage of Captaine Thomas James* (Franklin, Introduction).

In the eighteenth century the evil North began to be refigured through the aesthetic of the sublime. Following a classical treatise, *Peri Hypsous* (On the Sublime), now known to have been written by the second-century Greek rhetorician Longinus, that sought to understand and explain the sublime or elevated in art, eighteenth-century theorists sought to understand and explain the terror and mystery often felt by humankind in the presence of the natural world. They conflated the concept of the marvellous with the sublime, seizing on the notion of sublimity to discuss what had previously been figured as the preternatural in their discussions of nature.[1] Edmund Burke's explanation of the human response to the natural world in his *A Philosophical Enquiry into the Origin of our Ideas on the Sublime and the Beautiful* (1757) provided a psychological explanation for the phenomenon. Burke argued that "whatever is fitted in any sort to excite the ideas of pain and danger, that is to say, whatever is in any sort terrible, or is conversant about terrible objects, or operates in a manner analogous to terror, is a source of the *sublime*; that is, it is productive of the strongest emotion which the mind is capable of feeling" (36). In the nineteenth century, the representation of the natural world as sublime informed what has been termed the "sublime quest" (Bohls 26): British explorers sought to master areas of the globe hitherto unpenetrated by Europeans to prove the excellence of the masculine British subject and to prove the right of that subject to administer the Empire. The history of British incursions into the "evil north" was refigured as a centuries-long quest to master the sublime. For instance, Sir John Barrow, Second Secretary of the Admiralty and the man behind the Arctic expeditions of 1818, explicitly linked the early modern voyages firmly to the nineteenth-century endeavour in his *A Chronological History of Voyages into the Arctic Regions* (1818). Not only did he figure the nineteenth-century explorers as the inheritors and thus natural successors to the quest for the Passage, he chose the epigraph from Samuel Purchas's canonical collection of travel narratives, *Purchas his Pilgrimage* (1614), for his own: "How shall I admire your heroicke courage, ye marine worthies, beyond all names of worthiness!"

The idea of the North as the ultimate testing ground for the nation, entrenched as it was in the imagination of the British public, thus became a standard trope of nineteenth-century British literature. This trope has not gone unremarked; a number of critics have offered studies of the literature of northern exploration for the adult market, as well as investigations into the many "boys' books" of the period that deal with northern discovery and adventure.[2] But writers for the children's market engaged with the North in all genres. The survey that I offer here will examine some of these works and note the ways in which these examples of nineteenth-century juvenile literature are marked by the trope of the sublime quest to conquer the still-terrifying North.[3]

〜 The Imaginative Geography of Victorian Children's Literature of the North

"Tales for Boys and Girls" / Maria Hack's *Winter Evenings*

AS ANDREA WHITE HAS DEMONSTRATED, the "boys' book," the adventure story of later nineteenth-century writers such as R.M. Ballantyne and G.A. Henty, is both indebted to the literature of travel and exploration and shares its ideology of empire and its construction of the imperial subject (39).[4] In 1818, the year in which the British Admiralty embarked on its Arctic expeditions, Maria Hack produced the four-volume *Winter Evenings; or, Tales of Travellers*, a "book of travels" directed, unusually, to a child audience, and one of the first children's books of the period to engage with the representation of Canada's North, explicitly offering a nod to the Admiralty ventures (II.173). The "book of travels," or voyage collection, has its roots in the early modern collections of Richard Hakluyt and Samuel Purchas, and was ubiquitous in the eighteenth century. Unlike the travel narratives that record personal journeys through the British Isles or Europe, the voyage collections focus on accounts of travels to remote and dangerous parts of the globe. At the outset of the period, the voyage narrative was understood to be an important tool for the new philosopher, but as the century wore on, the emergent imperial project, the fascination with strange and distant lands and peoples, and a novel-reading public's desire for narrative instead of information fuelled the demand for the voyage collection. Like figures such as Edmund Halley and John Locke, Tobias Smollett, and Samuel Johnson, Oliver Goldsmith, and Christopher Smart, Hack edited her collection of voyages to coincide with a specific readerly interest, in her case, the instruction and delight of the juvenile reader (Franklin, *The Strange and Dangerous Voyage*, "Introduction"). As White points out, nineteenth-century educators were to value the narratives of travel and exploration and the adventure story or "boys' book" over the novel per se, as both genres were marked by their status as sources of "education and inspiration" (239).

Reprinted in three subsequent editions over the course of the century (1840, 1853, and 1857), *Winter Evenings* focuses firmly on the areas of the globe on which the British were casting an eye or in which they were actively engaged. *Winter Evenings* includes tales from *The Travels of Ali Bey* in the Middle East, *Golownin's Account of his Captivity* in Japan, Legh's *Journey in Egypt*, Bligh's *Voyage to the South Seas*, and Ulloa's *Journey across the Andes*, as well as devoting nearly two of its four volumes to the history of northern exploration. Hack, who was to become a prolific writer of juvenile literature and whose work would be praised for its "beautiful morality" (Mitchell), frames her offering to the children's market with an introduction that heralds her authorial intentions and discloses the didactic purpose of her juvenile "book of travels": "Let the actions, and enjoyments, and sufferings of men, form the subjects of the contemplation of children ... Examples of courage, of patience, of fortitude, of generosity and benevolence, and, above all, of reliance on the Supreme Disposer of events, on occasions of danger and distress, will have a natural tendency to strengthen and to elevate the character" (I.vi).

Winter Evenings opens with an epigraph from Mark Akenside's *Pleasures of the Imagination* (1744):

> By night,
> The village-matron, round the blazing hearth,
> Suspends the infant audience with her tales.—
>
> Witness the sprightly joy when ought unknown
> Strikes the quick sense, and wakes each active power
> To brisker measures:—the attentive gaze
> Of young astonishment; the sober zeal
> Of age, commenting on prodigious things.
>   Akenside (title page)

The epigraph signals the form of the narrative; Hack constructs it as she would several of her other bestselling juvenile works, as a series of dialogues between an extremely learned Mamma, "Mrs. B.," and her two curious and receptive children, Harry and Lucy. The text is placed firmly in the heart of the Victorian home, the woman's sphere, which is represented in both the epigraph and the setting of the stories by the image of the mother and children by the hearth. The annals of exploration offer any number of instructive tales, and Mrs. B.'s stories of northern exploration fully engage with the notion of the "sublime quest" in the service of fulfilling her duty to raise exemplary subjects for the Empire.

Mrs. B.'s North is very much the "wild, romantic country" (III.25), empty of people and full of danger, that the readers of the boys' books of northern adventure were to encounter and that previous generations of Britons knew as the "evil north." Mrs. B. recounts explorers' adventures with polar bears, walrus, and seals, near deaths through shipwreck and breaking ice, near escapes from disappearance in blizzards and on ice floes, describes the landscape of "mountains, rocks, ... precipices, [and] hills of ice" (III.25) and recites numerous lines of illustrative poetry such as these, from the account of Sir Hugh Willoughby's 1553 Arctic expedition in James Thomson's *Winter* (1726):

> Such was thee Briton's fate,
> As with first prow (what have not Britons dar'd?)
> He for the passage sought, attempted since
> So much in vain, and seeming to be shut
> By jealous Nature with eternal bars.—
> In these fell regions in Arzina [Lapland] caught,
> And to the stony deep his idle ship
> Immediate seal'd, he with his hapless crew,
> Each full exerted at his several task,
> Froze into statues; to the cordage glu'd
> The sailor, and the pilot to the helm. (III.53)

As this litany of English endeavours in the North provides "examples of courage, of patience, of fortitude, of generosity and benevolence, and, above all, of reliance on the Supreme Disposer of events, on occasions of danger and distress," and assists Mrs. B. in her project of "strengthen[ing] and ... elevat[ing] the character[s]" of her children, it also threatens to render not the Arctic, but the Arctic enterprise, sterile.[5] As young Lucy asks her mother, "Mamma, why do Europeans go to such a dismal country? does it produce anything worth fetching?" (III.85), Mrs. B. responds with remarks about fur, coal, and fish, but her main focus is to inculcate the idea that the Arctic enterprise is imperialist: "Knowledge is Power: it allows [us] to acquir[e] ... influence over our fellow-creatures, [and to] chang[e] and improve[e] the condition of the globe which we inhabit." She tells the children that "we have ... strong reasons for believing, that if the land in America were cleared and cultivated ... that the climate would be exceedingly improved":

> Cultivation is there gradually extending; and where the marshes have been drained, and the forests cleared, the climate has, in proportion, become more temperate; and not only more temperate, but more healthy: for, as the progress of cultivation has rendered the country more open and dry, many fatal diseases, which used to prevail, have been restrained and diminished. The wild beasts retire to less frequented regions; and health, plenty, and security surround the dwelling of enlightened man. (III.180–81)

The change and improvement that Mrs. B. proposes is a virtual erasure of the Arctic. If the end of the "sublime quest" is the mastery of the Arctic, the obliteration of the Arctic is perhaps its natural consequence.

At the other end of the century, the confidence and optimism that Mrs. B. shares with her fellow Britons had been severely dashed. At mid-century, the Franklin expedition, considered the best-manned and best-equipped force to have ever ventured into Arctic waters, was presumed lost; the search for Franklin's men and his two ships, the *Erebus* and *Terror*, was to last out the century. Hopes for their survival dimmed and then faded entirely as searchers turned up evidence of the failure of the expedition. The horror at their fate, including stories of a death march, starvation, and possible cannibalism, was universal, and the representation of the Arctic became even grimmer. The British public focused firmly on the bravery and fortitude that they imagined characterized the men of the Franklin Expedition and that they celebrated in the searchers and in subsequent explorers. As the second half of the century wore on and the conquering of the Arctic increasingly began to prove a chimaera, the "sublime quest" became a rich metaphor for the conquering of the self.

~ *The Nursery Rhyme / Robert Louis Stevenson's
"The Northwest Passage"*

ROBERT LOUIS STEVENSON, whose writing, including juvenile novels such as *Treasure Island* (1882) and *Kidnapped* (1883), famously deals with adventure and travel in an imperialist context, was not behindhand in exploring the theme of self-conquest in his book of nursery rhymes, *A Child's Garden of Verses* (1895). The nursery rhyme is directed toward the youngest children, and its didactic purposes are many, including the teaching of numbers, letters, time, the seasons, language, and so on, but it also inscribes the writer's and his or her culture's construction of the child subject. Denis Denisoff argues that the century's Romantic ideal of the child is countered by Stevenson's work:

> Stevenson stands out within this nineteenth-century context for his consideration of and respect for the template that imaginative play and exploration offered for aesthetics, philosophy, and politics. His poetry and other writings fuse the imagination—which his society habitually construed as childlike—with what were seen as adult issues. (231)

In *A Child's Garden of Verses*, Stevenson offers poems, such as "Whole Duty of Children" (9), "A Good Boy" (36), and "Good and Bad Children" (49), that treat of the virtues and rewards of self-conquest that would not be out of place in the period's evangelical and missionary publications for juveniles,[6] but he offers an unusually dark and moving representation of childhood fear overcome in "The Northwest Passage" (76–78). A tripartite poem in which conquering the fear of the dark, cold, and terrifying nightly trek a child must make from the warm and cozy parlour to the safe haven of the nursery is imagined as metonymic with the sublime quest of the Arctic explorer. The publication history of the poem reveals that it was first published in 1884 in a "lively, cosmopolitan review" (Greiman 53) for adult readers to which Stevenson was a frequent contributor, *The Magazine of Art*, and subtitled "A Childish Memory" (198). The subtitle directs the reader back to his or her experience of growing up in the early to middle years of Arctic exploration; however, if one considers that the Arctic was steadily being represented as unconquerable and that 1884 was the year in which the public was horrified to learn that American Adolphus Greeley's expedition to the Arctic Circle had resulted in the survival of only six of his twenty-five men from shipwreck, mutiny, starvation, and cannibalism, the nostalgia of the reader might well have been tempered with a darker valence. Relocated in *A Child's Garden of Verses*, the poem appears exactly as it did in *The Magazine of Art*, except that it is minus its subtitle, accompanied by Charles Robinson's illustrations, and made part of the apparatus of a volume of children's verse. Stevenson has thus collapsed the dual audience of the nursery rhyme—the child, and the parent who reads to the child—and the terrifying Arctic is exhibited in all its power.

Part I of "The Northwest Passage," "Good Night," depicts the child speaker steeling him or herself for the embarkation for the nursery by imagining him or herself in the company of explorers courageously departing for the cruel and brutal Arctic landscape of the Victorian imagination:

> When the bright lamp is
> carried in,
> The sunless hours again
> begin;
> O'er all without, in field
> and lane,
> The haunted night returns
> again.
>
> Now we behold the embers
> flee
> About the firelit hearth; and
> see
> Our pictures painted as we
> pass,
> Like pictures, on the
> window-glass.
>
> Must we to bed indeed?
> Well, then,
> Let us arise and go like men,
> And face with an undaunted
> tread
> The long black passage up
> to bed.
>
> Farewell, O brother, sister,
> sire!
> O pleasant party round the
> fire!

> The songs you sing, the
> tales you tell,
>   Till far to-morrow, fare ye
> well!

A regular iambic tetrameter underscores their steady purpose: as the "sunless hours" of the day begin and the "haunted night" returns, the children in the family circle around the hearth-fire must "arise and go like men, / And face with an undaunted tread / the long black passage up to bed" (76). The second section, "Shadow March," which details the actual journey, is accompanied by Charles Robinson's illustration of a grim-faced child who grasps the lowest banister in determination, eyes firmly fixed on his or her faraway goal (fig. 1):

> All round the house is the jet-black night;
> It stares through the window-pane;
> It crawls in the corners, hiding from the light,
> And it moves with the moving flame.
>
> Now my little heart goes a-beating like a drum,
> With the breath of the Bogie in my hair;
> And all round the candle the crooked shadows
> come
> And go marching along up the stair.
>
> The shadow of the balusters, the shadow of the
> lamp,
> The shadow of the child that goes to bed—
> All the wicked shadows coming, tramp, tramp,
> tramp,
> With the black night overhead.

FACING PAGE: *Fig. 1)* Shadow March (Charles Robinson, from "A Child's Garden of Verses," 77). Image courtesy of The Osborne Collection of Early Children's Books, Toronto Public Library.

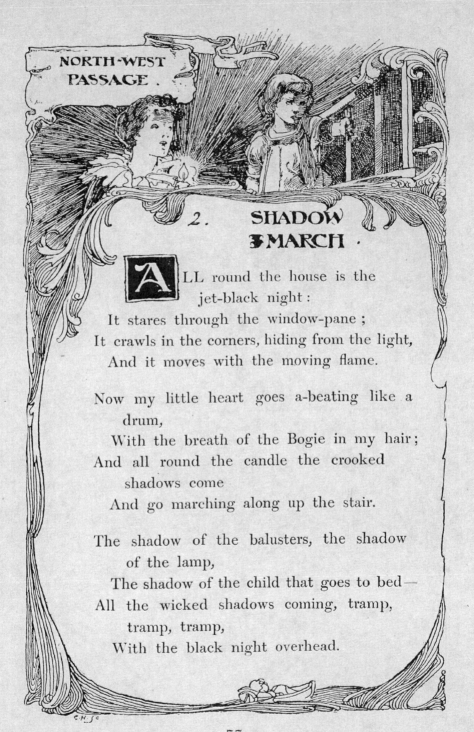

NORTH-WEST PASSAGE.

2. SHADOW MARCH

ALL round the house is the
    jet-black night:
It stares through the window-pane;
It crawls in the corners, hiding from the light,
    And it moves with the moving flame.

Now my little heart goes a-beating like a
    drum,
    With the breath of the Bogie in my hair;
And all round the candle the crooked
    shadows come
    And go marching along up the stair.

The shadow of the balusters, the shadow
    of the lamp,
    The shadow of the child that goes to bed—
All the wicked shadows coming, tramp,
    tramp, tramp,
    With the black night overhead.

"Shadow March," departing from the rhythm and metre of the first section, is composed in arrhythmic alternating tetra-meter and trimeter lines, suggesting the irregular heartbeat that results from the physical terror that the landscapes of the Passages inspire in the child and in the Arctic explorer: the erratic flickering of the "staring," "crawling" night that "moves" with the moving candle flame and brings the "crooked," "wicked" shadows and "the Bogie's breath," counterpointed by the drumming of the child's "little ... beating" heart" (77). The title of Part III, "In Port," details the triumph and the reward earned by the conquest of both the little one's and the explorer's fear:

> Last, to the chamber where I
> lie
> My fearful footsteps patter
> nigh,
> And come from out the cold
> and gloom
> Into my warm and cheerful
> room.
>
> There, safe arrived, we turn
> about
> To keep the coming shadows
> out,
> And close the happy door at
> last
> On all the perils that we
> past.
>
> Then, when mamma goes by
> to bed,
> She shall come in with tip-toe
> tread,
> And see me lying warm and
> fast
> And in the Land of Nod at
> last.

The "cold and gloom" is shut out when the "happy door" is "close[d]" on the "perils" that he or she has survived. The child arrives safe and warm at the "land of Nod," secure in the knowledge that Mamma will be in later to tuck him or her up (78). Both the adult and the child reader would be all too aware that the extended metaphor of the poem breaks down at this point: the Arctic explorer too often did not return to the safety of his home port or of any port at all. And, in fact, Stevenson neglects to attach any moral value to the child's journey: the end of this episode of self-conquest is solely the child's safety. We might question the latter part of the century's reading of the sublime quest here: is self-conquest a great enough reward for the explorer's travails in the North? And indeed, is it truly necessary for a little child to endure such terror to learn this lesson?

∾ "The Story Book" / George MacDonald's At the Back of the North Wind

THE NORTH OF THE SUBLIME QUEST was often anthropomorphized in the nineteenth century as female. As Britain was represented by Britannia and America by Lady Liberty, the North was figured as a *femme fatale*. One example, a cartoon by T.E. Donnison from late in the century entitled "The Siren of the Pole," features a scantily clad woman seated on an ice bench and leaning toward the viewer, wearing a come-hither look and dangling a laurel wreath from her fingers. Her hair, which on closer inspection proves to be the Northern Lights, floats freely and surrounds her head in an aureole. A skull and bones litter the ground around the bench, and bubbles emerge from the broken ice directly in front of her, suggesting that yet another of her swains has endeavoured to win her but has lost. Death awaits those who attempt to conquer the north.[7]

In George MacDonald's juvenile novel *At the Back of the North Wind* (1871), the author and his illustrator, Arthur Hughes, imagine the titular character, North Wind, as a Sirenesque figure, a capricious, demanding, dangerous woman. An allegory of the earthly journey to Heaven, the novel is based on the belief, ascribed in the novel to Herodotus (122), in the temperate zone of the open polar sea. Indeed a belief of antiquity, the idea of an open polar sea was also held by a respectable number of nineteenth-century theorists until the end of the century, when the polar ice pack had been thoroughly explored and the notion exploded.[8] In

MacDonald's allegory, the open polar sea represents Heaven, which lies at the "back of" North Wind, who represents Death.[9]

As the death that awaits Diamond, North Wind is as unpredictable as St. Paul says death must be.[10] At their first meeting, in Little Diamond's sleeping quarters in a stable, North Wind's voice, which is "gentle" but also "angry" and "loud" (47), leads Diamond to name her "Mr." North Wind (51). With "a beautiful laugh, large but very soft and musical," North Wind corrects Diamond, scolds him for his rudeness in addressing her as male, and forces their meeting:

> "Will you take your head out of the bed-clothes?" said the voice, just a little angrily.
>
> "No!" answered Diamond, half-peevish, half-frightened.
>
> The instant he said the word, a tremendous blast of wind crashed in a board of the wall, and swept the clothes off Diamond. He started up in terror. Leaning over him was the large beautiful pale face of a woman. Her dark eyes looked a little angry, for they had just begun to flash; but a quivering in her sweet upper lip made her look as if she were going to cry. What was the most strange was that away from her head streamed out her black hair in every direction, so that the darkness in the hay-loft looked as if it were made of her hair; but as Diamond gazed at her in speechless amazement, mingled with confidence—for the boy was entranced with her mighty beauty—her hair began to gather itself out of the darkness, and fell down all about her again, till her face looked out of the midst of it like a moon out of a cloud. From her eyes came all the light by which Diamond saw her face and hair; and that was all he did see of her yet. The wind was over and gone. (51–52)

Throughout the novel, there are no reasons given for North Wind's erratic, strange, and dangerous behaviour. One minute she blows sharp as a knife, the next so gently that Little Diamond feels warm; she announces her intention to drown a shipful of men at sea but at the same time carry Diamond in the "very core and formative centre" of the storm, in a nest of

---

FACING PAGE: *Fig. 2)* Cowering he clung with the other hand to the huge hand which held his arm, and fear invaded his heart (Arthur Hughes, from *At the Back of the North Wind*, Chapter VI). Image courtesy of The Osborne Collection of Early Children's Books, Toronto Public Library.

the "great billows of mist-muddy wind [that] were woven out of the cross-ing lines of North Wind's infinite hair, sweeping in endless intertwistings" (96); when she appears to him so huge that her "gigantic, powerful, but most lovely arm—with a hand whose fingers were nothing the less ladylike that they could have strangled a boa-constrictor, or choked a tigress of its prey," she tells him to "be a man. What is fearful to you is not the least fearful to me" (90; fig. 2). North Wind's function in the text is to personify the trials and tribulations that one must endure as one becomes fit for entrance into the Heavenly Kingdom. MacDonald's other works for chil-dren offer notably powerful female figures, for instance, Princess Irene's mysterious Great-Great-Great Grandmother in *The Princess and the Goblin* (1872). Irene's Grandmother changes, certainly, but only to steadily reveal her greater beauty and power, a model for Irene on earth as well as a guide to Heaven. North Wind's capriciousness is terrifying, even if Diamond does learn to love and trust her and is eventually borne away by her to Heaven. North Wind's allegorical power is drawn not from the century's engagement with the Angel in the House, like Grandmother's, but from its engagement with the North and the sublime quest.

ᗥ "The Picture Book" / Bertha and Flora Upton's
The Golliwogg's Polar Adventure

ALTHOUGH MODERN READERS may find the representation of North Wind a troubling one, her presence in nineteenth-century juvenile literature throws into relief the fact that one of the differences between literature of the North that is directed to an adult audience and literature of the North that is directed to a juvenile audience is the presence and participation of females. The project of representing the North for adult readers is largely undertaken by men, and women—other than Inuit women, who undergo a representation that is carefully divorced from that of British women—are largely excluded. But via both male and female writers, the imaginative geography of children's literature allows the presence and participation of girls and women. In the works examined here, for instance, Mrs. B. of *Win-ter Evenings* is informed and enthusiastic about northern exploration. She is the central figure of the collection and as noted above, her role as pre-ceptress is foregrounded in the epigraph to the volumes. Although Harry is encouraged in his urge to active manhood, and Mrs. B. approves Lucy's desire to "tend the home fires" (III.87), both children are firmly enjoined by Mrs. B. "to know and to feel interested in what is doing in the world. The more knowledge you gain, the more it will be in your power to be useful to yourselves and others" (III.176). In Stevenson's "Northwest Passage," the gender of the speaker is undetermined, and Robinson's illustrations and those of subsequent artists respect the voice of the poem: both boys and girls are depicted making the "haunted journey."[11] Some "boys' books" even feature active female characters: for example, in Charles Ede's *The Home Amid the Snow, or, Warm Hearts in Cold Regions: A Tale of Arctic Life* (1882), Silena, the daughter of a station master, "about sixteen years of age, tall, and elegantly formed, with all the golden beauty of the pure Scandinavian … [is] brought into frequent communication with the natives in her endeavor to assist and instruct them" (8). R.M. Ballantyne's 1858 *Ungava: A Tale of the Esquimaux* concerns the denizens of a Hudson's Bay Company trading post at Ungava Bay, including the factor's daughter Edith and his wife, Jessie, who is herself the child of an English fur trader and his English wife. And library collections of nineteenth-century children's books reveal that girls' names are as often inscribed in books about the North as those of boys, and that, whether school prizes or personal gifts, the givers of such books are often women—mothers, aunts, cousins, friends, and teachers.[12]

An example of a feminine North is found in Bertha and Florence Upton's picture book, *The Golliwogg's Polar Adventure* (1900), which features a quest undertaken by representatives of two groups who were normally barred from Arctic exploration: one black male and five white females, although they are all dolls. The *Polar Adventure* is the fifth of thirteen adventures that follow the life and times of Golliwog, Peggy, Sarah Jane, Meg, Weg, and The Midget. Inspired by Golliwog, the dolls undertake an expedition to the North Pole. They meet the typical challenges of the Arctic traveller in British literature, first having to procure food, equipment, stores, and clothing in preparation for their journey, and then experiencing the classic northern adventures: embarking from a grassy bank, they arrive in an ice channel; are stuck in ice; are rescued by an Inuit family who share their equipment with them; face, kill, and skin a polar bear; go seal hunting; get stranded on a floe; see the Northern Lights; climb great mountains of ice; upset a sledge and get pinned beneath it; are attacked by a walrus; and so on. The Golliwog is often more hindrance than help, initiating a racist stereotype that is exploited in Enid Blyton's Golliwog books and elsewhere in the culture, but the females are depicted as extremely capable of undertaking the challenges of Arctic exploration and adapting themselves to the landscape

and its people. One illustration depicts the dolls in intimate conference with the woman of the Inuit family. The picture may be read as a landscape stripped of sublimity, a female space that allows a warm and affectionate connection between the eldest doll, Peg, and the Inuit woman, and the younger dollies and the Inuit baby. The man of the Inuit family has gone to bring the dolls a sled, while they sit around the cooking fire. A homely scene is played out around the hearth, the scene of domesticity and metonymic with woman's sphere in the nineteenth century, as "Peg watched the firelight glow, / And smiled to see the children play / With 'Baby Eskimo.'" A friendship rooted in women's experience has begun (fig. 3):

ABOVE: *Fig. 3)* Peg watched the firelight glow, / And smiled to see the children play / With "Baby Eskimo" (Florence Upton, from *The Golliwogg's Polar Adventure*, n.p.). Image courtesy of The Osborne Collection of Early Children's Books, Toronto Public Library.

[Peg] promised to come back again
When they had found the Pole,
Then thanked them for such kindly help
With all her heart and soul. (21)

The dolls do discover the North Pole. In the final verse of the book they congratulate themselves:

"Farewell, North Pole! We're going back
To let the children know
A problem old is newly solved
By Golliwogg and Co." (62)

Britons did not, in fact, discover the North Pole or navigate the Northwest Passage. Their centuries-long attempt to conquer the North ended in disaster and disgrace, as did their expedition to the South Pole. The heroes of other countries had bested those of Britain. In 1926, A.A. Milne recorded this exchange in *Winnie the Pooh*:

You remember how [Pooh] discovered the North Pole; well, he was so proud of this that he asked Christopher Robin if there were any other Poles such as a Bear of Little Brain might discover.
"There's a South Pole," said Christopher Robin, "and I expect there's an East Pole and a West Pole, though people don't like talking about them." (135)

In fact, it seems that after the turn of the century, grown-up people, at least, "didn't like talking about" the North Pole, or the South Pole, or the Northwest Passage, either. In literature for adults, a silence descended on a topic that had engrossed the country since the Middle Ages. The British representation of the North underwent a shift as Canadian writers and artists, such as Ralph Connor and the Group of Seven, figured a beneficent animism for the North, and the stories of British maritime heroes and their exploits were relegated to specialist texts on the history of exploration. But British writers of juvenile literature were not about to concede the battle. They have produced such worlds as the White Witch's Narnia; the hotbed of witches and wizards that is Hogwarts; and the breathtakingly, horrifyingly evil Bolvangar that is destroyed by Lyra Bellacqua.[13] In the realm of faerie, at least, British literature continues to be fascinated with the imaginary North.

Notes

My sincere thanks to Norah Franklin and to David Humbert for reading and commenting on the various drafts of this paper.

1)   See, for example, Tamworth Reresby, "A Miscellany of Ingenious Thoughts and Reflections" (1721), in *The Sublime: A Reader in British Eighteenth-century Aesthetic Theory,* ed. Andrew Ashfield and Peter de Bolla (Cambridge: Cambridge UP, 1996), 43–44. See also Sir Richard Blackmore, *Essays Upon Several Subjects* (1716), in Ashfield and de Bolla, 40–41; and Thomas Stackhouse, from Chapter XVIII, "Of Sublimity or Loftiness," in *Reflections on the Nature and Property of Languages* (1731), in Ashfield and de Bolla, 50.

2)   See, for example, Stan Atherton, "Escape to the Arctic: R.M. Ballantyne's Canadian Stories," *Canadian Children's Literature / Littérature Canadienne pour la Jeunesse* 1, no. 1 (Spring 1975): 29–34; Robert G. David, *The Arctic in the British Imagination 1818–1914* (Manchester: Manchester UP, 2000); Jennifer Hill, *White Horizon* (Albany: SUNY P, 2008).

3)   Scandinavia is also represented in the period's children's literature, primarily in folk and fairy tales. One of the most popular and most illustrated tales of the imaginary North in the nineteenth century, and beyond, is Hans Christian Andersen's "The Snow Queen" (1844), most recently in its very loose adaptation in Disney's *Frozen*. See Joanne Findon's "The Imaginary North in Eileen Kernaghan's *The Snow Queen*" in this collection.

4)   Andrea White, "Adventure Fiction: A Special Case," In *Joseph Conrad and the Adventure Tradition* (Cambridge: Cambridge UP, 1993), ch. 2, 39–61.

5)   See Jen Hill, "Ends of the Earth, Ends of the Empire: R.M. Ballantyne's Arctic Adventures," ch. 6, 151–73, in *White Horizon: The Arctic in the Nineteenth-Century British Imagination* (New York: SUNY P, 2009), for a discussion of this problem in Ballantyne's *Ungava* (1858), *The World of Ice* (1858), and *Giant of the North* (1887).

6)   For the relationship of the poetry of *A Child's Garden of Verses* to the literature of the nineteenth-century juvenile missionary magazines, see Ann C. Colley, *Robert Louis Stevenson and the Colonial Imagination* (Aldershot: Ashgate, 2004), ch. 2, 179–202.

7)   Other examples include Edward Robert Hughes's painting, *Heart of Snow,* undated but approximately late nineteenth century, and the cartoons "The Queen of Swords" (1875), "Waiting to be Won" (1875), "A Cold Reception" (1876), and "The Sleeping Beauty of the North" (1876), all in *Punch*. English versions of Hans Christian Andersen's "The Snow Queen," of which there were many in the period, offered children visual and verbal representations of a cold and cruel woman metonymic with the north.

8) For instance, the evidence for the existence of an open polar sea is argued in Daines Barrington's *The Possibility of Approaching the North Pole Asserted. By The Hon. D. Barrington. A New Edition. With an Appendix containing Papers on the same Subject and on a North West Passage. By Colonel Beaufoy, F.R.S.* (London: T. and J. Allman, 1818).

9) MacDonald, an Aberdonian, was heir to a long tradition of writing about Scotland as a place of evil. Scotland is, of course, north of England, and northern Scotland in particular was represented through reference to Hell in medieval and early modern literature, including Shakespeare's northern Hell, complete with witches, in *Macbeth* (1623). As late as 1814, Sir Walter Scott made use of the trope in *Waverley*, employing Miltonic allusion to a northern hell to evoke a Satanic rebelliousness in the (northern) Highlands during the Jacobite Rebellion of 1745. Most recently, J.K. Rowling has set the *Harry Potter* series' Hogwarts School of Witchcraft and Wizardry in the Scottish Highlands. See Arthur H. Williamson, "Scots, Indians, and Empire: The Scottish Politics of Civilization 1519-1609," *Past and Present* 150 (February 1996): 46-56.

10) Thessalonians 5:1-3:

   1. But of the times and the seasons, brethren, ye have no need that I write unto you.

   2. For yourselves know perfectly that the day of the Lord so cometh as a thief in the night.

   3. For when they shall say, Peace and safety; then sudden destruction cometh upon them, as travail upon a woman with child; and they shall not escape.

11) See, for example, illustrations to the poem by Bessie Collins Pease in *A Child's Garden of Verses* (New York: Dodge, 1905), 55-61, in particular 59, in which a little girl's enormous hairbow casts one of the "crooked ... wicked" shadows.

12) I have examined copies of books in numerous libraries in Canada, the United States, and the United Kingdom, including the Toronto Public Library's Osborne Collection of Early Children's Books, UCLA's Charles E. Young Library, the British Library, and the Bodleian Library.

13) Respectively, C.S. Lewis, *The Lion, the Witch and the Wardrobe*; J.K. Rowling, the *Harry Potter* series; and Philip Pullman, *The Golden Compass*. See Cory Sampson's essay in this collection, "Pullman and Imperialism: Navigating the Geographic Imagination in *The Golden Compass*."

3) Envisioning Ireland

Landscape and Longing in Children's Literature

❦ MARGOT HILLEL

The savage [*sic*] loves his native shore,
Though rude the soil, and chill the air;
Then well may Erin's sons adore
Their isle which nature formed so fair.
What flood reflects a shore so sweet
As Shannon great, or pastoral Bann? (Orr in *The Child's Irish Songbook*)

═════════

This poem, titled "The Irishman," which extols both the Irish landscape and its people, is by James Orr, a late-eighteenth- and early-nineteenth-century Irish poet. It is reproduced in a collection called *The Child's Irish Songbook*, published sometime later in the nineteenth century.[1] Orr, who was sometimes called the Bard of Ballycarry, was not really a poet for children, and the inclusion of his work in such a book reinforces the sometimes nationalistic nature of literature given to children. This verse encapsulates a number of the discourses that appear in some of the literature for children I will discuss here. In this short piece we have discourses of patriotism, of love of the land, of them and us, and of identification with the landscape. The rhetorical question simply reinforces the poet's view of the unsurpassed beauty of Ireland.

An examination of how children grow up having particular ideas reinforced through their literature makes an important contribution to understanding the political scene of Ireland. Literature can be used to manipulate viewpoints in readers. Indeed, literature is infused, although sometimes quite subtly, with the ideologies of the authors, and an analysis of literature over a wide historic range elucidates the persistence of the tropes. The examination of Irish children's literature covered in this chapter—because of space constraints, a small selection—helps us understand Irish attitudes towards the past, present, and future geography of their land. Importantly, many of the constructions are found on both sides of the religious divide, with Catholic and Protestant writers using literature to construct a form of geographic and sometimes nostalgic patriotism.

In the foreword to Nancy Watson's *The Politics and Poetics of Irish Children's Literature,* Declan Kiberd writes that "Watson is not exaggerating when she suggests that the re-examination by our younger contemporary historians of painful themes in Irish history, from the Great Famine to the Easter rebellion, was rehearsed first of all in children's literature" (vii). It is also true to say that some of these themes were explored in children's books in the nineteenth and early twentieth centuries, depictions set against a background of the geography of Ireland. Ken Taylor has argued that "one of our deepest needs is for a sense of identity and belonging and a common denominator in this is human attachment to landscape and how we find identity in landscape and place" (Taylor 1). Ireland was frequently depicted in literature as a land of great natural beauty, a beauty unparalleled elsewhere and a beauty much missed by the Irish diaspora scattered throughout the world, especially after the Great Famine. Reminiscence and narrative are thus intertwined, a discourse, as Muller-Funk puts it, of memory, reminding, and forgetting—and indeed, of imagining (Muller-Funk 207–8).

Terence Brown, discussing nineteenth-century adult Irish literature, writes that "the sense of depth of history so essential to national *amour propre* meant drawing on Gaelic antiquity" (Brown 23). Reverence for a Gaelic past was also evident in some nineteenth-century children's literature. Author Marie Bayne, for example, discussing in her book *Tales of Ireland for Irish Children* a "golden age" of Ireland, tells her readers that the stories of great heroes from the past make clear that

> their country was very dear to them. We see too, that though, in many ways, the land was unlike what it is now, yet the same soft air blew over it, and the same kind waters washed its shores. The same red berries grew on the quicken [rowan] tree, the same white blossom decked the thorn, and the same little shamrock threaded its way among the grass and wove a green mantle for Erin. (Bayne 6)

Asking the children who read the stories of heroes and of the imagined space in which they lived "to fix in their minds and hearts a picture of the ... Island of the Gael," she writes, "let us always remember that what a nation has been it may be again. Some day, we trust, the Golden Age will dawn once more in Ireland; and all Irish children who live nobly and well are doing their part to bring that day about" (Bayne 84).[2] The discourses of heroism, story, and patriotism and indeed notions of "Ireland" are thus interlinked in her work.

"Let Erin remember the days of old," wrote Thomas Moore in his poem of the same name, thus echoing the sentiments expressed by Bayne. The fisherman walking in the evening on the banks of Lough Neagh, the largest lake in what is now Northern Ireland, imagines Moore,

> sees the round towers of other days
> In the wave beneath him shining:
> Thus shall memory often, in dreams sublime,
> Catch a glimpse of the days that are over;
> Thus, sighing, look through the waves of time,
> For the long-faded glories they cover ... (Moore in Mulholland 1891)

Once again, past and present are intertwined and the imagery used by the poet functions as a succinct and evocative method of linking landscape and love of Ireland. Describing the country as an "emerald gem," Moore encodes in that phrase both the vibrancy of the colour of the fields of rural Ireland and its perceived value, a jewel of the Western world. Memories here are depicted as a kind of folk memory, with the glorious past forming a kind of present. The "round towers" suggest the medieval past when such towers were built and that remain, for example, in Glendalough near Dublin.

There is something of a blurring between the intended audiences of some of these writings. Moore was a prolific and much-loved poet and songwriter of the nineteenth century, but he wrote largely for adults, not young readers. Some of his work, however, including this one, was included in Rosa Mulholland's edited book *Gems for the Young from Favourite Poets,* published in 1891.[3]

The importance of traditional stories is praised in many books, such as Standish O'Grady's *In the Gates of the North*, and the revival and translation of ancient stories allowed "them to exist once more in the timeless spirituality of the nation's continuous being," writes Terence Brown (23). O'Grady himself was very important in the Irish Literary Revival, a movement that encouraged nationalism and a revival of interest in Ireland's Gaelic heritage and that was later embraced by Anglo-Irish writers such as W.B. Yeats and Lady Gregory (Kiberd 1995, 107). Lady Gregory published a version of the Cuchulain story, *Cuchulain of Muitehemne*, in 1902. Furthermore, these stories were important to both Protestant and Catholic Irish. Padraic Frehan discusses the importance of these stories in Catholic schools (Frehan 72–73).

Cuchulain and his tales of valour appear in stories for children and adults, and the story that he held the "gates of the North against a host of invaders" is a commonly told one. According to ancient Irish legend, when Cuchulain was wounded in battle, he tied himself to a pillar so that he could face his enemies when he died. It was only when a raven landed on his shoulder that his enemies dared approach him. The past lives on in the present with this story, for there is a sculpture in honour of the 1916 rising in the General Post Office in O'Connell Street in Dublin that depicts the death of Cuchulain.

The beauty of the Irish landscape, so dear to these heroes and to those coming after them, was extolled in prose and poetry, which often constructed an iconography of an idealized landscape. As D.B. Mah points out with reference to Australia, "landscape imagery has been tied to the mythology of national identity for a variety of reasons, [including] the psychology of the people's relation to the land, their possession of it and a sense of belonging" (Mah 19). Such reasons can also be seen to underlie the discourse of the Irish landscape found in some of the books for young readers that I want to consider here. The books chosen particularly demonstrate the tropes discerned in the readings. Space has required that I make a choice, as many more books could have been chosen that would also demonstrate the same attitudes towards Irish geography as those discussed.

Nowhere, the readers of many of these books were told, was a country as beautiful as their native land:

> By Killarney's lakes and fells,
> Em'rald isles and winding bays,
> Mountain paths, and woodland dells
> Mem'ry ever fondly strays;
> Bounteous nature loves all lands;
> Beauty wanders ev'rywhere
> Footprints leaves on many strands,
> But her home is surely there. (Falconer in *Child's Irish Song Book*)

Another author, Kathleen Tynan, concurs with this description of Killarney: "The beauty of Wicklow," she writes, "[with] its wonderful woods, its deep glens, its placid waters, its glorious mountains, is only less than the beauty of Killarney, which is an earthly paradise" (Tynan 38). Tynan's book, titled rather sentimentally *Peeps at Many Lands—Ireland*, is a kind of travelogue for children, divided into sections: Arrival, Dublin, The Irish Country, The Irish People, South of Dublin, The North, Cork and Thereabouts, Galway, Donegal of the Stranger, Irish Traits, and Way. The book extols the beauties of the Irish landscape; the author is less enthusiastic about man-made structures, regarding Irish farmhouses as "seldom even pretty" and Irish villages as marring the landscape (Tynan 26). The picturesque joy of the landscape, however, begins with the arrival on the "lovely coastline" of Ireland, where the air is "soft as a caress" (Tynan 4). There is, in descriptions like this, a recurring trope of the gentle, nurturing landscape. The land itself becomes almost a character in the book, constructed as a feminized ideal, within which the human characters develop their sense of space and place. The geographical narrative contributes to the reader's understanding of Irish culture, self-identity, and the way the past has contributed to the present.

There is a strong nexus between geography and patriotism in such books, and Tynan makes it quite explicit: "no good Irishman will concede England a beauty, natural or otherwise, which Ireland does not possess" (Tynan 26). Somewhat contentiously, however, in terms of Irish politics, while describing North-East Ulster, she claims there is "nothing Irish about it except the country itself" (46). This might be symptomatic of what Colette Epple calls Tynan's tendency to walk "a thin line between criticising Britain for its societal prejudices and offending her British audience." Epple argues that Tynan needed to keep her British audience happy because the money she earned through her publications supplemented her husband's modest income as a magistrate (Epple 39). (Tynan was an Irishwoman who married an English lawyer, subsequently living for most of her life in England.)

Rosa Mulholland, another prolific Irish woman writer, published more than forty novels and also poetry, children's stories, and some history and biography. It is sometimes difficult to determine the age of intended readership of earlier children's books as notions of children's capabilities in reading change, and the interests that children have may also change. Some of the books discussed from the nineteenth century might well have been for a slightly younger audience than that which we would now designate as young adult. On the other hand, as Heather Scutter points out, the boundaries between adult fiction and that for young adults were also blurred in the nineteenth century, and she instances the "boys' own adventure" genre, which "implied and encompassed readers all the way from young teenagers to old (but boyish of course) adults" (Scutter 3). Furthermore, Declan Kiberd describes, "the vastly increased amounts of knowledge to be mastered at school also had the effect of extending the years of childhood beyond primary level and well into the teens, so that by the end of the nineteenth century it was lasting up until the age of seventeen or eighteen, when formal schooling came to an end" (Kiberd 2004, 55). This too, will have had an effect on the sorts of material given to young readers.

Mulholland's books often included a young woman who might well be the focalizer of the story and whose concerns and feelings were thus reflected in the narrative; this appealed to older girls, the likely audience of her books. Mulholland was Roman Catholic, and her novels took as their subject the Catholic upper classes, a new departure in Irish fiction at that time. Her later work grew more nationalist in tone and she never shirked tackling major social concerns, such as emigration, illegitimacy, and eviction. Her book *The Return of Mary O'Murrough,* published in 1907, is one such that tackles the issue of emigration. A review of the book in the *Irish Monthly* in December 1907 described it as a book in which the changing fortunes of Mary O'Murrough "form a deeply interesting and instructive drama which is worked out with lifelike reality" and that is "the very latest in a wonderful series of stories, chiefly of Irish life, that we owe to the genius and patriotism of Lady Gilbert" (*Irish Monthly* 701).

Mulholland's patriotism is manifested in part in this book by the geographic narrative, as she extolls the beauties of her homeland. The book tells the story of Mary O'Murrough, who has had to emigrate to America after losing her house. The threat of impending migration hangs over her village. The description of the area in which she lives is detailed and explicit, constructing for the reader a clear picture of the sorrow of parting with the very land itself:

> The priest's walk home was through what might be called the most beautiful bit of Ireland, if other visions did not rise before one to dispute that statement. Nowhere is there a more continuous stretch of absolute loveliness and striking grandeur, made up of mountain and valley, lake and river, and scattered woodland. That mingled tenderness and sternness of expression which is the great charm of Irish scenery is hardly more impressive anywhere than here; and, for colouring, the grave greys and violets, the solemn purples deepening to black, of the mountain crags and sides, the fantastic fringes of orange and tawny brown, the sprightly greens of the fields and pastures that bring their golden irises and star-daisies to the wayfarer's feet, all these have a peculiar brilliance and softness in the dreamy and luminous Southern atmosphere. (Mulholland, *Return*, 68)

In an idealized construction, Mulholland has her character Mary return to this beautiful landscape a changed person, marked by her life of toil in America. Her dislocation from Ireland is written on her body, and she is not recognized by the other villagers. The romanticized space of the constructed geography in the novel is, however, transformative. In a representation of femininity that ignores the hard labour needed to win a living from the land, the priest expresses his hope that Mary might be somewhat restored to "the flower-aspect so sweetly important to a woman, the comely look which the eye of another delighteth to look upon" (167). His wish is granted, and Mary is "much improved," with her "sweet, pale face" signifying her once more as a true daughter of the loveliness of Ireland (224).

Inhabitants of this beautiful land are not always human; they can also be what K.M. Briggs calls the "personnel of fairyland." Oscar

Wilde's mother, Lady Wilde, a respected folklorist of the nineteenth century, describes, in her 1888 collection, the importance at that time of the continuing oral tradition in Ireland, with its amalgam of the stories of saints and fairies. There was, she claimed, an ongoing and "instinctive belief" in the existence of "unseen agencies that influence all human life." Such a collection would not be possible a few years later, she writes, because of widespread emigration "and, in the vast working-world of America, with all the new influences of light and progress, the young generation, though still loving the land of their fathers, will scarcely find leisure to dream over the fairy-haunted hills and lakes and raths of ancient Ireland" (Briggs xii). Writing in the nineteenth century, Mr and Mrs S.C. Hall, described by William Trevor as "the most devoted of the Victorians who wrote about Ireland" (86), said that "every lake, mountain, ruin of church or castle, rath and boreen, has its legendary tale; the Fairies people every wild spot; the Banshee is the follower of every old family; Phookas and Cluricannes are—if not to be seen, to be heard of, in every solitary glen" (quoted in Trevor 92).

These are not the jokey leprechauns of the tourist brochures of today, signified by green top hats and a shamrock; in the stories gathered by Lady Wilde, Crofton Croker, W.B. Yeats, and others, they are present for good or ill and it is a foolish person who discounts their occurrence in the landscape. Lady Wilde's collection contains a number of stories of farmers who decide to build a house or barn on the greenest spot of land—a spot that is a fairy rath and against which their neighbours warn them. The fairies always take revenge, killing cows, making the farmer himself ill, or even, in the worst cases, causing a child to sicken and die (Wilde 46). Pooka, the Irish Puck, haunts wild places, and fairy hills contain such inhabitants as the Dana O'Shee, who on May eve open the door from fairyland to the mortal world (Briggs 207, 199). The love of the land and a strong sense of place was shared by its human and its fairy inhabitants.

This was God's own country, young readers were told; little wonder then that emigrant literature constructed this idealized landscape as part of a discourse of "home," "homeland," and memory. Whatever this imagined ideal, for some of the nineteenth century Ireland was, of course, also a place of poverty and misery for many people—the nurturing land seemed to have turned on its occupants as the potato crops repeatedly failed (Grianna 153). The importance of that crop is suggested in the chorus of "The Potato Digger's Song." That song emphasizes the importance of the potato in the Irish diet and the welfare of the Irish family where there was "many a mouth to fill." A young man who is courting is told he must get on with his harvest as "love in this country lives mostly still on potatoes":

Work hand and foot,
Work spade and hand,
Work spade and hand
Through the brown, dry mould.
The blessed fruit
That grows at the root
Is the real gold
Of Ireland. (Irwin in Mulholland, *Gems for the Young*)

The use of the word *blessed* signifies both its God-given status and its importance to the farmers and their families. In addition, the song illustrates a pre-industrial age, with the harvesting being done by hand with families digging together and the relationship between man and land a more direct one.

Not surprisingly, given the importance of the potato and the devastation caused by the failure of the crops in the mid-nineteenth century, the Famine figures prominently in Irish consciousness and history. Past and present again meet in modern fiction that looks back to this era. The political overtones of the Famine loomed large in the mind of one child writing at the time of the 150th anniversary of the Famine:

I was a sailor on the Fortuna,
An emigrant ship ...
The Crew were harsh and bitter ...
At home the English landlords
Would eat as much as they liked
While we were starving outside. (Eve Gleeson in Coughlan 21)[4]

The Famine was the single biggest reason for Irish emigration: "Altogether, about a million people in Ireland are reliably estimated to have died of starvation and epidemic disease between 1846 and 1851, and some two million emigrated in a period of a little more than a decade (1845–55)" (Donnelly 1). This in itself changed the geography of Ireland, leaving villages abandoned and agricultural practices changing:

Land holdings became larger, as the tendency to subdivide the family farm declined. From now on, the farm was given to one son and the others often had little choice but to emigrate. The Famine also changed centuries-old agricultural practices, hastening the end of the division of family estates into tiny lots capable of sustaining life only with a potato crop. The famine affected the poorest classes – the cottiers and labourers – most of all, the cottier class being almost wiped out.[5]

Thus the landscape the emigrants left behind had already changed, and that which they remembered so fondly "in exile" was often a romanticized one.

Rosa Mulholland, in *The Return of Mary O'Murrough*, has a character tell of an area of Killarney, that imagined paradise we saw described in those earlier quotations, where five hundred families had been put off the land and the land turned over to cattle (104). The evictions too, were sometimes captured in poetry as with William Allingham's "Tenants at Will," quoted by Willam Trevor:

> Threescore well-arm'd police pursue their way;
> Each tall and bearded man a rifle swings,
> And under each greatcoat a bayonet clings … (Trevor 103)

One survivor of the Famine described mothers lying "in their beds with the children beside them … they used to lie there until one after another they died of hunger," and others going "out in the fields on all fours and eat[ing]their fill of grass and weeds" (Grianna 154).

Migrants left such scenes of misery for countries like Canada, America, and Australia, which they imagined—sometimes erroneously (Griffin 186)— as places of potential plenty and prosperity:

> I'm biddin' you a long farewell,
>   My Mary—kind and true!
> But I'll not forget you, darling!
>   In the land I'm goin' to;
> They say there's bread and work for all,
>   And the sun shines always there—
> But I'll not forget old Ireland,
>   Were it fifty times as fair! (Helen Selina, Lady Dufferin,
> in Mulholland *Gems for the Young*)

Importantly too, the poem suggests an ongoing commitment to Ireland, a significant characteristic of the Irish that set them apart from almost all other ethnic groups of immigrants to America, as William Griffin notes (Griffin 186).

Another poem laments the loss of friends and family, recognizing the distance between Australia and Ireland as half a world away:

Och, I wish that Maurice and Mary dear
Were singing beside us this soft day!
Of course they're far better off than here;
But whether they're happier who can say?
I've heard when it's morn with us, 'tis night
With them on the far Australian shore;—
Well, Heaven be about them with visions bright,
And send them childer and money galore ...
(T.C Irwin, "The Potato Digger's Song" in Mulholland, *Gems for the Young*)[6]

The question in the fourth line is a rhetorical one, with an implied answer suggesting that migrants are unlikely to be happier and that material wealth is no substitute for home. The use of the description a "soft day" (a particularly Irish description for a day with a gentle drizzle or mist) suggests a contrast with a harsher land far away.

Children's books, poetry, and song constructed specific kinds of attitudes towards the landscape. It was this love of country constructed, in part, through literature that was reflected in the emigrant stories with their strong longing for the homeland. Nostalgia for the land left behind was a notable discourse in many emigrant tales, the geography imagined through the lens of homesickness. These stories became "a narrative of exile and loss" (Higgins and Kiberd 11). In *An Irish Cousin,* by Catherine MacSorley, for example, Gerald, an orphan, lives with his grandfather and devoted Irish "nanny" in a house in a remote part of Ireland. Eventually his grandfather dies, having asked Gerald's aunt and uncle to be guardians. His will cannot be found, however, and the house reverts to a nephew in Australia. With his aunt's family, Gerald goes to England, where he is dreadfully homesick. And once again this longing is encoded in a description of landscape:

So July drew to a close ... and Gerald knew that very soon the bog would be putting on its August glory of purple heather and golden furze, and a longing he could not have put into words came over him, to see it again ... to feel the keen air blow across it; to smell the peat smoke from the cottages, and the myrtle and heather from the bog, all mingled into a fresh, wild sweetness. (MacSorley, *Irish Cousin*, 115)

The description here invokes the sense of smell, of sight, and even of touch to construct place and the nostalgia for all that is associated with that place. Ireland is colour and light, space and freedom, unlike the confined and constrained space of England.

So strong is the attachment to Ireland that when Bridget, the Irish servant returns with the family to their homeland, she says it is like "'life from the dead' ... as she walked down the avenue, and watched the sun setting behind the hills, and the turf cribs coming back across the bog" (118). Particular scents can be strongly evocative of place and conjure up memories of times of happiness. Prior to leaving for England, Bridget had bought some turf from a turf-seller in Dublin so that she could take it with her to England and have the feel and smell of home with her when she was far away (96). Yet even this was no substitute for the reality of life in Ireland, so strongly evoked by her exclamation of her view of the land bringing her back from the dead. The Irish landscape is here constructed as being itself life-giving.

Mulholland, in *Mary O'Murrough*, has one of the intending migrants illustrate this discourse of loss; she expresses her misery "at the thought of not seeing the fields, the turf-stacks, crossing the threshold of her mother's cabin and hearing the lark sing perhaps ever again" (88). James Monaghan's poem "Song of an Exile" is a litany of things past, of images of home, of the pain and joy of such memories. It appeared in a journal called *Young Ireland*. This weekly journal, which cost a penny, called itself "An Irish Magazine of Entertainment and Instruction." It contained poetry, stories, games, and serials. Monaghan's poem reads, in part:

Oh! Oft I wandered long ago
The Deel's green banks along
To watch its placid waters flow
And hear the wild birds' song ...

Ah me! Long years have passed away
Since when in dear Rockview

I loitered many a fleeting day
With playmates fond and true ...

I often think how sad 'twould be
If here by Hudson's wave
I should by cruel fate's decree
Descend into the grave ... (Monaghan 73)

The Deel, a river in County Mayo in northeastern Ireland, functions in this imagined pastoral depiction as a metonym of "home," emblematic of Ireland as a whole. The first-person focalization enhances the construction of exile and nostalgia.

Perhaps the ultimate loss for the "exile" is dying far from home, away from one's friends. The need for one's burial to be part of the geographical space of homeland is an important one for many people—expressed strongly in this poem. The reality that many of those who migrated were unlikely to return home is acknowledged by one poet in the *Child's Irish Song Book*:

But deep in Canadian woods we've met,
And we never may see again
The dear old isle where are hearts are set,
And our first fond hopes remain! (Sullivan in *The Child's Irish Songbook*)

Like Lady Dufferin's imagined exile in the poem quoted earlier, this one too is strongly nostalgic for home and makes it clear that "home is where the heart is." The expression "dear old isle" encodes a loving familiarity for the place where the exiles were born and that, as such, is forever the place of fondest memories.

A book by "M.E.T." titled *Exiled from Erin: A Story of Irish Peasant Life* has been described as a "poignant story of misery and heartache" (Loeber and Loeber 386). It is the story of an Irish peasant family, especially of two of the brothers. Fergus wants to marry Nora, his first cousin, but it is forbidden by the church. Nora and her family immigrate to America, where she eventually marries. Fergus and his brother Art have, in the meantime, left home to find work in England, where there is a great deal of unrest and wages are not paid properly. Fergus and Art hear that Nora has married and travel to America to make sure she is happy. Art goes to work in a silver mine and, when Fergus follows, he finds people fleeing from an "Indian fight," among them Art. Fergus nurses his wounded brother and dreams of their return home. Art tells Fergus he is blind and will never again see home; for Art the beloved geography of Ireland can be revisited only in memory and imagination. While in Canada, they meet an Irish family called Mooreland who display more compassion and righteousness, we are told, than the English the brothers meet. Unfortunately the family all die and are unable to return to Erin, even to be buried there. On this the author states: "Thus does the dust of our faithful Irish people go to swell the soil of a hard and unsympathetic land" (211). Encoded in such a statement is a depiction of a softer land in Ireland that somewhat belies the need for people to emigrate but that emphasizes the trope of landscape as denoting home, with all that implies. Furthermore, it looks towards the future in which the "hard and unsympathetic land" has been both literally and figuratively enriched by the bodies of the Irish.

In their article "Culture and Exile: The Global Irish," Michael Higgins and Declan Kiberd comment on the "longing for a Sligo sod felt in a London street by the young [W.B.] Yeats" (13). Yeats himself described how he would "remember Sligo with tears" when he went to London, and according to Declan Kiberd, "Sligo became a place sacred to the youth ... It was some old race instinct [Yeats] recalled" (Kiberd 1995, 102). Higgins and Kiberd further argue that "before emigrating, a person might be known as a Kerry woman or a Wicklow man. In the precincts of London or Boston, however[,] such persons learn[ed] what it means to be Irish" (9). Many migrant stories written for children also established a kind of iconography of the Ireland the migrants had left behind, which remained a fixed entity, emblematic of "their" Ireland. Such a viewpoint constructs an imaginative geography, which, unlike the reality, does not really change and remains forever idyllic.

In most of these stories it is the countryside that is lamented, not the towns. This reflects the fact that most of the migrants came from the countryside. The "threat" of departure hung heavily over villages, as Mulholland writes in her novel *The Return of Mary O'Murrough*: "The coming autumn movement of a wholesale emigration was already casting its shadow before it [and in each] little dwelling [that should have been a] very home and secure haven of happiness ... there reigned the woe of impending separation" (69).

Mrs. J. Sadlier was a prolific mid-nineteenth-century author of books such as *Alice Riordan, or the Blind Man's Daughter; New Lights, or Life in Galway; The Blakes and Flanagans; a Tale Illustrative of Irish Life in America; The Confederate Chieftains, a Tale of the Irish Rebellion of 1641; Bessy Conway, or the Irish Girl in America;* and *The Daughter of Tyrconnell.*[7] Many of these, as can be seen from the titles, were migrant tales. She herself was a migrant, from Ireland to North America. Her books are often didactic, valorizing "good old-fashioned Catholics" who support a particular pride in Ireland. In *Bessy Conway,* Sadlier warns of the dangers of America, particularly to "the Irish Catholic girls who earn a precarious living at service in America," for there is "no class more exposed to evil influences." The book follows the generic "home-away-home" narrative identified by Perry Nodelman and Mavis Reimer (quoted in Reimer and Bradford 200): Bessy resists all temptations, stays true to the faith, and returns home just in time to prevent her family from being turned out by the bailiffs. While in America, she meets numerous Irish people, many of whom long to return. One is Paul, described as a "dwarf" who sings a song of mingled joy and sorrow:

> More dear than the roses by all Italy yields,
> Are the red-breasted daisies that spangled the fields,
> The shamrock, the hawthorn, the white-blossomed sloe,
> Oh! My heart's in old Ireland wherever I go. (174)

Poetry of this kind functioned as a readily accessible narrative form, one that established a set of images of Ireland that lexically signified home. Landscape, Margaret Drabble has written, can form "a living link between what we were and what we have become" (quoted in Taylor 2). This construction is embodied in Paul's poem, in which the first-person narrative emphasizes the "then and now" of his sense of place. A particular vision of Ireland is encoded in the imagery. It is to this land of the well-remembered flowers, which may not be as exotic as those of other countries, but which have the charm and comfort of familiarity, that Paul wishes to return. Once there, he vows:

> Then I sigh and I vow that e'er I get home,
> No more from my dear little cottage I'll roam,
> The harp shall resound and the goblet shall flow,
> For my heart's in old Ireland wherever I go. (174)

The countryside is missed even by those who are forced by circumstances to move to a city even in Ireland itself. In MacSorley's early-twentieth-century *Nora: An Irish Story*, the impoverished gentry family must move to Dublin, where the father dies of a broken heart, "away from the hills and the bog, and the fresh air and the wide, open spaces" (96). For his wife, "like all genuine country people ... living in town was like being in prison" (109). His daughter speaks lyrically about the bogs as "such wide open space, and rich dark colours [and in August] all crimson with heather and shaded with all kinds of beautiful tints and colours" (20).

The city becomes a metaphorical prison, and the only way to regain freedom is to escape—that is, to return to the countryside. The image of the bogs thus functions as a metonym for liberty and for the natural world, often constructed as the proper place for children and young people to grow up, suggesting an imagined ideal childhood on the land. Nora has been dislocated from her country, and it is the memories of that landscape and her people that constitute her real home. The focalizing of the description through the character of the daughter allows the reader to understand—and indeed perhaps share—the nostalgia for place that she expresses so strongly. This is a geographical construction of "home," and as Patricia Holland has argued, home is important for children as a place of nurturing and for performing childhood (Holland 57).

Ken Taylor has argued: "Landscape is not simply what we see, but a way of seeing: we see it with our eye but interpret it with our mind and ascribe values to landscape for intangible—spiritual—reasons" (Taylor 1). It is an important part of how we construct identity. "Seldom are the desires of an Irish heart gratified in a foreign land," muses the author of *Exiled from Erin* (M.E.T. 201). As we have seen, one of those desires is for visions of the geographical space called Ireland. Many of the authors I have discussed here were inspired by such visions, and from their imaginations came descriptions of characters shaped by the landscape around them. In these authors' works, Ireland itself becomes a character presented to us with love, hope, nostalgia, longing, and patriotism.

Notes

1) Despite the name, the book was a collection of song lyrics without the music. The term "savage" is clearly unacceptable to a modern audience. I have included this quotation because it exemplifies the sort of material given to young readers, extolling the wonders of Irish geography and culture and a justifiable pride in both.

2) Padraic Ferehan notes that Bayne's writings were included in Irish schoolbooks for two decades (296).

3) Interestingly, in *The Child's Irish Songbook*, there is a note at the bottom of the back page: "In this collection there are none of Moore's Melodies, as it is taken for granted that these are or ought to be already well known in every Irish family."

4) The Principal of the one of the schools involved in this project wrote: "In this, the 150th anniversary year of the famine, our young students [average age 10] have researched and written about the factors that caused this tragic happening. They have become conscious of the famine as a struggle between rich and poor, between farmer and labourer, between corrupt administrators and their victims. They have become aware of the scale of the tragedy and the terrible spectre of evictions and mass emigration which have consequences for Irish society even in these days" (Coughlan 41).

5) In Ireland, this was a peasant renting and farming a smallholding under cottier tenure (the land was let annually in small portions directly to the labourers). See Lurgan Ancestry, "The Great Famine," http://www.lurganancestry.com/famine.htm. See also *The Oxford English Dictionary*.

6) This poem also appeared in the *South Australian Weekly Chronicle* on 4 May 1867. This is an indication of the widespread nature of some of this material and of the Irish diaspora who read it.

7) Although the intended audience for many of the books discussed here may not seem to fit into the category of children, the term can be extended to the upper end of that range, to what we would now refer to as Young Adult novels. The protagonists in many of these books, often young women, would be in their late teens, of an age of the readership. In the case of Mrs. Sadlier's books, for example, this made the "lessons" being taught particularly apposite. At Trinity College Dublin, her books are housed in the Pollard Collection of Children's Books (see, for example, https://nccb.tcd.ie/catalog/6108x6407).

4) From Vanity to World's Fair

The Landscape of John Bunyan's Allegory in Frances Hodgson Burnett's *Two Little Pilgrims' Progress*

❧ SHANNON MURRAY

———

As Frances Hodgson Burnett's *Two Little Pilgrims' Progress* opens, the twin orphans, Meg and Robin, have been enduring what they consider to be a miserable life with their dour Aunt Matilda. That life changes when two things happen at once: Meg finds and reads a copy of John Bunyan's *The Pilgrim's Progress*, and Robin overhears two farmhands talking about the Columbian World Exposition. As they talk, the children superimpose Bunyan's Celestial City on the very worldly world's fair and make a plan: they will save up, boil some eggs, run away, and see the White City for themselves.

In many ways, the novel is pretty conventional fare: orphans, a journey, various trials and tribulations, and the discovery of a real—that is, wealthy—home at the end, all woven together over the framework of John Bunyan's *The Pilgrim's Progress*. What makes it remarkable is the purposeful erasure of the real landscape of late-nineteenth-century America—not just erasure, in fact, but a moral rejection of it—in favour of a newly created landscape formed from the combination of the imaginary one in Bunyan's allegory and the manufactured one of the World's Fair. The novel leaves us with three landscapes superimposed on one another or in competition: the rejected rural one, the constructed landscape of the fair itself, and the complex and perhaps accidentally ironic one adapted from Bunyan's allegory.

In a book constructed along the lines of *The Pilgrim's Progress*, one might expect a real interest in the journey itself, in how the little pilgrims work their way through the physical landscape and what they learn from those travels. A simple illustration on the title page of the first edition suggests that the illustrator, Reginald Birch, thought so: it is a picture of the two children, backs to the reader, facing a long road ahead. Stories constructed as pilgrimages obviously have a goal in mind, but almost inevitably, it is the journey that matters to the narrative, with less ink spilled once the goal is reached. In Bunyan's allegory, the point of the journey is the arrival at the Celestial City, but the point of the narrative is to show *how* Christian got to that destination and what he learned along the way. As a rule, a Yellow Brick Road will always matter more than an Emerald City. Even though Burnett's *Two Little Pilgrims' Progress* signals both in its title and in its frontispiece a pilgrimage and a journey narrative, the book itself subverts that expectation. This is a journey with a detailed starting point, a remarkably quick and easy journey, and then a loving and detailed exploration of the goal.

The starting point is a farm in rural Illinois. When their parents die, Meg and Robin are raised by their paternal aunt, Matilda. Hers, according to the narrator, is a narrow life that the children describe as "hideous and exasperating and sordid" (2). Readers would be forgiven if they had trouble discovering any evidence to support that harsh judgement. Matilda is no Aunt Petunia or Miss Minchin. She is not cruel or insulting, and she doles out neither physical nor psychological abuse; indeed, she encourages the children to have money of their own, so they enjoy some financial independence. Their lives are not ones of deprivation or active misery; they have what they need, at least physically. Here is what is wrong with her, according to the narrator:

> Mrs. Matilda Jenkins was a renowned female farmer of Illinois, and she was far too energetic a manager and business woman to have time to spend on children. She had an enormous farm and managed it with a

success and ability which made her celebrated in agricultural papers ...
She cared for nothing but crops and new threshing machines and fertil-
izers. (2-3)

There is a hint here that there might be something suspect about a successful *female* farmer; Meg calls her a man in woman's clothes, whereas
her beloved but unsuccessful father was more like a "woman in man's
clothes" (3). Mark Noonan in a recent article makes an argument for
Burnett as a "progressive feminist" and contends that her early work is misunderstood (368), but it is hard to miss the judgment levelled at Matilda
that she is inappropriately masculine and more interested in business
than in a happy home life. Matilda's real fault as a guardian is uninten-
tional intellectual and emotional neglect, but what she represents is made
to appear even worse. She is a poor role model for these children because
she is devoted to a life that neither builds nor imagines: she reaps and
sows. When Matilda eventually does travel to the fair, she is not impressed
by its beauty or by its signs of human promise: she comes to gather practical ideas from the agricultural exhibit to improve her own farm. The land
and life on it take on a sinister quality out of proportion to what actually
happens to the children there, though while in her care they see themselves
as living not like people but "like pigs who are comfortable" (Burnett *Two*,
27). Rather than being praised for success and industry, Matilda is presented as poor soil for the children to thrive in; what's more, her kind of
industry runs counter to what the novel considers "truly American." It is an
oddly industrial and anti-pastoral idea of the American Dream, one supported by the kind of wonders to be found at the Columbian Exposition.
When the children eventually do run away, they are able to escape unde-
tected partly because of the narrow focus of a farming community: "'They
are not thinking of us' said Robin. 'They are thinking about crops'" (76).

This world, then, is an interesting contrast to what Jane Suzanne Carroll identifies as one of the most prominent *topoi* of Western literature: the
green world. In her article on death and landscape in children's literature,
she argues that "the green topos, from its earliest inception in English
literature, is expressly associated with the cycle of life, death and renewal"
("Death," 75). Green spaces—woodlands, fields, open spaces, and farms—
offer locations that allow children in children's literature to escape and to
engage with the natural instead of the human world. Burnett's presentation of the land and particularly of the farm is remarkably anti-pastoral.

For these children, the natural landscape is the oppressive space. It is only in the city, and particularly in the highly artificial and temporary city of a World's Fair, that the children reach their full potential and have a chance for true happiness.

So if farmers and farming are socially and intellectually stifling, what is the alternative that Burnett is proposing? Here's an early hint, from her narrator's description of the boy, Robin:

> He was a young human being, born so full of energy and enterprise that the dull, prosaic emptiness of his life in Aunt Matilda's world had been more horrible to him than he has been old enough to realize ... The truth was that in this small, boyish body was imprisoned the force and ability which in manhood build great schemes and not only build but carry them out. In him was imprisoned one of the great business men, inventors, or political powers of the new century. (Burnett *Two*, 7)

Energy, enterprise, force, ability, great schemes: none of these, the book suggests, are possible on the farm. Rejecting their circumscribed life, the children quietly save and plan and set out for the World's Fair, assuring themselves that Aunt Matilda won't be at all unhappy when they have gone—and indeed, she barely notices.

Off they go to Chicago, and perhaps it is not surprising that the landscape Matilda represents—any natural or agricultural land that the train would have passed through—is virtually absent on their journey. This is not a Huck Finn experience of the river. The twins camp out once (no Crusoe-like detail here), hike four miles to the railway depot, and get on a waiting train. Their brand of roughing it is deciding not to book a sleeping car. In fact, a reviewer in the *Dial* criticized the novel precisely for presenting a pilgrimage without struggle: the children "meet no difficulties whatever. They have no fear. They never get lost or confused, and find only courtesy and kindness" (qtd. in Thwaite, *Waiting* 151). Once they are on the train, all sense of potential danger is gone, and the countryside they are rushing through blurs together. Rather than any details about the sweep and beauty of the American Midwest, we get this: "The country went hurrying past them, making curious sudden revelation and giving half-hints in its haste; prairie and field, farmhouse and wood and village all wore a strange, exciting, vanishing aspect" (Burnett *Two*, 79). This is ellipsis, not travelogue.

The landscape conveniently vanishes because Burnett and her child protagonists are not interested in what exists, what is already there; more interesting is what humans create, either through technology (great schemes) or through imagination (great stories). Those are the two strengths these children represent, in a tellingly gendered dichotomy: Robin invents the schemes and Meg the stories. Both forms of invention will be united in the fair itself. But before we follow them to the fair, a word about the genesis of the novel.

Hodgson Burnett, of course, is now best known for her children's books, *Little Lord Fauntleroy* (1886), *A Little Princess* (1905), and *The Secret Garden* (1911), but she published more than fifty novels, thirty-two of them before *Two Little Pilgrims' Progress*. That means that she was already a well-known and financially successful author, though as her biographers point out, she was acutely aware of the need to keep writing if money were to keep flowing. Her biographer Gretchen Gerzina writes that at the time, she was "one of the world's most popular living women writers" (166), so she was one of those invited to contribute her works to the Woman's Building library at the World's Fair. She was less than pleased with the invitation: she called it "one of those endless demands that one should send some of oneself to some Womans Department of Something in the Worlds Fair [*sic*]. I have grown so tired of Woman with a capital W though I suppose it is rank heresy to say so ... Nevertheless if every body is sending books I must send mine" (qtd. in Gerzina 166).[1] At about the same time, she was invited by Eugene Field to submit a story for a collection that was to celebrate the fair's opening. Her idea was to write a revised version of *The Pilgrim's Progress*, but as Gerzina suggests, the story grew out of proportion to the request (and the proposed collection was never published). The now novel-length work was published on its own two years later.

Burnett's biographer Ann Thwaite argues that "Frances must, at [that] point, have gone to the Chicago Fair herself, for her *Two Little Pilgrims' Progress* could surely not have been written otherwise" (*Waiting* 149). While it is possible that she did go, I do not think it a necessary inference. In the same year as the fair, a large variety of souvenir guidebooks—some very large indeed, with hundreds of photographs, maps, and charts—were published, with names like *The World's Fair, being a Pictorial History of The Columbian Exposition, Containing the Complete History of the World-Renowned Exposition at Chicago; Captivating Description of the Magnificent Exhibits, such as Works of Art, Textile Fabrics, Machinery, Natural Products, the Latest*

Inventions, Discoveries, Etc. Etc., with A Description of Chicago, its Wonderful Buildings, Parks, Etc.[2] The degree of detail about the fair in her novel could easily have been gleaned from one or another of those guidebooks. She may or may not have attended what was one of the most extraordinary and talked about events in America in the nineteenth century.[3] And while the fact that the book began as a commission might seem to undercut her sincerity in praising the fair—if she becomes more public relations officer than artist—that need not colour our reading of her work either; whether or not she was paid to do it, she could well have been taken with the extraordinary marvel in Chicago. Everyone was.

That fair had two main purposes. The first was to celebrate the 400th anniversary of Columbus's arrival in North America, hence its official name: the World's Columbian Exposition. But the other, more nationalistic purpose was to establish Chicago and the United States as industrial, cultural, and technological leaders in the world. The Paris Exposition of a few years earlier—the one that gave us the Eiffel Tower—had both shown off French dominance and shown up the disorganized and, by comparison, puny American efforts in their national pavilion (Larsen 14-15). This is the fair that saw the first Ferris wheel (with George Ferris's train-sized cars), the first presentations of shredded wheat, Quaker Oats, Cracker Jack, Juicy Fruit gum, Pabst Blue Ribbon beer, and Aunt Jemima's pancake mix (Larsen 247). This fair also, much to Thomas Edison's chagrin, chose AC over DC power, and, as Larsen points out, "helped change the future of electricity" (131).

The grounds of the fair were extensive, and Daniel Burnham's plans were ambitious. "Make no little plans," he wrote. "They have no magic to stir one's blood" (xii). Central canals, lakes, and lagoons, complete with boardwalks and bridges, ran through the grounds, and full-scale domed buildings, all painted a gleaming white, filled the parcel of land on the shore of Lake Michigan. Once opened, it must have been exhausting to work one's way through: 199.7 acres in total, 30.5 acres of exhibits, and enough food outlets to feed 60,000 people at once (Smith, Vendl, and Vendl 29-30). This Fair was four times the acreage of any previous fair, and it boasted twice the exhibition space within the buildings (19). Its intention was to overwhelm the visitor with the scope, depth, and variety of human achievements, but especially of American achievements. As the official guide to the fair proclaimed, this fair was "[primarily] to illustrate American progress," though "the United States appropriately extends the hand of fellowship and hospitality to all other nations already represented on her soil" (qtd. in Smith, *Colorado* 19). So Chicago's fair had a very strongly nationalistic purpose, and, I'll argue, so does Burnett's novel. Of course, like all such expositions, the fair was meant to show where human beings stood: where is technology, where is science, where is industry at that moment? What does the present, but more importantly, what might the future look like?

No wonder, then, that it is that World's Fair that Robin and Meg long for. Here is the language the narrator uses to describe it:

> And this wonder working in the world beyond them—this huge, beautiful marvel, planned by the human brain and carried out by mere human hands; this great thing with which all the world seemed to be throbbing, and which seemed to set no limit to the power of human wills and minds—this filled them with a passion and restlessness and yearning greater than they had ever known before. (26)

Stirring words. The children react to this terrestrial City Beautiful as if it were the home they have never known, just as Bunyan's Christian responds to the idea of the Celestial City.[4] A connection to place is very strong in this novel, but that connection is with the urban landscape, not the rural, and Burnett uses her powers of description to celebrate the bustle of Chicago, not the peace of the land:

> The pavements were moving masses of human beings, the centres of the streets were pandemoniums of wagons and vans, street cars, hotel omnibuses, and carriages. The brilliant morning sunlight dazzled the children's eyes; the roar of wheels and the clamour of car bells, the clattering of horses' feet, of cries and shouts and passing voices, mingled in a volume of sound that deafened them. The great tidal wave of human life and work and pleasure almost took them off their feet. (84)

What she describes is excess of sensation in which much of what is experienced is controlled or created by humans, who themselves become part of the scenery. The children love it, but repeatedly, we're reminded that this is not a longing general to all the world's children: it is bred in the bones of right-thinking young Americans.

Precisely what soil produces such children in America and not, say, in the Old World is never specified. An understood, unquestioned chauvinism runs through the novel (and through *Little Lord Fauntleroy* as well) that makes innovation and industry natural to children of America. It is important to note that for Burnett, America was an adopted country. Born in Manchester in 1849, she moved with her family to Tennessee in 1865, and throughout her adult writing life lived by turns in England and America, with time in Italy as well (Bixler 1–7). Though *A Little Princess* and *The Secret Garden* were both set in England, in her earlier "beautiful child" story *Little Lord Fauntleroy* she demonstrates that same preference for the plain, fresh American child who can show the stifling English aristocrat a thing or two, by gum.

We get this contrast between complacency and American energy when Meg and Robin meet a kind German woman in a Chicago bakery. She is shocked at first to find that they have travelled to Chicago by themselves, then adds, "'but your 'Merican childrens is queer ones.'" Here is how the narrator describes her judgment of the children: "She had been a peasant in her own country, and had lived in a village among rosy, stout, and bucolic little Peters and Gretchens, who were not given to enterprise, and the American child was a revelation to her" (107). And if American children are generally separated from their European cousins by enterprise, Meg and Robin are more extraordinary still, with an "attraction all American children had not; they look so well able to take care of themselves, and had such good manners and no air of self-importance at all" (107). I think it is remarkable that in this novel the word "bucolic" is twice used as a pejorative, to denote a passive, complacent contrast to the best of America. Again, any association with the land, no matter how beautiful or pastoral, is a threat to the innovation and industry that the novel prizes.

Replacing that pastoral is the highly artificial landscape of the fair. Once the children have arrived at the gates of the exhibition grounds, the narrative becomes static, their travels and possible trials behind them. What follow are chapters that lovingly describe what the children find as they return daily to explore new sights and new pavilions. Here the landscape is all lovingly constructed with human hands. Even when the natural world surfaces, it is in an unnatural setting. For example, they visit the Palace of the Flowers, which held up a great crystal of light glowing against the dark blue of the sky, towers and domes were crowned diadem, thousands of tools hung among the masses of leads, or reflected themselves sparkling in the darkness of the lagoons, fountains of molten jewels sprung up in flames and changed. The city beautiful stood out whiter and more spirit like than ever in the pure radiance of those garlands of clearest flame. (179)

In chapter after chapter, the description is overwrought and exaggerated, emphasizing that even when the natural world appears, it is controlled by those who built the fair. The "sweep of columns, statue-crowned," is lapped by the "waters of blue lagoons ... as if a homage to their beauty" (195). Note that the beauty here is in the columns and colonnades, not in the blue waters. The fact that all this beauty is designed to be temporary, to be consumed and then discarded, is not a flaw; in fact it appears to be part of its charm. How wonderful that all this could be built so quickly and for so short a time. The fair, the narrator says, is "radiant and unearthly in its beauty." We get references to the gondoliers in the fake Venice, to the Midway (the first in American history, complete with the first Ferris wheel), to the Chinese theatre, to the Arabian Nights entertainment, and to a variety of global costumed exhibits.

It is interesting, though, that most of the buildings and exhibits are not given their proper names. Instead, as part of Meg's "fairy stories," they are given new names, ones that give the unearthly beauty of the fair an even more extraordinary character. Only the Agricultural Exhibit gets to keep its proper name, because that is where the children encounter their pragmatic Aunt Matilda once again. Had it been published when it was supposed to, the novel might well have encouraged even more pilgrimages to the fair. But interestingly, this ideal manufactured landscape is still not enough on its own, and if the organizers of the fair had wanted a full description and review of the wonders of the experience, they might have been disappointed. The American landscape pales in comparison to the world of the fair, but the world of the fair is itself secondary to the creative colouring that Meg's imagination gives it.

For that colouring, Meg relies on some folk tales but especially on *The Pilgrim's Progress,* which supplies the third landscape in this book. As I mentioned earlier, it is Bunyan's allegory that prompts the journey in the first place. Meg had found a copy, which she read and reread in the barn loft, eventually declaring to her brother, "Oh Robin! ... I don't want to hear of the people down there. I've been reading *The Pilgrim's Progress,* and I do wish—I do wish there was a city beautiful."[5] Robin's response to this is "there is going to be one" (9). He tells her about the World's Fair in Chicago, which they both immediately picture as the Celestial City on earth, and from that point, the two plot their escape from their own City of Destruction to this Promised Land. As they make their plans, they imagine repeatedly what stage in Christian's journey they have reached. They identify their own The Hill Difficulty and the Slough of Despond and write their journey to Chicago over Christian's.

From then on, *The Pilgrim's Progress* becomes their personal road map for their journey from the farm to Chicago, and even before they leave, they see their own difficulties and obstacles as types of Bunyan's metaphorical landscape. The narrator tells us that "they had never read that old, worn *Pilgrim's Progress* as they did in those days. They kept it in the trough near the treasure and always had it on hand to refer to. In it they seemed to find parallels for everything" (54). Aunt Matilda's farm world is the City of Destruction, and their Hill Difficulty is represented by having a goal and no way to attain it. Meg tells her brother: "That day you said you would not let it go by you … that was the day we reached the Wicket Gate." So the real landscape of Matilda's farm is ignored, even rejected: in its place on the short journey—as well as in the preparation phase—is Bunyan's landscape.

Even more interesting on their journey, though, is that they are so self-aware as readers. They know explicitly how they are reading *The Pilgrim's Progress*; as the narrator says, "somehow one could scarcely tell where one ended and the others began, they were so much alike, the three cities—Christian's, Meg's, and the fair, ephemeral one that the ending of the 19th century had built upon the blue lake's side" (53). Meg is sure that the Celestial City and the World's Fair "must be alike," so she combs through the book to look for telling descriptions. Note that there are actually three cities altogether, not just two: Christian's Celestial City, the world's fair, and the third, Meg's imaginative conflation of the two, a kind of heaven on earth, Bunyan's allegory superimposed on Mr. Burnham's dream.

Of course, Burnett is not alone in redirecting *The Pilgrim's Progress* for a young audience.[6] There have been dozens, even hundreds of these adaptations, revisions, reworkings, and significant allusions to *The Pilgrim's Progress* in children's literature since the end of the eighteenth century. Some are simple abridgements; others use only words of one syllable, change theology to suit Anglican children, or completely recast the hero as a child or children, as in Enid Blyton's *Land of Far Beyond*.[7] The number and variety of versions for children is an indication of how important and ubiquitous Bunyan's allegory had become in English nursery libraries. The book's own journey seems to have gone something like this: children appropriated and read books like *Gulliver's Travels, Robinson Crusoe,* and *The Pilgrim's Progress* because they offered exciting adventures and perhaps because the work written specifically for them was less appealing. Take James Janeway's *A Token for Children: Being an Exact Account of the Conversion, Holy and Exemplary Lives and Joyful Deaths of Several Young Children* (1672) as an example.[8] As the importance of Bunyan's allegory grew, as it came to be considered both a spiritual guidebook and a "classic" of Englishness, so did the number of ways parents and publishers contrived to put the book in children's hands. The result was that for almost two centuries one could assume that every literate English-speaking child would know some version of *The Pilgrim's Progress.* Louisa May Alcott could confidently use it as her structural framework in *Little Women,* and L.M. Montgomery could have her *Emily of New Moon* confess to liking the book, because both could assume that their readers would know what that meant.[9] And the adaptations continue: over the past twenty-five years we have seen pop-up versions, contemporary retellings, graphic novels, and even a video game, all to make *The Pilgrim's Progress* accessible to children.

In referencing Bunyan's seventeenth-century allegory, then, Burnett is certainly doing nothing new. She is using a common touchstone, a book that on the one hand she could expect would lend an air of familiarity to the story she was to tell, and on the other hand, might elevate her own text and by extension elevate the World's Fair by association with Bunyan's Celestial City. The fact that Burnett uses Bunyan is not extraordinary: *how* she does it is.

First, the landscape of *The Pilgrim's Progress* gets more attention than the real landscape either of rural Illinois or of Chicago and the fair. The children map their own struggles onto Bunyan's allegorical map, just as Bunyan might have expected his readers to do. If the World's Fair trumps Matilda's farmland, so the imaginary worldscape of Bunyan's allegory trumps the World's Fair. That is all Meg's doing as she overlays the fair with her imaginative colouring. When they arrive, for example, she has a particular way she wants to enter: "They knew that there were gates of entrance here and there, through which thousands poured each day; but

Meg had a fancy of her own, founded, of course, upon that other progress of the Pilgrim's. 'Robin,' she said, 'oh, we must go in by the water, just like those other pilgrims who came to town'" (92). So the fair takes its meaning from Bunyan's allegory, and Meg adds to that meaning as she goes, imagining, for example, a genie for each of the pavilions.

Compared to Christian's journey, though, the children's is easy. As I suggested before, imaginative pilgrimages invite stories of struggle and triumph; these children feel no struggle. Christian's journey took him through sloughs, up hills, across fields, and before fierce foes. Meg and Robin, by contrast, take a train and meet a rich old man who wants to adopt them. Is this just a lapse in attention for Burnett, who is more concerned at this point with celebrating the fair than with creating the kind of conflict that might result in a good yarn? Or is she deliberately undercutting the conventions of the pilgrimage, first signalling and then rejecting the idea of struggle, because the kind of life she wants to paint—the hard-working life of ideal, beautiful American children—has no real struggles? In America, in the kind of America in which this fair could exist, such obstacles melt away. This last reading possibility makes the effect more deliberate and perhaps more interesting, though it does not make it a better book.

But there's a third, larger issue raised by Burnett's use of *The Pilgrim's Progress*: it is not a religious use, and it may be ironic. Meg herself draws attention to the fact that hers is a secular reading of Bunyan's Christian work when she, a little embarrassed, explains her method of connecting the Celestial City and World's Fair to the rich saviour, John Holt: "Rob and I talk to each other and invent things about it, just as we talked about this. We just *have* to, you see. Perhaps we say things that would seem very funny to religious people—I don't think we're religious—but—but we do *like* it" (167). The narrator goes so far as to suggest that "the place she and Robin had built to take refuge in was a very real thing. It had many modern improvements upon the vagueness of harps and crowns" (166). Bunyan's heaven is vague; the fair offers modern improvements on heaven. Earlier, Rob says that this is not a world in which they can wait for miracles to happen: "'The world is all different,' said Robin. 'You have to do your miracle yourself'" (27). They may rely on *The Pilgrim's Progress* for directions, for ways to read the steps of their journey, but they will make those steps on their own, through their own work and ingenuity and with thoughts of this world alone.

That secular reading of *The Pilgrim's Progress* may explain the odd conflation of the World's Fair and the Celestial City. There are, of course, three cities in Bunyan's allegory. The City of Destruction from which Christian comes and the Celestial City towards which his journey leads are the first and last, but on the way from one to the other, he and his companion go through Vanity Fair. And surely Vanity Fair is a better parallel for the Columbian World's Exposition than heaven is. Vanity Fair is established by Beelzebub, Apollyon, and Legion,

> and as in other Fairs of less moment, there are the several Rows and Streets, under their proper names, where such and such Wares are vended: So here likewise, you have the proper Places, Rows, Streets ... where the Wares of this Fair are soonest to be found: Here is the Brittain Row, the French Row, the Italian Row, the Spanish Row, the German Row, where several sorts of Vanities are to be sold. (70)

The Chicago World's Fair is explicitly a celebration of the heights of human accomplishment, and it is as temporal and temporary as they come: a city built to be torn down again in a year. The Columbian World's Exposition looks more like one of those other "Fairs of less moment," a typological reflection of Vanity Fair, not of the Celestial City.

I'll confess that I'm still not sure how to read Burnett's use of Bunyan. I am not content with a reading that dismisses the novel and its allusions merely as an extended public relations piece for which she got paid, because there is too much else involved in her notions about American entrepreneurship and imagination—and there are too few specifics about the fair itself. I cannot imagine that she blindly missed the better parallel between the evil and unjust Vanity Fair and the one she celebrates; nor does she seem to wish to undermine the accomplishments of the Chicago fair by criticizing its essential vanity. What I am left with is the possibility that she recognizes the worldliness of her fair and is content with that worldliness because that is now what matters—not the next world and its vague landscapes, but *this* world, and in an effort to elevate human and especially American accomplishment, she will substitute a temporal city for a heavenly one as a reasonable goal. Francis Molson argues that Burnett's use of Bunyan is a culmination of the secularization of *The Pilgrim's Progress* in America in the nineteenth century, one that shifted the pilgrim's goal from an eternal heaven to domestic happiness, a home on this earth (58).

So perhaps the fair is not the ultimate goal, but just a stage that leads the children to their final home with the wealthy widower, John Holt.[10]

Two Little Pilgrims' Progress is a celebratory book: it celebrates the Chicago World's Fair, wealth, American enterprise, plucky orphans, and above all the power of the imagination to transform real worlds into better, more meaningful imaginary landscapes. It celebrates happy endings, whether they occur in Heaven or in rich men's homes, with industrialists standing in for God. It demonstrates both the awe that America felt for the fair and the changing responses to Bunyan's *The Pilgrim's Progress*. And it rejects the life of the real American landscape and those who would, either in "bucolic" Alps or on Illinois farms, bind themselves to that land. In its final words, it is unabashedly, unironically devoted to the truth of what is promised in fairy tales:

> Perhaps, as Meg said often to John Holt, theirs was a fairy story—and why not? There are beautiful things in the world, there are men and women and children with brave and gentle hearts; there are those who work well and give to others the thing they have to give, and are glad in the giving. There are birds in the sky and flowers in the woods, and Spring comes every year. (191)

―――――――

Notes

1) Burnett may have thought little of the idea of a Woman's exhibit (capital "W"), but when it was finished the building housed an extraordinary library—extraordinary both for its vision of the design of a public library and for its opening up of the canon in its invitations to submit. In their history of the building's library, Sarah Wadsworth and Wayne A. Wiegand argue that the library played with a "much broader view of culture that contested the genteel traditions and canonical texts the dominant patriarchy promoted" (Wadsworth and Wiegand 25).

2) See also Hubert Howe Bancroft's *The Book of the Fair* or the two books published by Rand, McNally: the *Sketch Book* and the *Handbook to the World's Columbian Exposition*, large-format tomes with many pictures and painstaking if hagiographical descriptions of the buildings, grounds, and exhibits.

3) David Silkenat identifies the fair as instrumental in shaping Chicago's working class and points to other historians who have explored the fair's central place in American late-nineteenth-century cultural history. It was, to quote one, "the greatest tourist attraction in American history" (Donald Miller, qtd. in Silkenat 268).

4) The choice of "City Beautiful" is a conflation of the Celestial City, Bunyan's Heaven, and the House Beautiful, a human habitation that Christian encounters early in his journey. It is inhabited by Prudence, Piety, and Charity. The combination means that Meg can make her final destination an earthly one.

5) Part of Matilda's spiritual poverty, by the way, is evidenced by the fact that in her "bare, cold house there was not a book to be seen," though their poor parents had managed to surround them in their cozy house with "an atmosphere of books, by buying cheap ones they could afford and borrowing the expensive ones from friends and circulating libraries" (3).

6) Bunyan did write a book aimed specifically at a child audience: his *Book for Boys and Girls or, Country Rhymes for Children*, published in 1686, is perhaps the first collection of original verse for children in English. It seems that his *Pilgrim's Progress* was intended not for a young audience but for a general readership.

7) Mary Godolphin wrote the first of at least three monosyllabic versions, with only proper names retaining more than one syllable; the Rev. Neale was concerned that children not be led astray by Bunyan's idiosyncratic Christian doctrine, so he shoe-horned the work into an Anglo-Catholic framework. Mary Sherwood wrote *The Infant's Progress*, replacing the grown-up Christian with a child pilgrim. Many more are documented in David Smith's *John Bunyan in America* and in Ruth MacDonald's *Christian's Children*.

8) Accounts of the early centuries of children's reading in English can be found in Gillian Avery's "The Beginnings of Children's Reading to c 1700" and in F.J. Harvey Darton's *Children's Books in England*.

9) As Elaine Showalter points out, Alcott was merely having her characters do what she and her sisters did when they were children: playing pilgrims; and in *Emily of New Moon*, Emily Bird Starr's lines about being proud of liking *The Pilgrim's Progress* are almost word for word from Montgomery's own journals. Both women, then, were not simply imaging child readers of Bunyan: they themselves had been child readers of Bunyan.

10) Readers of Erik Larson's *The Devil in the White City: Murder, Magic, and Madness at the Fair That Changed America* might well feel some unease when the children follow this old man back to his hotel room. We now know that the World's Fair was the hunting ground of America's first well-documented serial killer, H.H. Holmes. The fair was an attractive target for him precisely because people came and went and might not be missed. Fortunately for Meg and Robin, theirs is not that kind of story.

5) Old World, New World, Other World

Overcoming Prosaic Landscape with *The Golden Pine Cone*

❦ LINDA KNOWLES

This Golden Earring has great magic. It has taken you both into a new world—if you only knew.

—Clark, *The Golden Pine Cone*, 24

───────

S ince the days when the country was dismissed as "a few acres of snow," Canada's immense and unsettling geography has been one feature guaranteed to attract attention from the outside world. In the 1930s Susan Buchan, wife of the author John Buchan (Governor General of Canada, 1935-40), declared that the wilderness gave Canada her greatest charm:

> It is possible to live in one of her great cities and to forget how near the wild country is to your home. I remember we once went to a skiing camp, and ... I heard someone say, "There are no houses between us and

the North Pole," and then someone else remarked that a bear had come down from the woods and devoured part of the contents of the camp store cupboard. We were not many miles from the … capital of Ottawa.

The wild is always there, somewhere near. (Gatenby v)

Marjorie McDowell, writing in *The Literary History of Canada*, asserts that "if children's books from Canada have enjoyed an international vogue beyond their strictly literary deserts, it is because the Canadian terrain has spread itself so enticingly behind a child's dream" (624). However, the way Canadians chose to interpret the "child's dream" of the Canadian terrain significantly affected the development and the character of Canadian fiction for children.

Alice's Adventures in Wonderland (1865) was first published just two years before Confederation, yet Canadian fiction for children over the next eighty-three years was marked by a tendency towards realism that contrasted strikingly with the broad stream of fantasy in Britain. In England, Tom swam with the Water Babies (1863), Alice stepped through the Looking Glass (1871), and E. Nesbit's Five Children were given wishes by a Psammead (1902); meanwhile, children in Canadian fiction had very different adventures. Until quite recently, Canadian children's literature has been dominated by adventure, historical fiction, and coming-of-age novels. From *The Canadian Crusoes* onward, Canadian fictional children, orphaned, kidnapped by the Native people,[1] lost in the wilderness, alienated from their parents, struggled to find their way home. We have *Anne of Green Gables* but no Canadian *Alice;* we have indomitable, hard-working, enduring children but no fairies.

The Canadian Crusoes (1852), one of the first Canadian children's books, appeared just at the beginning of the great flowering of fantasy that—as Humphrey Carpenter points out—"took root most quickly and deeply in England. Other European countries produced only a tiny handful of memorable children's fantasies before 1914" (16). *The Wizard of Oz,* the first significant North American fantasy, did not appear until 1900. In this context, the late development of fantasy in Canada does not appear so unusual, but it is still remarkable that even when it did begin to appear, the genre was slow to take hold. After 1950 there was a sudden flurry of fantasy with books such as *The Golden Pine Cone, The Return of the Viking,* and *The Secret World of Og,* but this was followed by a series of stories (*Lost in the*

Barrens, The Long Return, The Incredible Journey) that reverted to the theme established by Mrs. Traill in the 1850s. If survival, as Margaret Atwood asserted, is the central theme of Canadian literature, its presence in children's fiction is unmistakable.

We are all familiar with Tolkien's description of the storyteller as a sub-creator who makes a Secondary World: "Inside it, what he relates is 'true': it accords with the laws of that world. You therefore believe it, while you are, as it were, inside" ("On Fairy Stories," 37). Northrop Frye makes a similar point when he describes fiction as a "conscious mythology" that "creates an autonomous world," but he goes on to employ a different emphasis by saying that "it gives us an imaginative perspective on the actual one" ("Conclusion," in *Literary History of Canada*, 837). If this is true, and all fiction takes the reader into a secondary or autonomous world, with an imaginative perspective on the actual one, then the fiction created for children, the worlds chosen to present to them, can be particularly illuminating.

John R. Sorfleet points out that "the history of children's literature in Canada is the developing story of how Canadians have come to terms with the place in which we live" and that "allying ourselves with the reality of our land ... is the secret to Canada's strength and potential" (222). What effect has this "coming to terms" had on the imaginative perspective offered to Canadian children, and how does that perspective relate to others in the broader canon of children's literature? What is the significance of the fact that Canadian children were, for all those years, offered a prosaic geography, a landscape with scarcely a hint of fantasy? What was it about that geography that excluded fantasy? "Where," an academic once asked me, "is a sense of awe and wonder, and where is the Lady Galadriel?"

∾ New World / A Country without Myth

THE SIMPLE ANSWER to Canada's lack of fantasy literature can be found in the early conviction that Canada's geography is inherently prosaic, that the new country was devoid of mythical or supernatural associations. The classic example is provided by Charles Sangster in *The St. Lawrence and the Saguenay*:

> Oh! Stately bluffs! As well seek to efface
> The light of the bless'd stars, as to obtain
> From thy sealed, granite lips, tradition or refrain! (LXXXV, ll. 7–9)

This belief had its positive side, in that it asserted that early settler Canadians met nature face to face, without the aid or intercession of mythical beings:

> ... No Nymphic trains appear,
> To charm the pale Ideal Worshipper
> Of Beauty; nor Neriads from the deeps below;
> No hideous Gnomes, to fill the breast with fear:
> But crystal streams through endless landscapes flow,
> And o'er the clustering Isles the softest breezes blow. (VII, ll. 4–9)

but Sangster's words echo those of Catherine Parr Traill in 1836 when, willingly sacrificing poetry to pragmatism, she boasted that even the Highlanders and Irish had no time for superstition in their rush to build a new world:

> Fancy would starve for lack of marvellous food to keep her alive in the backwoods. We have neither fay nor fairy, ghost nor bogle, satyr or woodnymph ... I heard a friend exclaim ... "It is the most unpoetical of all lands; there is no scope for imagination; here all is new." (*Backwoods*, 153–55)

This attitude meant that when these authors began to write for children they had no sense of First Nations' tradition of fantasy to support them.

But then, what about *Anne of Green Gables*? It is tempting to think that L.M. Montgomery had Traill in mind when she put the phrase "scope for imagination" in Anne Shirley's mouth (11). Renaming her prosaic surroundings, populating them with fairies and dryads, and terrifying herself with her own haunted wood, Anne showed what could be done about a country without myth. Similarly, John Hunter Duvar, another Prince Edward Island resident, devised a neat solution to the problem with his comic poem *Emigration of the Fairies* (1857), in which a fairy band is washed across the Atlantic to Canada on a floating island of seaweed. And anyone familiar with Edith Fowke's *Folklore of Canada* will know—despite Mrs. Traill's assertions to the contrary—there were plenty of fairies smuggled into the country in the baggage of the Scots settlers of Cape Breton Island. Yet, these migrants did not find their way so easily into published writing for children. Why didn't others follow Montgomery's example, or Duvar's?

Perhaps besides the lack of "scope for imagination," there was a philosophical bias against fantasy when promoting recognizably Canadian books to children. Mrs. Traill's *Canadian Crusoes*, which influenced novels as recently as Farley Mowat's *Lost in the Barrens* (1956),[2] had a distinctly pragmatic origin, as Agnes Strickland's preface to her sister's work attests:

> Our writer has striven to interest children, or rather young people approaching the age of adolescence, in the natural history of this country, simply by showing them how it is possible for children to make the best of it when thrown into a state of destitution as forlorn as the wanderers on the Rice Lake Plains ... [It] is well if those young minds are prepared with some knowledge of what they are to find in the adopted country; the animals, the flowers, the fruits, and even the minuter blessings which a bountiful Creator has poured forth over that wide land. (3)

Marilynne V. Black and Ronald Jobe's 2005 survey of Canadian picture books shows that this bias persisted into the twenty-first century:

> To evolve a national identity, youngsters need to develop a sense of place, a feeling of "This is where I belong." It is crucial, therefore, that they see their communities, regions and country reflected accurately and authentically in literature. Also, it is equally important for children to gain a sense of their nation's past and the impact of the land on our history. (35–40)

In a new world where young people like Madeleine de Verchères found themselves defending forts against hostile tribes, and where they could be lost in the forest and never found again, the sense of an alien world was ever-present, and this could not help but make its mark on the imagination. Northrop Frye put this most clearly in his Conclusion to the *Literary History of Canada*:

> The mystique of Canadianism ... came so suddenly after the pioneer period that it was still full of wilderness. To feel "Canadian" was to feel part of a no-man's-land with huge rivers, lakes, and islands that very few Canadians had ever seen. "From sea to sea and from the river unto the ends of the earth"—if Canada is not an island, the phrasing is still in the etymological sense isolating. One wonders if any other national

consciousness has so large an amount of the unknown, the unrealized, the humanly undigested, so built into it. (826)

It is as if the stern demands of Canadian geography, the wild, unclaimed, unsettled wilderness, presented authors with a Perilous Realm that displaced the Realm of Faerie. In such surroundings, did Canadian children need fantasy? In fiction, if not in fact, Canadians were too busy fighting bears to imagine fighting ogres. As adventure stories became the accepted response to the New World, and the wilderness took the place of the Perilous Realm, Canadian children were offered a prosaic geography where they might engage in struggles for survival like miniature adults, an imaginative geography that gave them a sense of place but, however exciting the events portrayed in it, lacked any sense of magic or the numinous. Once again, our geography, both real and imaginative, was against us.

This state of affairs may have satisfied nationalist educators, but as Bruno Bettelheim observed, "a fare of realistic stories only is barren":

> Outlawing realistic stories for children would be as foolish as banning fairy tales; there is an important place for each in the life of the child … When realistic stories are combined with ample and psychologically correct exposure to fairy tales, then the child receives information which speaks to both parts of his budding personality—the rational and the emotional. (54)

Fantasy allows children to try out, in imagination, emotional experiences that they will encounter as they grow up: "The child intuitively comprehends … that fairy tales depict in imaginary and symbolic form the essential steps in growing up and achieving independence" (73). Fantasy literature could, in this way, allow them to try out in imagination what Robertson Davies described as "large spiritual adventures" but, as he also observed, "Canada is not the place where you are encouraged to have large spiritual adventures" (Monk 13). Moreover, creating the literature of a new nation left little room for creatures associated with the Old World such as elves and fairies. For Canadian children, as for New Zealander Margaret Mahy, "magical displacement" had to come from the Old World,[3] which, for most Canadians, until later in the twentieth century, meant Britain, but even Britain was not as rich in fairies as might be supposed.

WHEN WE SPEAK OF THE NEW WORLD as devoid of fairies, we are speaking as if fairies actually existed. In fact, *belief* in fairies has long been in decline, and (with apologies to Arthur Conan Doyle, J.M. Barrie, and Tinker Bell) fairies are as scarce in the Old World as in the New:

> "The men in green all forsook England a hundred years ago," said I, speaking as seriously as he had done. "And not even in Hay Lane or the fields about it could you find a trace of them. I don't think either summer or harvest, or winter moon, will ever shine on their revels more." (Brontë, 40–47)

> You see, children know such a lot now, they soon don't believe in fairies, and every time a child says, "I don't believe in fairies," there is a fairy somewhere that falls down dead. (Barrie, *Peter Pan 35*)

There is, however, a striking difference between the two sorts of absence, a difference that Hugh MacLennan observed in the emptiness he found in the Old World and the New: "Above the sixtieth parallel in Canada you feel that nobody but God has ever been there before you, but in a deserted Highland glen you feel that everyone who ever mattered is dead and gone" (7).

In *Puck of Pook's Hill* we meet this second sort of emptiness as Kipling relates the tale of two children who accidentally conjure up the fairy Puck by playing *A Midsummer Night's Dream* three times in a fairy ring on Midsummer Eve. Unlike Canada, with its lack of mythical associations, Kipling's England resonates with ancient magic, and Pook's Hill is a corruption of an older name:

> Pook's Hill—Puck's Hill—Puck's Hill—Pook's Hill! It's as plain as the nose on my face ... If Merlin himself had helped you, you couldn't have managed better! You've broken the Hills—you've broken the Hills! It hasn't happened in a thousand years. (8)

As Puck explains, however, the world has changed:

> Unluckily the Hills are empty now, and all the People of the Hills are gone. I'm the only one left. I'm Puck, the oldest Old Thing in England. (8)

Kipling dates the fairies' absence to the Reformation:

> This Reformations tarrified the Pharisees same as the reaper goin' round
> a last stand o' wheat tarrifies rabbits ... they says, "Fair or foul, we must
> flit out o' this, for Merry England's done with, an' we're reckoned among
> the Images." (267)

Others relate the dearth of fairies to the Industrial Revolution and the
advent of the Machine Age. Whatever the reason, fairies disappeared
from the Old World landscape, and though fairy tales became the object
of scholarly research, as literature they were relegated to the realm of the
nursery. As Tolkien pointed out: "Adults are allowed to collect and study
anything, even old theatre programmes or paper bags," but an adult who
enjoyed fairy tales, as opposed to collecting and studying them, had to do
so in the company of children ("On Fairy Stories," 34).

Nevertheless, it is just at this time, when science and rationalism were
rendering the geography of the Old World as prosaic as the New, that children's literature began to flower and fantasy to become a key feature of
its practice. Humphrey Carpenter's *Secret Gardens: A Study of the Golden Age
of Children's Literature* gives a very clear analysis of this development, so it
is not necessary to go into detail here, except to point out that one of the
remarkable effects of the rise of children's fantasy in Britain was the decision, most clearly seen in the works of E. Nesbit, to make contemporary
middle-class children—not fairy tale princes, princesses, or peasants—the
protagonists of the stories. It then became necessary to create a world in
which magic could exist without violating the plausibility of the real world
of these characters:

> The natural world has its laws, and no man must interfere with them in
> the way of presentment any more than in the way of use; but they them-
> selves may suggest laws of other kinds, and man may, if he pleases, invent
> a little world of his own, with its own laws; for there is that in him which
> delights in calling up new forms—which is the nearest, perhaps, he can
> come to creation. (MacDonald, 1893)

Though each invented world has its own laws, locating magical tales
in a plausible setting has been done in several distinct ways. One method
is to place the story long ago and far away, in a time when magical crea-

tures could be said to have existed, as in George MacDonald's *The Princess and the Goblin* (1872) or, as a modern example, in Cressida Cowell's *How to Train Your Dragon* (2003), of which she says, "The past is another land, and we cannot go to visit. So, if I say there were dragons, and men who rode upon their backs, who alive has been there and can tell me that I'm wrong?" (quoted in an exhibition at Seven Stories, Newcastle upon Tyne, 27 October 2012–4 September 2013). Another method is to place the action in the present day but in a world isolated from the real one, such as the Land of *Oz,* or the Neverland of *Peter Pan.* These worlds can be reached by a journey (however magically undertaken or unlikely), but other worlds may be reached only through a portal, as in the Alice books, or *The Lion, the Witch and the Wardrobe* (1950). Such worlds can be kept distinct from the real world and easily sustain their own interior logic. In a third method the magical events may take place in the present day and in the real world, with the aid of a magical device or creature that has survived from the age of magic, such as the Psammead in Nesbit's *Five Children and It* (1902) and *The Story of the Amulet* (1906), or Puck in Kipling's *Puck of Pook's Hill.*[4] None of these approaches, however, provides that "sense of place" demanded by writers and critics intent upon establishing a sense of national identity. When, at last, a Canadian writer dared to invent another world, it was entered by means of a talisman, a magical object (the eponymous golden pine cone), but Canada's first fantasy for children locates the other world in the present day and with the same geography as the real one.

∾ Other World / The Golden Pine Cone

THE PRECEDING DISCUSSION showed how writers in the New World and the Old responded to a lack of fairies in quite different ways; one group embraced realism, the other created fantasy. Catherine Anthony Clark, however, managed to overcome the traditional Canadian resistance to fantasy by combining First Nations myth with realistic features drawn from the traditional "survival" tales of Canadian children's fiction.

Catherine Anthony Clark may have been one of those children for whom the Canadian terrain was a dream. According to Harbour Publishing's "About the Author" note, "when she was growing up in England, Catherine Anthony Clark always knew that she wanted to live in Canada" (191). She persuaded her father to immigrate to British Columbia in 1914, and there she married a rancher, Leonard Clark, and had two children before publishing *The Golden Pine Cone* in 1950, the year *The Lion, the Witch and the Wardrobe* was published and four years before the appearance of *The Lord of the Rings*. Clark's fantasy, like Lewis's, has talking beasts, but the people of her other world are drawn from Native American cultures, and it is this feature of her fantasy that attracted the attention of critics:

> Catherine Anthony Clark looked to the myths of First Nations peoples in writing *The Golden Pine Cone*, which contains one of the few fantasy otherworlds that is decidedly not run by humans. (Berry)

> The mountainous, lake-filled Kootenays which are the settings for her stories are not, obviously, the natural habitat of the rarefied and delicate fairies of Europe. Her fairies must be less of gossamer and more of buckskin. (Selby 40)

> [The children] encounter magical beings who are either spirits in the form of natives, or creatures from traditional west-coast mythology, or invented creatures of the British Columbia landscape. (Johansen 16)

If First Nation myth seems suddenly the appropriate place to look for creatures of Canadian fantasy, why had they not been used before? Certainly there had been many years of anthropological exploration of Aboriginal myth, but the stories of Aboriginal myth, unless heavily revised, were considered too lacking in narrative and too erotic for children (Edwards and Saltman 199; Fowke, *Folklore of Canada* 10). Nevertheless, in the years since the publication of *Canadian Crusoes*, interest in and understanding of First Nations myths and legends had grown. The poems of Pauline Johnson, Isabella Valancy Crawford, and Duncan Campbell Scott used Native cultural images of flint, feather, drum, and canoe to establish their Canadian character. Not long after Clark arrived in Canada, First Nations myth was made more accessible and familiar by the publication of *Canadian Wonder Tales* (1918), a collection of folklore from Native and immigrant peoples, and, later, Emily Carr's *Klee Wyck* (1946). Carr's documentation in paint of West Coast Aboriginal art, and productions by the National Film Board of Canada, contributed to the growing influence of First Nations culture in Canadians' sense of national identity. It is unsurprising therefore that Catherine Anthony Clark should have used elements of this culture to mark her fantasy as identifiably Canadian. In doing so, however, she did not restrict herself to one Native culture. As Mary Jane Miller points out, "Clark drew elements from several different First Nations for her portrayal but did not perpetuate the familiar stereotypes" (81); indeed, she incorporated many other elements of European folk and fairy tales into her work.

J. Kieran Kealy says that "initially she may have tried to formulate a Canadian mythology which brought together the Indian and European traditions" (8), but I believe this is to overstate the case. I believe that the eclectic mix of European and North American figures in Clark's fantasy is simply a blending together of all the elements familiar to a child growing up in British Columbia, and that her chief distinction as a fantasy writer is not the adoption of Native American myth. The other world she creates is not separated from the real world, in the manner of Carroll or Lewis; rather, it allows the real and the imaginary, the natural and the fantasy worlds to interpenetrate.

The revolutionary quality of Clark's fantasy is at first difficult to see. Indeed, her tale follows many of the precedents set by *The Canadian Crusoes,* beginning with a realistic description of a prosaic life "not very long ago and not very far away" (8), on the banks of Kootenay Lake surrounded by the sort of natural loveliness Mrs. Traill believed compensated for the lack of history and legend:

In spring, the ground was covered with pale pink, scented twinflower and there were clumps of yellow violets among the pippissawa and sometimes a lady's-slipper orchid. But now, in September, there were no flowers ... (14)

The protagonists are modern children who go to school by bus and "spend ... money in the ten-cent store" (13). Like Traill's children, Clark's are competent and resourceful: eleven-year-old Bren "had an axe of his own and was building himself a play-cabin" in the time he had to spare from splitting stove wood, fetching cows, or other chores, and the Mother taught Lucy to "churn butter and bake and many other things." Bren does not believe in fairies, "though Lucy would like to" (18).

The finding of the Golden Pine Cone is undramatic. At first it is only "a most peculiar thing." Though the romantic Lucy's first thought is that "A Tree Spirit must have dropped it" (14), the practical Bren asks, "Have you ever seen a tree Spirit?" (15). Neither child recognizes, at first, that it enables them to understand the speech of birds and animals:

A Harsh voice called behind them, "They've found it—they've found it." But when the children turned in fright there was nothing to be seen but two blue jays in a fir tree.

"It must have been the jays shrieking," said Bren, very relieved. Lucy agreed. It had sounded much like a jay's cry. (16)

Their many duties put the pine cone out of the children's heads, and it is fully two weeks before they discover its power and learn that the "peculiar thing" is an earring pendant that belongs to Tekontha,[5] the Kootenay's spirit guardian. It must be hidden from the renegade giant Nasookin, who wants to possess it because it contains a piece of Tekontha's magic. Clark's children thus embark on a quest similar to that of *The Lord of the Rings,* to return the pine cone to its owner, accompanied by Ooshka, the malamute (part dog and part wolf) that has been given to them by an old prospector. Ooshka is under threat of punishment by Tekontha for crimes committed by the wolf side of his nature, and they hope his help on the quest will gain him Tekontha's pardon.

As in Narnia, time in the Other World flows differently: the children's adventure takes many months, but when they return home they find that it is the same day they left, though not before their parents have been searching for them for half an hour. Unlike Narnia, however, the Other World of *The Golden Pine Cone* is not an alien world with different stars, but, more like the Old England of Kipling's *Puck of Pook's Hill,* a mythologizing of the world the children already know and love. As Sheila Egoff notes in her introduction to *The Golden Pine Cone*

> The children ... are not swept ... into a wholly unrecognizable land. Clark's children are exposed to events that seem only somewhat larger than life and to a land that remains familiar to them. (Clark 10)

The Other World to which the Golden Pine Cone gives access is freely modelled on tales of the Spirit World such as those found in *Canadian Wonder Tales.* I am not concerned here with the authenticity of Clark's interpretation of the First Nations Spirit World, which in any case is only briefly suggested and, as Berry says, "makes use of mythology only on a basic, straightforward level"; but it is clear that she takes her idea of the Spirit World's time from Native belief systems:

Mythic time describes that era when the world and its inhabitants were very different than they are now. However, myths, or what Boas called folktales, can also be set in historical time, when the earth—its landscape, its flora and fauna, and its people—was no different than it is today. Even though the mythical age is past, mythical beings and even mythical lands still exist somewhere in the present, and in these tales, mythical beings often interact with humans. (*Handbook of Native American Mythology*, 41)

Using the concept of the Spirit World allows Clark to transform the real world in much the same way that the diamond in Tyltyl's hat in Maeterlinck's *The Blue Bird* (1911) brought the "souls" of everyday objects to life so that Bread, Milk, and Water become human-like companions on the children's journey, and the cat and dog are given the power of speech. The Spirit World of *The Golden Pine Cone* differs from that of *The Blue Bird*, however, in that it departs entirely from the tame domesticity of the children's home life and is expressed through elements of wild nature and First Nations myth.

In the prosaic world live the human characters, unaware of and untouched by the Other World:

"Your parents are still outside the Other World and nothing of it can hurt them. But you and Bren have entered Tekontha's kingdom—like me. You will see things you never dreamed of." (26)

In Tekontha's world, "where the Little People live and where the Indians [*sic*] have snow-white tepees and blankets spun with gold" (121), there is an eclectic range of characters. Some are described as spirits, but others are reminiscent of creatures from traditional European folklore and fairy tales. Many of the spirits, such as the Ice Folk and the Flower-flits, are visible personifications of frost and dew, akin to the familiar folk of fairy tales or the fairies of Disney's animation of the Nutcracker Suite in *Fantasia* (1940):

By ones and twos and then by dozens tiny creatures with glass bodies crept out from the chinks in the logs, and bands of them were soon at work smoothing drops of water into glossy sheets and sliding down tiny icicles growing from the slow-dripping roof logs … As they worked, their bodies lit up with tiny flames that wavered and glowed like star fire. (27)

The Pearlies that live under the lake and guard the treasures that fall into it have a life cycle that is described as realistically as that of creatures described in *The Water Babies*. Compare, for example:

When Pearlies begin to get old, after fifty years, their skins get thicker and thicker till they are covered with rock-scales, and when the lime reaches their hearts, then they die. The dead Pearlie cracks in two and out floats a pair of bubbles. These get bigger and bigger, for they are baby Pearlies … The bubble-stuff of their bodies changes to pearl, and then to lime when they are old. Then that is the end of those Pearlies. So it goes. (57)

with

And then the caddises grew quite tame, and used to tell [Tom] strange stories about the way they built their houses, and changed their skins, and turned at last into winged flies; till Tom began to long to change his skin, and have wings like them some day. (Kingsley)

Here, and in other descriptions of the Spirit World creatures, we find echoes of the natural history Mrs. Traill included for the edification of readers of *The Canadian Crusoes*. The Muskrat, Head Goose, Ooshka the Malamute, and other creatures from Tekontha's world all prove to be useful guides to the Other World:

"What are those bright little things along the logs? Are they fairies?"
Muskrat chuckled. "Certainly not. They don't amount to much, except for splitting rocks at the seam and loosening landslides. They are just the Ice People. When the nights grow cold in the fall, then the Ice Folk come out of the hills and get to work making crystals and sheets of ice. Very stupid! They work all night and the sun undoes it all next morning." (28–29)

These nuggets of "natural history" contribute to the sense of realism that underpins the fantasy. This is a real world, just one that human beings ordinarily cannot perceive.

The people of the Spirit World embody human faults and virtues, but they also express natural forces. Nasookin, who is filled with rage and boastful energy, first appears as a crashing in the woods like the sound and fury of a storm (21). He takes the children prisoner (again, a parallel with *The Canadian Crusoes,* in which Catherine is taken prisoner) and forces them to work for him. Lucy, who is the younger and more timid of the children, finds the courage to stand up to their captor in defence of her companion—"If you kill the Head Goose then I will kill you Nasookin" (104)—and in turn becomes a pet of the giant, who calls her "little squirrel" and says she has "the heart of a fighting beast" (107). Nasookin proudly sings the song of his life, beating its rhythm on a drum. Only Lucy understands the pathos of the giant's boastful song:

And it was always a sad song of wasted youth and of strength that was always doing, with nothing done; of love that was spoiled and soured by a blood crime. However he boasted and sang, it was always sad ... (95)

The adult reader, though not the child, might here detect a faint echo of Duncan Campbell Scott's "Powassan's Drum":

Is it a memory of hated things dead
That he beats—famished—
Or a menace of hated things to come
That he beats—parched with anger
And famished with hatred—? (Ross 115)

Onamara, Nasookin's estranged wife, who has had her heart torn out and replaced with a lump of crystal, lives on an island that rises from the bottom of the lake on moonlit nights. Her heart lies at the bottom of the lake, and it becomes the children's task to recover it from the Pearlies. Cold and unfeeling, perpetually seeking amusement from her magical toys, Onamara seems derived from a European fairy tale like Hauff's "Heart of Stone."[6] The children soon tire of her selfish idleness and her constant refrain of "I have to be amused" (50), but she gives them the magical deerskin clothes that will never get torn or dirty and will keep them warm through their long winter in the Spirit World (52).

Occupying a middle space between the two worlds, and appearing almost in the centre of the book, is Bill Buffer, the creator of the Golden Pine Cone, an old prospector who has chosen to live in the Spirit World: "I left the man-world of my own will" (119). Others like him, who have left the prosaic human world to work for Tekontha, include the Squareheads:

> They are men, really, who have lived their lives like wild animals and so have somehow come under Tekontha's power … Every few years, a Square-head can spend the winter down in the man-world … If they behave well, they win their freedom back, if not, they wait another two years and try again. Some have given up trying … they got their name from their square fur hats. (57–58)

Figure 1 illustrates the relationship of these characters to their respective worlds. In the Prosaic World live the human adults: Bren and Lucy's parents and neighbours, and the "Old Man" who leaves Ooshka with them. In the Spirit World live the supernatural or mythical characters such as

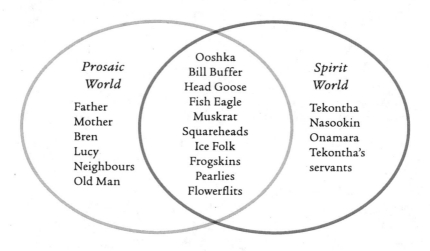

Fig. 1) Distribution of characters between the worlds

Tekontha, Nasookin, and Onamara, as well as the Lake Snake and the Ice Witch. The birds and animals and nature spirits, representing nature itself, exist in both worlds but can be perceived in their spirit guise only by the children with the aid of the pine cone, or by those whose transgressions have put them in the power of Tekontha.

In Tekontha, who embodies strict but not implacable justice, Catherine Anthony Clark created a tutelary spirit that is recognizably Canadian but that embodies the sense of the numinous previously lacking in Canadian children's fiction. It is in Tekontha that we can identify a being with a power equivalent to the Lady Galadriel. Compare, for example, the following descriptions of Tekontha

> And then the figure of a woman veiled in mist stood at their centre.
>
> Lucy watched awestruck. This must be Tekontha. The skeins of vapour eddying around her showed the outline of her slender figure and her thin feet in foot-gloves with golden tassels. Her face was hidden though they had come to a halt thirty feet from her.
>
> The music had faded away and there was such a silence that Ben could feel his heart thumping. Then a voice spoke that was sweet, yet it thrilled their veins … (162)

> Then the mist about her parted and they saw she was more lovely than any woman on land or sea, with eyes as strange and beckoning as running water and hair that flowed to her knees, as black as charcoal. (164)

with these descriptions of Galadriel:

> They were clad wholly in white; and the hair of the Lady was of deep gold, and the hair of the Lord Celeborn was of silver long and bright; but no sign of age was upon them, unless it were in the depths of their eyes; for these were keen as lances in the starlight, and yet profound, the wells of deep memory. (Tolkien, *Lord*, 373)

> She held them with her eyes, and in silence looked searchingly at each of them in turn. None save Legolas and Aragorn could long endure her glance. Sam quickly blushed and hung his head. (376)

Tekontha, unlike Galadriel, rules alone, but in other respects they are very similar. Both are beautiful, with long hair, though one is dark, the other golden, and both have eyes that hold the viewer captive.

I have not found any source for Tekontha in First Nations myth, but a similar figure is found in the Lakota tale of the White Buffalo Calf Woman, bringer of the sacred pipe and its ceremonies. In one version of the myth she is described in terms very similar to that of both Tekontha and Galadriel:

> Coming out of the growing heat haze was something bright, that seemed to go on two legs, not four. In a while they could see that it was a very beautiful woman in shining white buckskin. As the woman came closer, they could see that her buckskin was wonderfully decorated with sacred designs in rainbow-coloured porcupine quills. She carried a bundle on her back, and a fan of fragrant sage leaves in her hand. Her jet-black hair was loose, except for a single strand tied with buffalo fur. Her eyes were full of light and power, and the young men were transfixed. ("Living Myths")

There are many versions of this myth, but all agree that White Buffalo Calf Woman has great power, enveloping in dust and turning to bones the young man who has lustful thoughts about her.[7]

Before they can return to their own world, Bren and Lucy must plead with Tekontha to release Ooshka from the wolf part of his nature so that he can be an ordinary dog. As Bill Buffer explains to the children: "Anything is better than living in two worlds and belonging to neither" (125). Though Tekontha at first resists their appeal, Bren makes a spirited and effective account of their efforts:

> We give you your golden earring; it was Ooshka who thought how to return it. We bring you the heart of Nasookin's Princess from the Pearl Folk; it was Ooshka who helped against the Snake. We had to go from one thing to another; we had to do things the hard way. We have tried to earn our dog's freedom. (164)

Tekontha at last relents:

"Listen, Ooshka—malamute—dog with two natures! You killed with cruelty. You refused to bear your just punishment ... But these children love you, and you love them. As a dog you have been faithful and courageous. Behold, the wolf in you is dead. You are free, as these children are free." (167)

K.V. Johansen complains that the children "have adventures with no real danger or cost to themselves ... They do not come home changed or wiser; the lessons are not learned, but are demonstrated solely for the reader's benefit" (16). Lucy's defiance of Nasookin and Bren's killing of the Lake Snake rather contradict that assertion, but in any case, Johansen ignores the possibility that the true lesson of *The Golden Pine Cone,* intended for the reader, is not a moral lesson but an imaginative one.

∾ *Prosaic Geography Transformed*

THERE ISN'T THE SPACE HERE to explore the possible source for Tekontha, nor for the other elements of Native myth that Catherine Anthony Clark uses, nor to explore the clues that identify the real places on which the tale is based. In any case, the child reader is not necessarily concerned about that identification. As Tolkien observed:

> If a story says "he climbed a hill and saw a river in the valley below" ... every hearer of the words will have his own picture, and it will be made out of all the hills and rivers and dales he has ever seen, but especially out of The Hill, The River, The Valley which were for him the first embodiment of the word. ("On Fairy Stories" 78)

I was a first-generation child-reader of *The Golden Pine Cone,* and I have to confess that when I read that book, I saw my own lake. No mountains surrounded it (as I knew they should), but there was an island that I imagined to be Onamara's and I imagined the Pearlies living beneath the water and taking possession of the things I lost in it.

C.S. Lewis said that the boy reading a fairy tale "does not despise real woods because he has read of enchanted woods: the reading makes all real woods a little enchanted" (Egoff 215). John R. Sorfleet was angry when he found "twittering penguins" in a story set in Canada because "[he] realized ... the world of the schoolboy's annuals, the world in all the English and American children's books [he] was reading, *was not [his] world ...* The writes *didn't know [his] country*" (215). I read *The Golden Pine Cone* long before I discovered Narnia, and in it I found what I most passionately desired: a world that was recognizably mine, my prosaic, everyday world, but that world transfigured, become numinous, my real woods a little enchanted and full of scope for imagination. Flawed and awkward as Clark's writing may occasionally be, her intermingling of realism and fantasy provided a quality of enchantment that I would not find again until I read Alan Garner's *The Owl Service*, another tale in which two worlds mingle.

From Catherine Parr Traill onwards, Canadian writers accepted that Canada was a country devoid of myth, a prosaic geography in which fairies were as alien as Sorfleet's penguins. With *The Golden Pine Cone*, Catherine Anthony Clark transformed this prosaic geography, "broke the hills," and began to teach their "sealed, granite lips" to "speak tradition and refrain."

———————

Notes

1) For a comprehensive analysis of the practice of kidnapping by First Nations, see Axtell 1975.

2) *Canadian Crusoes* was also published as *Lost in the Backwoods*. *Lost in the Barrens* also appeared as *Two Against the North*.

3) Quoted in "Margaret Mahy Obituary," *The Guardian*, 26 July 2012.

4) The animal fable and talking and dressed animals belong to a different level of fantasy, one that can overlap all three of these categories, and again, it is significant that Canada's contribution to the animal story is the realistic animal biography as told by Charles G.D. Roberts and Ernest Thompson Seton.

5) I have been unable to find any source for the name "Tekontha." It may be derived from Tekonsha, a place near Chicago.

6) In German, *"Das kalte Herz." Märchen almanach auf das Jahr 1826* (Fairytale Almanac of 1826).

7) Unlike Tekontha, White Buffalo Calf Woman is a shape-shifter, transforming four times into a buffalo calf as she disappears into the distance. There are several useful sources of the myth recounted by Native Americans on video. See, for example, "The Legend of White Buffalo Calf Woman," an interview with Bill Means, at http://www.youtube.com/watch?v=ezNKgRbnVPY.

6) Healing Relationships with the Natural Environment by Reclaiming Indigenous Space in Aaron Paquette's *Lightfinder*

❧ PETRA FACHINGER

We need to observe history (the path you come from), understand the present (the top of the mountain), and look at possibilities of the future (the things you can see) with the idea of spirit and resonance, a participation of the world that indigenous people have always had. (Gregory Cajete, The Banff Centre, 4 September 2014)

Extraction and assimilation go together. Colonialism and capitalism are based on extracting and assimilating. My land is seen as a resource. My relatives in the plant and animal worlds are seen as resources. My culture and knowledge is a resource. My body is a resource and my children are a resource because they are the potential to grow, maintain, and uphold the extraction-assimilation system. (Simpson 75)

―――――

In the bestselling 2015 Burt Award–winning novel *Lightfinder* (2014),[1] Aaron Paquette uses the genre of young adult dystopia to explore the connections between the degradation of the natural environment as a result of human greed and the effects of intergenerational trauma on Indigenous families and communities. As the two Elders, Kokum Georgia and Auntie Martha, tell Aisling at the beginning of their quest, the "end of the world" has begun (Paquette 45). Kokum explains that "this is a world war, not against nations and countries, but truly, against the worst kind of darkness" (66).[2] The novel presents the intricately linked journeys of its two protagonists in alternating chapters, focusing on the coming-of-age stories of two Cree teenage siblings, one of whom feeds the "good wolf" and the other the "evil wolf" (Taylor, "Prologue," in *Night Wanderer*) in the struggle for the survival of Indigenous identity, culture, and knowledge and the future of life on earth. Both Aisling (meaning "dream" or "vision" in Irish Gaelic) and Eric (derived from Old Norse meaning "king" or "ruler") find themselves displaced from family, community, and white society and disconnected from the lands of their cultural heritage. Whereas Aisling is guided by Kokum Georgia and her Auntie Martha, as well as other human, animal, and mythical protectors, Eric, who has been left to his own devices, becomes the victim of dark forces. He gradually loses his humanity on his way to see Raven, The Devourer of the Light (158), in the company of Raven's son Cor.

Using elements of the end-of-the-world narrative, Paquette situates his novel in specific regional and local spaces and places and refers to some nation-specific traditional stories such as the prophecy of White Buffalo Calf Woman. The novel is set in Alberta, and the siblings' father, who dies in an explosion at the beginning of the novel, used to work as a safety inspector on the oil patch in Fort McMurray (Paquette 2). Massive resource extraction at the Oil Sands has destroyed the natural environment of the unceded territory of the Lubicon Cree, who have long relied on those lands for hunting, fishing, and trapping. This is one of the causes of the uprootedness of the novel's characters.[3] The ecocide of the Oil Sands, in which the siblings' father is to some degree complicit, is linked to colonization and the genocide of Indigenous peoples.[4] Although the text thus situates the events in a particular geographic space, it also emphasizes the importance of transnational connections by including the Indigenous Australian characters Matari, a young warrior and activist whom Kokum met on her global travels in search of like-minded people, and the Elder Inkata, who helps to save Aisling's life when the dark forces attack her. The chapters focusing on Aisling foreground the beauty of the Alberta landscape and the diversity of its fauna and flora, whereas the chapters discussing Eric's journey include descriptions of a ravaged landscape. Matari evokes the "imaginative geography" of the Western Desert of Australia. His character underscores Kokum's vision of kinship as well as one of the foundational principles of Indigenous Knowledge: everything is related.

The reference to Indigenous Australian culture has several functions in Paquette's novel. Apocalypse or dystopia is a genre that Indigenous Australian writers, including those of Young Adult fiction, have repeatedly drawn on. As Australian literary scholar Roslyn Weaver observes, in doing so these authors "rewrite Australian history as apocalypse to represent the impact of white colonization on Indigenous peoples. The disaster scenarios of apocalypse can allow minority groups to invent a new world in which to challenge and change dominant cultural constructions of widely differing agendas" (Weaver 136). Stereotypically, Australia's interior has been portrayed as desert and *terra nullius* by white writers in much the same way as have the Canadian prairies, ignoring the presence of Indigenous peoples. Apocalyptic themes are therefore an important tool for Indigenous writers to "re-inscribe the unwritten future with themselves as a significant part of the landscape" to challenge the fiction that the land is empty (142). *Lightfinder* has a particular affinity with Ambelin Kwaymullina's (Palyku) young adult Tribe series, *The Interrogation of Ashala Wolf* (2012), *The Disappearance of Ember Crow* (2013), and *The Foretelling of Georgie Spider* (2015), which takes place three hundred years after an ecological catastrophe that is the result of climate change and exploitation of natural resources. As Kwaymullina points out:

> I am often told that it is unusual to be both Indigenous and a speculative fiction writer. But many of the ideas which populate speculative fiction books—notions of time travel, astral projection, speaking the languages of animals or trees—are part of Indigenous cultures. One of the aspects of my own novels that is regularly interpreted as being pure fantasy, that of an ancient creation spirit who sung the world into being, is for me simply part of my reality. (Kwaymullina 27)

Ashala and Aisling, both sixteen years old, share the same gifts, or "abilities," as they are called in *Ashala Wolf's Interrogation*: they are sleep-walkers and dreamers, that is, they are not limited by the time and space constraints of the "real" world. They are able to time travel and to move unencumbered between distant places, and they both fight to save the environment and to create a new society. The notion of "Dreamtime" is of particular importance in this context.

By linking the Indigenous worlds of North America and Australia in this way, the novel suggests that the health of the natural environment needs to become a central issue worldwide and that Indigenous people need to unite in the struggle to save it. To underscore the significance of kinship, the novel highlights commonalities among various Indigenous oral traditions, such as the legend of the cannibal spirit of the Wendigo, who is driven by insatiable greed, and the prophecy of White Buffalo Calf Woman, whose return heralds the advent of an age of peace and harmony. This connection can also be read as one of the decolonizing strategies of the novel in that it reimagines kinship in the aftermath of the residential schools in both countries. Julia Emberley's argument that the "Indigenous uncanny opposes the formation of systems of power that can harm or destroy kinship relations between peoples, animals, the land, and the genealogical histories of the ancestors" (Emberley 213) can also be made for "Native dystopia."

∾ Indigenous Futurism and "Transgenre"

IN THE SAME WAY that Anishinaabe writer and narrator Drew Hayden Taylor tells a "Native Gothic" story in his Young Adult novel *The Night Wanderer: A Native Gothic Novel* (2007), Paquette, who is of Cherokee, Cree, and Norwegian descent, tells a "Native dystopian" story in *Lightfinder*. In doing so, he participates in Indigenous futurism and rewrites the familiar genre of the dystopian story by incorporating elements of Indigenous traditional stories.[5] Moreover, he adapts the Native coming-of-age narrative for his non-Indigenous readership by showing young readers that his Indigenous adolescent characters have much in common with them, but that they are also culturally different in that they can draw on an additional kind of "energy."[6] According to scholar and writer Lynette James, Indigenous futurist Young Adult narratives "make use of many story structures ... including common sf or YA genre tropes such as the hero(ine)'s quest or the protagonist's coming-of-age. This is not simply attention to commercial conventions, but an organic outgrowth of the oral storytelling traditions in which the authors have been steeped" (James 156). And Dene writer Lindsey Catherine Cornum explains that "Indigenous Futurism is in part about imagining and cultivating relationships to land/space and each other" and "illuminates the vast network of complex connections that link everything in the world, and then further on out, to everything in other world" (Cornum n.pag.). Importantly, within the spectrum of Indigenous futurisms, Paquette adopts elements of what Anishinaabe scholar Grace L. Dillon refers to as "Native slipstream." According to Dillon, Native slipstream, "a species of speculative fiction within the sf realm, infuses stories with time travel, alternate realities and multiverses, and alternative histories. As its name implies, Native slipstream views time as pasts, presents, and futures that flow together like currents in a navigable stream. It thus replicates non-linear thinking about space-time" (Dillon 3). In this way, both place and time are envisioned as Indigenous. Immanent catastrophe, like the end of the world in *Lightfinder*, is linked to intergenerational trauma, loss of Indigenous knowledges, and colonial violence.

In mapping the "imaginative geography" of *Lightfinder*, Paquette mixes genres and narrative traditions and makes direct and indirect reference to both classic texts of English literature (*Alice's Adventures in Wonderland*, *Frankenstein*, *The Lord of the Rings*) and pop culture (*Star Wars*, *The Hunger Games*, *Harry Potter*). As it borrows from the dystopian novel, the coming-of-age novel, science fiction, the adventure novel, the vampire story, and the captivity narrative, *Lightfinder* can be read as an example of what Chickasaw scholar Jodi A. Byrd has called "American Indian transgeneric fictions," one of whose most important features is its potential to disrupt generic expectations across cultural contexts in order to write back to "colonialist narratives that attempt to trap Indigenous Peoples in a dead past" (Byrd 357). As Byrd observes,

> as American Indian authors test the limits of genre's laws and transgress the boundaries of taxonomy, they produce texts that might be understood as transgenres, texts that experiment, refuse categorization, and that genre-bend narrative fiction into poetry, traditional stories into science fiction, fantasy into the historical, and horror into the epistolary. After all, genre is socially constituted, produced at the site of

interpretation, and is a system of identification. And popular genre says as much about the cultural moment that produces it as it does about definitions. (Byrd 247)

In the spirit of transgression, "transgeneric fiction" also creatively draws on texts from the British literary canon in an effort to defamiliarize them and decolonize their colonial implications. Forty-five pages into the novel, Aisling confesses to her Auntie Martha and Kokum Georgia: "I feel like I tripped and fell down a rabbit hole but I haven't even hit the ground yet" (*Lightfinder* 45). This reference to Lewis Carroll's *Alice's Adventures in Wonderland* (1865) serves to contrast Aisling's situation with Alice's. Aisling has just been notified of her father's sudden death, and she has not yet had time to properly mourn him. Eric, her twelve-year-old brother, has run away from home after insulting his mother and blaming her for his father's death, and nobody knows where he is. Most significantly, Aisling is about to embark on a quest in search for him, which involves fighting Raven and his helpers and saving Earth. While going down a rabbit hole opens the door to a world of fantasy for Alice—Alice's adventures in Wonderland are all make-believe—Aisling's "adventures" are the result of colonization, human disrespect for the natural environment, and greed. And whereas Alice, an upper-middle-class Victorian girl, lives a sheltered life, one she finds "boring," Aisling, who has an alcoholic mother on the Rez, lives all by herself in the city to have access to a better high school. In contrast to Alice's dream world, where curious creatures do strange and curious things, the animals and mythical figures that Aisling meets on her journey have their origins in Creation stories such as the giant turtle on whose back Earth rests. Alice's quest is to locate the beautiful garden in Wonderland; Aisling's is to save the world from destruction. At the end of Carroll's book, we find ourselves abruptly returned to the "real world," while Paquette's novel ends in "Dreamtime." Most significantly, while Alice's "adventures" are a product of her imagination, Aisling's are real. The beautiful garden turns out to be a place that has ultimately no meaning or consequence for Alice, whereas Crowsnest Mountain, where Raven resides, becomes the site of the critical battle between good and evil, in which Kokum sacrifices her life to save that of her grandson. The site of the battle inside the mountain is described as an inferno and as a world out of control. As Aisling observes on their approach: "The Earthblood was flowing out of the mountain on which they stood but it was flowing the wrong way. Not down and out of the valley, but up toward what appeared to be a spinning whirlpool, sucked into darkness" (220). Here as elsewhere the novel's "imaginative geography" mirrors the ideas of Blackfoot scholar Leroy Little Bear about the similarities and differences between Indigenous knowledge and Western science. According to Little Bear, the Native paradigm comprises four principles: everything is in constant flux; all creation consists of energy waves/spirit; everything is animate; and everything is related. As he explains, any Native approach to renewal is thus necessarily holistic.

⤳ Intergenerational Trauma and the Birth of a New Generation of Elders

IN "HOME AND NATIVE LAND: A Study of Canadian Aboriginal Picture Books by Aboriginal Authors," Doris Wolf and Paul DePasquale observe that most of the picture books, which make up 75 percent of all narratives portraying the lives of Aboriginal children and youth in Canada, have an intergenerational focus and attempt to "sanitize the living conditions of Aboriginal youths" (103) because of the "pressures to produce ... positive images of Nativeness" (DePasquale and Wolf 92). The authors suggest that only a few of these books "go beyond offering positive depictions of the family as their sole decolonizing strategy" (96). In contrast, *Lightfinder* uses a number of decolonizing strategies, including Indigenous futurism, "transgenre," and the defamiliarization of literary sources, to underscore the importance of the "imaginative geographies" of environmental health and Indigenous space in which Indigenous peoples would be able to connect locally and globally in self-determination and would not have to choose between "tradition and necessity" (*Lightfinder* 51). Aisling's pursuit of a better academic education away from home, which in her opinion is a "necessity," deprives her of family and community, and Eric, who has a gift for drawing, will not be able to study at the Banff Centre for the Arts because, as he puts it, he is from the Rez (192), a reality that places him at a significant social disadvantage. In addition to actively resisting racial stereotyping of Indigenous peoples, the novel draws attention to social issues such as poverty, lack of funding for Indigenous education, substance abuse, and the high rate of youth suicide. But it also suggests renewal within the Indigenous community. By portraying the limitations of both the middle generation, those most severely affected by intergenerational trauma caused by the residential school legacy, and those of the Elders, who have not been able to keep up with the challenges of a changing world, the novel tackles the issue of how to deal with the tragedies of the past without standing in the way of the young generation's future.[7] Young people's broken relationship to the land as a result of colonization is one of the many relationships in need of healing.

Lightfinder opens with Aisling, who lives alone in her father's city apartment, waking up from a dream in which a mysterious voice utters the words "Save me" (1). As we learn later, it is Earth herself who is making this plea. When Aisling inexplicably collapses at school and is taken to a nearby hospital—the hospital staff suspects that she is on drugs because she is an Indigenous teenager—she is informed that her father had a fatal accident. In the novel's opening lines, the crisis of the future and the crisis of the present are thus intimately linked. As Emberley explains, "the use of physical catastrophes to implant a tangible sense of loss is a very common technique in Indigenous storytelling. However, there is just as often a shadow event that, by its very nature, is less tangible and, thus, more difficult to comprehend, such as 'social illnesses' ... Such storytelling is about the restoration of human relations as the very basis of what it means to be human, contrary to an ideology of individualism that runs through the more dominant and contemporary driving forces of 'Western' society" (Emberley 215). The premature death of Aisling's father puts the rift between generations in focus. Alone in the city, Aisling ponders that her Kokum or her mother might have been able to figure out what the dream means, "but both of them were out of her life" (*Lightfinder* 1). Aisling has special gifts and was born with "a second set of eyes" (63); this enables her to see the veins of Mother Earth as well as "earthlight" (65). She can also hear Mother Earth telling her she has been waiting for her to help "stop the ending of the songs" (94). Earth explains to Aisling that Raven and his dark forces have returned to destroy her: "He digs and poisons. He is ending life, ending the songs. He must be stopped before we are all silenced in order to feed his empty soul. He can never be filled and yet he tries" (95). Raven's insatiable appetite is a reference to that of the Wendigo, the cannibal spirit of the North. Although her parents and Kokum have long been aware of Aisling's gifts, they neglect to talk to her about the importance of this inheritance. Oral transmission of cultural knowledge has been almost completely interrupted in this family. The lack of transmission of Indigenous knowledge from one generation to the next goes hand in hand with the loss of Indigenous languages, and the alienation from traditional food, which the novel refers to humorously when Aisling suspects that her dreams might have been affected by consuming "too much pizza and Diet Coke" (1). All she can remember of her cultural heritage is a couple of songs that Kokum used to sing to her when she was a child and her appreciation for her Auntie Martha's bannock.

However, the family members are connected through their special abilities. All members of Aisling's family have spiritual gifts, which are, as the text explains, genetically passed down from one generation to the next (64). Aisling's mother has the gift of dreaming the future (65), an ability that overwhelms her because she is unable to prevent tragedies from happening: she has turned to alcohol to drown her fears and lack of control over future events. Kokum has the gift of making everybody "relax and feel safe" (187). For this, she is being criticized by Buffalo Calf Woman—another traditional character in the novel, who provides Aisling with important information—as it has made her "far too complacent" (187). Eric, like Aisling, has the gift of seeing "Earthblood," and he is being sought by Raven, who is unable to do so, as an accomplice in the destruction of the world. Both siblings are able to "suck" energy from the veins of Mother Earth in a quasi-vampiric manner. But whereas Aisling does so to heal, to preserve, and to foster connections, Eric taps into the veins of the Earth to gain more power and control. In doing so, he becomes more and more like the Wendigo. The narrative thus makes it clear how important the teaching of Indigenous knowledge is to guide young people in using their abilities respectfully for the greater good of the community. During their travels in pursuit of Eric, and ultimately Raven, Kokum and Auntie Martha initiate Aisling, but they admit that they are not prepared for Raven's return. The novel repeatedly describes Kokum as tired, lacking energy, and not recognizing the portents of evil. At a crucial turn of events, both Kokum and Auntie Martha are fast asleep while Aisling battles yet another of Raven's helpers, her non-Indigenous schoolmate Jake, who has invited himself along on the journey and has pretended to care about Aisling's well-being.

After a particularly intense dream about Mother Earth, Aisling's body is marked as that of an Elder. She wakes up with two white strands of hair flowing from her temples. Her Kokum explains to her that she has been given a gift, "a mark of favour that hasn't been seen in generations, not since the days of the buffalo" (102). With the help of a healing song that her grandmother teaches her and a present from one of the mythic characters, Walking Man, Aisling is able to make plants grow, whereas Eric is given a bloodstone by Raven's son Cor, whose task it is either to make Eric a willing accomplice on their journey across Alberta or to kill him.[8] According to Kokum, bloodstones are very dangerous because everything they touch "becomes stronger. More emotions than anything else" (152). As she explains, the "last time anyone even saw one was back in the days of Louis Riel and the Red River battles. It fell from person to person and made them crazy. One man even claimed to have become Wendigo and he slaughtered his entire family" (152). To maintain its power, the stone needs to be fed blood. In critical moments, Eric uses a knife to make himself bleed to gain more strength from the bloodstone, which empowers him to knock one of his bullying classmates unconscious and to overwhelm a police officer who attempts to detain him.

Unlike Aisling, the Indigenous Australian Matari, who is only slightly older than she is, has many traditional skills and great knowledge of Indigenous science. The group comes to rely heavily on Matari and his boomerang and bullroarer for survival. At the beginning of their quest, he uses a "very rare and very old Anangu tracking technique" (43) to determine in which direction they should move to catch up with Eric. He also has the special "ability to see and know things that [are] hidden to others" (92). He explains to Aisling:

> Every year, our family and many others gather in the outback near Ayer's Rock—or Uluru, that's the real name—for the Dreamtime, and we share songs a lot like the one your Gran just sang for you. Our family always has the strongest dreams. The forms, the animals are different—the Raven is a Vulture or a Shadow in our telling—but the tales, the visions are near the same. After all, the danger's the same for all of us, unna? (91)

Matari saves her twice from attackers sent by Raven to kill her and nurses her back to life after Raven tears away half of her leg. Her healing occurs in "Dreamtime." According to Kevin Gilbert,

> the Dreaming is the first formation, the beginning of the creative process of mobile life/spirit upon and within the land. It is the days of creation when the Great Essence, the Spiritual Entity and minion spirits

formed the Aboriginal version of the "Garden of Eden" and recorded that creation and the laws abounding upon the tjuringas. (Gilbert xix)[9]

The "Dreamtime" itself is an "imaginative geography" in that, as Matari observes, "in the Dreaming you can reach a world of consensus, a world that makes sense" (*Lightfinder* 209). But Aisling needs to "stay in both worlds to allow the healing to complete itself" (178), which seems to imply that in an ideal world, Canada's Indigenous youth would be able to connect to their own "Dreaming" to facilitate healing. Aisling eventually returns to her Kokum and Auntie with a prosthetic leg. As she finds out later, her Kokum sacrificed her arm so that she could walk (207). The symbolism of this sacrifice is significant: the Elder generation makes sacrifices in passing its power and wisdom along. Before Kokum enters the final battle to rescue Eric, she admonishes Aisling: "You *must* take up my work, girl. And do a better job of it than I have" (235). At the end of the novel, Matari returns to reunite with Aisling, and the couple assumes the role of the new Elders in the mutual struggle to save the planet. With Kokum's return to the Earth and the transition from one generation of Elders to the next, death and birth touch and begin a new cycle. In the words of the narrator of Syilx writer Jeannette Armstrong's *Whispering in Shadows*: "I said that I would / give my flesh back / but instead my flesh / will offer me up / and feed the earth / and she will / love me" (296).

The representation of a young woman, often trained by her grand-mother, assuming the role of new Elder has a tradition in Indigenous Young Adult fiction in North America. I am thinking here of Anishinaabe writer and educator Ruby Slipperjack's novel *Little Voice* (2001), which describes the journey that Little Voice or Ray takes to become a medicine woman under the tutelage of her grandmother, who is a skilled and well-respected healer and midwife. Omakayas, the sole survivor of the smallpox that ravaged her family and community in Anishinaabe author Louise Erdrich's *The Birchbark House* (1999) and its sequels *The Game of Silence* (2005) and *The Porcupine Year* (2008), is a powerful dreamer, healer, and clairvoyant, who can hear the medicine spirit voices, particularly those of the bear clan, talk to her. On her journey, she has many mentors, including her adoptive grandmother and several animals, to guide her. In all of these novels the Elders do not seem to have any shortcomings, whereas Taylor's *The Night Wanderer*, to which I have referred earlier, addresses the need for a new generation of Elders persuasively.[10] Although she attended residential school,

Granny Ruth—who speaks Anishinaabemowin fluently and can read the land like no other—spends a good part of the day resting in her favourite chair. As the text explains, "Granny Ruth had seen her world go from growing up in a house where only Anishinabe was spoken and the outhouse crawled with spiders and flies, to a school where the teachers tried to beat all the Indian ways out of her" (*Night Wanderer* 159–60). Like Aisling's parents, sixteen-year-old Tiffany's are divorced, and their mothers are absent from the girls' lives. Tiffany's father is addicted to Hollywood TV sitcoms while Aisling's is preoccupied with work. In this novel, in which Taylor "culturally appropriate[s] a European legend and Indigenize[s] it" (Taylor, "The Night Wanderer"), cannibalism and greed are associated with the Europeans through the figure of the vampire. Vampirism is portrayed as a contagious European disease with which Pierre L'Errant was infected hundreds of years ago. Pierre, who is both a "good Elder" and the vampire/Wendigo, prevents Tiffany from committing suicide after a particularly upsetting fight with her father and her breakup with her abusive white boyfriend. Pierre is able to save her by teaching her about Anishinaabe history and reconnecting her to the land. When he tells Tiffany his own story in the guise of a vampire story at the end of the novel, he explains to her that he needs to die because the world has been changing while he is unable to do so (207). As discussed earlier, Paquette uses "slipstream" to show that despite ongoing colonialism the circle will remain unbroken as long as the new generation has the opportunity to learn from the Elders and is able to adapt this knowledge to ever-changing social and environmental circumstances.

∾ Relations versus Corporate Logic of Growth and Consumption

Lightfinder critiques capitalism and consumerism along with colonialism and the exploitation of the natural environment, all of which it shows to be closely connected. Cor, who carries an inexhaustible amount of cash, manipulates Eric by giving him a cellphone and other material goods and by treating him to junk food, to which Eric is addicted. When Eric finally faces Raven, he is surprised to find "a man seated behind a wooden desk," wearing "corduroy pants and a light-grey dress shirt under a sweater vest" (227), looking like a captain of industry on his day off. The narrator notes Eric's surprise: "Eric stood in stunned silence. This was not at all what he'd expected. Fire breathing dragons? A nest of ravens ready to claw him to pieces? Sure, but not this" (227). But Kokum sees through Raven's veneer. As she warns her companions: "He already toppled Turtle Mountain in our world. If he can topple it here he will have made a wound so severe he could call on the full force of the Power [the Earthblood]" (221). She realizes that Raven has almost succeeded. When Aisling and Kokum hold hands singing a traditional song in defiance of Raven, he observes, acknowledging the song's power, that Kokum's family's "bloodlines have grown stronger" (230). Kokum retorts:

> More than you know, old Raven, sitting here like a spider. You have been busy casting your webs, manipulating governments, playing corporations like pawns. Oh yes, you've put mankind to work for you, led humanity to foul his own home with the power of greed and fear. But it has made you blind to the individual, to the community. But then, you never could understand that in the first place, could you? That's what you fear, the bonds of love. The ability of our species to overcome our instincts and act with the good of our children and grandchildren as our guide. (230)

Not at all impressed by Kokum's speech, Raven claims that he has already "destroyed humanity's love of the earth" by giving them "comfort, luxuries, 4K televisions" (231). Kokum replies by telling him "we are rising. One by one" (231). The pronoun "we" seems to be referring to Indigenous communities and nations, an assumption that is confirmed by Raven's response: "Your young grandson, the future of your oh-so great nation, has become my willing partner" (231). Metaphorically speaking, what happens in this critical battle at the end of the novel is a showdown between the principles of corporate culture and consumerism on the one hand and kinship relations and Indigenous sovereignty on the other. As Cherokee writer and scholar Daniel Heath Justice in his discussion of kinship observes:

> Indigenous nationhood is more than simple political independence or the exercise of a distinctive cultural identity; it's also an understanding of a common social interdependence within the community, the tribal web of kinship rights and responsibilities that link the People, the land, and the cosmos together in an ongoing and dynamic system of mutually affecting relationships. It isn't predicated on essentialist notions of unchangeability; indeed, such notions are rooted in primitivist Eurowestern

discourses that locate indigenous peoples outside the flow and influences of time. (Justice 151)

On the final stretch of their journey, Kokum explains to Aisling that she, like her mother, is a member of the raven clan while her father belonged to the bear clan. Kokum herself, Auntie Martha, and Eric are foxes. The three women shapeshift into these forms to prepare for a confrontation with Raven. According to Abenaki writer Joseph Bruchac,

> the line between human and animals is so lightly drawn in American Indian cultures that it ceases to exist at certain points. The ideas of totem and clan offer some clue to this connection. *Totem* is a word that comes from *ototeman*, an Anishinabe word roughly meaning 'one's relative.' There is a mystical connection between a human and his or her totem. (Bruchac 160)

Eric temporarily transforms into a fox on his journey across Alberta, when suddenly, in Alice-like fashion, he falls into a different kind of reality, a "summer land" where he meets his future mate the fox spirit Skia. With his clan affiliation thus clearly established, Eric comes to the rescue of his family in the end, although he was temporarily let astray by the dark forces:

> Eric couldn't believe it. He never suspected anyone would come so far to try to help him. He was a fool for thinking he was on his own, that no one cared. But things had gone too far for him to turn back. Just knowing he had led his own family into danger was proof enough for him that he had to stop things now. (*Lightfinder* 228–29)

Both siblings experience psychological and spiritual rebirth at the novel's conclusion: Aisling, the dreamer turned activist, will be the new Elder, who will "help other people understand what they can do so they can help to heal the Earth" (240), while Eric, who earlier in the novel explained "I'm from the Rez. I just can't imagine going anywhere" (192), will be living in a utopian "summer land" for the rest of his life in the shape of a fox. Auntie Martha and Skia teach him the rules of proper fox behaviour, and Aisling and Matari promise to visit him often. Here as elsewhere, the novel asks to be read metaphorically. The respective outcomes of the siblings' quests suggest that Indigenous people need to overcome the effects of continuing colonization and move on to take leadership in healing their own communities and the planet.

The scene of the showdown between "good" and "evil" in *Lightfinder* is firmly anchored in the realm of fantasy and defies logic. One could also refer to it as "irrational" or "nonsensical," the latter being one of the characteristics often used to describe the illogical events in *Alice's Adventures in Wonderland*. *Lightfinder* also lends itself to be read alongside Métis author Warren Cariou's short narrative "An Athabasca Story," another Native fictional response to the environmental and social injustices created by the Alberta Oil Sands.[11] In his essay "Tarhands: A Messy Manifesto," Cariou warns readers that his "paper sets aside any attempt to make reasoned arguments about conservation or regulation, and instead embraces irrationality as the last possible mode of engagement with a contemporary public that will no longer listen to reason" ("Tarhands" 17). Posing the question whether Canadians should "embrace irrationality in response to" the Oil Sands dilemma, literary scholar Jon Gordon suggests that "An Athabasca Story" offers "an alternative form of (ir)rationality" (Gordon 107) in that "the utopian ideology of the community created through bitumen extraction is negated through Cariou's fictionalization of that community's treatment of Elder Brother" (96). According to Gordon, Elder Brother in the story encounters a world

> in which the rules governing normative behaviour are very different from what he is used to, a world in which acquisitiveness has replaced sharing and selfishness has replaced community. Elder Brother's actions, and his ignorance, in "An Athabasca Story," serve to remind readers of the normative rules of Cree culture and to point to an alternative idea of what the world might be. (97)

Gordon concludes that we might

> understand Cariou's call for an "irrational response" to bitumen extraction as an attempt to expose the flaws in the "rational" and "common sense" logic of capitalism, a move to "uncommon sense. This "logic" of capital is countered through the "irrationality" of relatedness. (107)

In a similar way, *Lightfinder* calls into question the rationale for drilling the Oil Sands. Paquette concludes his novel in the alternative world of "Dreamtime," whose logic contrasts with that of capitalism and consumption. As Matari observes, "in the Dreaming you can reach a world of consensus, a world that makes sense" (*Lightfinder* 209).

∾ *Conclusion*

IN HER INTRODUCTION to *Walking the Clouds*, Dillon points out that "Native apocalyptic storytelling ... shows the ruptures, the scars, and the trauma in its effort ultimately to provide healing and a return to bimaa-diziwin. This is the path to a sovereignty embedded in self-determination" (Dillon 9). She observes that all forms of Indigenous futurism are narratives of *biskaabiiyang*, an Anishinaabemowin word "connoting the process of 'returning to ourselves,' which involves discovering how personally one is affected by colonization, discarding the emotional and psychological baggage carried from its impact, and recovering ancestral traditions in order to adapt in our post-Native Apocalypse world" (10). All Indigenous characters in *Lightfinder* "return to themselves" and to "the Earth," albeit in different ways. Most notably, Aisling as the new Elder complements the gifts that were passed down genetically and culturally to her with Indigenous knowledge and Indigenous scientific literacy to save the planet. The novel ends on a hopeful note, although the last chapter refers to Raven's potential return some time in the future. On their way to Crowsnest Mountain, Kokum, pointing at Bow River Valley, which today is part of Banff National Park, observes,

> "This is the place White Buffalo Calf Woman brought the last great herd of free, wild buffalo before the white man could slaughter the last of them. They entered her Gateway and were saved. When the time is right, they say the mountain will shake and people will know: the time for the buffalo's return is at hand." (210)

The novel thus draws attention to the special relationship between Indigenous peoples of the Plains and the buffalo—the myth of the disappearance of the buffalo often goes hand in hand with that of the "Vanishing Indian"—and shows that visions of the future arise out of Indigenous knowledge and prophecy rather than out of Western scientific insights.

By reclaiming "space," that is, by reconnecting with tradition, language, and the land, Indigenous people are able to heal not only themselves but also humanity's relationship with the natural environment. As Paquette reminds readers, Indigenous peoples in Canada are alive and well and are in a good position to guide the country on issues of social and environmental justice. In writing this novel, he also reclaims a literary genre that is often wrongly considered exclusively European. *Lightfinder*, which is firmly rooted in contemporary Canadian geography and Indigenous reality, is no Native *Harry Potter*, but rather an enthralling example of "imaged earth writing" with a political message.

<hr />

Notes

1) I approach *Lightfinder* and other Indigenous texts discussed in this chapter from the limited perspective of a white settler scholar.

2) In many dystopian societies portrayed in fiction, from Margaret Atwood's *Maddaddam Trilogy* to Jean Hegland's *Into the Forest* to Suzanne Collins's *The Hunger Games* to Algonquin playwright Yvette Nolan's *The Unplugging*, people live in fear and societies disintegrate because of environmental disasters that make the planet uninhabitable.

3) See John Goddard's *Last Stand of the Lubicon Cree* for an analysis of the Lubicon Nation's ongoing struggle against the human rights violations committed by the Canadian government and the oil industry. Dene and Métis as well as dozens of other First Nations communities have been affected as well.

4) Her father's association with the Oil Sands can be read as a reference to the economic boom of Fort McKay, which has come at significant environmental cost for the Cree, Dene, and Métis communities in the area as the Athabasca River is now too polluted to drink from, and the fish are too contaminated to eat.

5) See Dillon's introduction to the Indigenous futurisms reader edited by her. Dillon uses the term "Native Apocalypse," instead of "Native dystopia," as one of the subgenres of Indigenous futurism.

6) Like other teenagers, Aisling is self-conscious about her appearance and is frustrated that she is unable to recharge her cellphone out in the bush.

7) In appreciation of her family's sense of humour, Aisling observes: "The humour of their family and of their community always helped them to get through the hard

times. And goodness knew they had seen hard times. The words 'Residential School' flashed through her mind and she shuddered" (64).

8) She is thus quite a contrasting figure to the Anishinaabe character Gabriel Quinn in Thomas King's environmental novel *The Back of the Turtle* (2014), who, as the brains behind the powerful defoliant GreenSweep, is complicit in a major environmental disaster and the death of an entire Indigenous coastal community in British Columbia.

9) According to the *Encyclopaedia Britannica*, *tjuringa* in Indigenous Australian tradition is "a mythical being and a ritual object, usually made of wood or stone, which is a representation or manifestation of such a being."

10) I am indebted here to Donna Ellwood Flett's perceptive reading of Taylor's novel. As she writes, "tied directly and intimately to these aspects is the Gothic spectre in the novel, Pierre L'Errant: he is the vampire villain who passes his cultural and historical knowledge to the young, contemporary, heroine Tiffany, thereby transforming to hero at the same time as he transforms the role of Elder as the carrier of knowledge from older male to younger female. These transformations offer some unusual commentary on dealing with the sins of the past and imply a rethinking of the role of Elders in today's growing, youthful Aboriginal population" (Flett 25).

11) See also Dogrib author Richard Van Camp's story "On the Wings of this Prayer," the short film *Mihko* by Cree filmmaker Myron A. Lameman, and Cariou's documentary *Land of Oil and Water: Aboriginal Voices on Life in the Oil Sands*.

7) History, Hills, and Lowlands

In Conversation with Janet Lunn

❧ AÏDA HUDSON

═══════════

J anet Lunn was a beloved children's author whose passion for history shone throughout many of her novels, picture books, and non-fiction. Her historical novels *Shadow in Hawthorn Bay* (1986), *The Root Cellar* (1981), and *The Hollow Tree* (1997) picture Canadian, Scottish, and American geographies that revolve to a greater or lesser extent around Hawthorn Bay. This trilogy fictionalizes Prince Edward County on the shores of Lake Ontario, whose first settlers included the family of the man she married, Richard Lunn. It was in an old County farmhouse that their five children spent much of their childhood and where Janet started writing full-time. Set in the early 1800s, *Shadow in Hawthorn Bay* moves from the moors of Scotland to the Lowlands of Upper Canada and follows a young Scottish girl who has second sight. In the second novel of the trilogy, the root cellar is the portal for time travel between the present and past, between the present-day farmhouse pleasantly haunted by a ghost and the American

Civil War in the 1860s. *The Hollow Tree* begins in 1777 during the American Revolution and follows the journey of a young heroine and pacifist through New England to Hawthorn Bay.

Janet's other writerly pursuits include *The Story of Canada*, which she wrote with Christopher Moore, a beautifully illustrated and highly readable history for the young that has seen many editions, the latest being in 2016. She wrote book reviews for the *Kingston Whig-Standard* in the 1950s, a time when no one in Canada was reviewing children's books. In 1967 her history of the county, which she wrote with her husband, simply titled *The County*, was published. In the mid-1970s she became the first children's book reviewer in Canada for Clarke Irwin. She was also the first children's author to head the Writer's Union of Canada. Her many honours include two Governor General's Awards, the Vicky Metcalf Award for her body of work, the Order of Ontario, and the Order of Canada. Although history was her main focus whether in the fiction or the non-fiction she wrote, her love of the land and her cherishing of family also run through her writing. Her early years in Vermont and her life in Prince Edward County helped shape her literary geographies, particularly in *Shadow in Hawthorn Bay*. Janet Swoboda Lunn was born in Texas in 1928. She passed away in Ottawa in June 2017.

∞

As a historical novelist, what do you think the function of geography is in historical fiction?

I think that geography has a lot to do with how people are in life and in fiction. Hill people are not like Plains people or like Sea people or River people (who are not like each other, either). And stories are told about and out of the sensibilities of these different people. We are moulded, in part, by where we are born and raised. I don't think Kevin Major could have written W.O. Mitchell's stories, nor could Mitchell have written Major's. The geography is too different and so is the culture. But, then, culture and geography come in one package.

As for geography in specifically historical novels, it probably matters most in immigrant stories. My own interest in writing historical fiction was sparked by wanting to find out how the big, dramatic moments in history have impacted ordinary lives. One—I think only one—of my books came to be because of geography. That was *Shadow in Hawthorn Bay*. I read in one of Catherine Parr Traill's books how happy she was, when coming to Upper Canada, to find (and it was in her words as I recall them) that "the Scots and Irish peasants had left their foolish superstitions behind them." I remember sitting holding that book in my lap for long time thinking, "Lord that must have been painful."

I had read a lot about the early Scottish settlers and how terrified they were of the deep forests, because the highlands of Scotland had been treed centuries earlier (there are far more trees in the highlands now than there were two hundred years ago—plantations). And those mountains, great, sheer, looming shapes of grey mist, were so very different from the clear, bright clearings surrounded by dark forests in lowland Upper Canada and Nova Scotia. I thought about those differences and that those mythical creatures in those treeless, misty mountains could never have been imagined in all our bright light nor could they survive here. So, perhaps, in this book, the geography is as much what it's about as the history. Maybe more.

In my other two historical novels, and in *Dear Canada, A Rebel's Daughter*, that is not the case. In *The Hollow Tree*, I wanted to write about a girl caught in the American Revolutionary War who had no interest in it. I did use the geography I knew and loved, the village in Vermont where I lived when I was a child, and Prince Edward County, Ontario, where my husband was born and grew up and where I spent so many years. Prince Edward County was settled by United Empire Loyalists; their descendants were my neighbours. With my husband, I wrote a history of the county for the centennial year, 1967. When I was reading those Loyalist land grants in the Ontario archives, there were the names I knew so well, names I saw on mailboxes, some on my own road. So, when I came to write Phoebe Olcott's story, I guess it's obvious that I would take her from my old home in Vermont to the Loyalist settlement where I was living.

Which was your village?

My village was called Norwich, but I didn't call it that in *The Hollow Tree*. I called it Orland village, but it is Norwich as I think it would have been in 1777. Hanover, across the river in the story, is very much the real Hanover. In fact, in the book, some of the names of the background characters (the blacksmith, the tavern keeper) are their real names. (Later in the book, I used a real historical character as more than background, a man named Justice Sherwood. He's a minor but important character.)

I chose Norwich, Hanover, and Prince Edward County because they are important to me; Phoebe could have lived in almost any New England village and ended up in any United Empire Loyalist settlement along Lake Ontario. It was the Revolutionary War, not those specific places, that drove the story.

It wasn't geography that drove *The Root Cellar*, either. It was my own house. That's it on the wall.

That's it, is it?

Scott Cameron painted that picture for the deluxe edition of the book as the house probably looked in 1865. By the time we had the house, it had no porch and the bricks were in such bad repair, we covered them with yellow stucco.

Is it in Hillier?

It's in Hillier Township outside Hillier village on South Pleasant Bay Road beside Pleasant Bay. The house is so dilapidated in the present, but so vibrant in Will's and Susan's time!

Yes, dilapidated is sure the way it was when we moved in. I described the house in the book absolutely as we found it. It also reflects a state of mind. A child confused and a house in disrepair are analogies for one another. I didn't think that while I was writing, but, later, I realized that I must have made the subconscious connection. I'm sure I just thought, "Oh, yes. That works." And we did have a ghost.

How does that tie in with the fact that shadows and ghosts play an important role in your book?

The ghost in our house was where the story started. After Richard saw her, I knew I had to write a story with her in a central role. I have always been intrigued by ghosts. Ghosts and shadows. "Shadows." When I say that word, it sounds negative, but I don't mean that at all. I have always lived with the ghosts of the past.

My earliest memories are of our house in Vermont. It was an old farm-house built in the late 1790s. When I was small, I imagined all sorts of ghosts there. My sister Martha was a good storyteller; she was four years older than I, and I don't suppose I was more than four when she told me that the reason the brook that ran behind our house was called Blood Brook was because of all the battles between the Native Americans and the settlers. I lay awake long into many nights imagining those terrifying battles; I could almost hear the shrieks and shouts. Actually, there were never any battles in that area. Years later, when we buried our mother in the Norwich cemetery, I found that the cemetery was full of people named Blood. But, until then, I believed every word of my sister's story. Probably the first people who owned our house were called Blood.

Let's go back to The Root Cellar. *Why is the root cellar your time travel portal?*

Alas, our Hillier house did not have a root cellar. (I'm sure it still hasn't one.) One afternoon, while I was working on this book, friends from the city came to see us and we were showing them around our four acres. Out behind the house, I stumbled over rough ground and fell and broke my foot. My mind was only half on what I was doing, in any case, because, when I am in the middle of writing anything, my head is more off in the story than it is in the immediate three-dimensional world. I had been fussing and fussing how Rose was going to go into the past. I think it was only a few days after my fall that my two worlds came together in one of those ah-ha! moments that we all love. I remembered telling our friends that the one thing we didn't have that most of the other nineteenth-century farm-houses had was a root cellar and I suddenly thought, "There *is* a root cellar. There has to be. It's an obvious place. Down underneath."

Oh, I was happy when that came to me! It's true. Most old farmhouses do have root cellars. Our house must once have had one, because the house dates back to about 1820. It had obviously disappeared. So my Hawthorn Bay house got its root cellar. I didn't have to make up the fireplace and the bake oven in the kitchen; they were there. (There's a Scott Cameron painting of those, too.)

There are so many magical passages in this book. There is a two-page description (48–49) of the bay and the trees and flowers in spring as Rose and Will sit there in the boat. The trilliums and other wildflowers are in bloom, the bluebirds and orioles are singing, and Will is playing his song on a flute. Whenever that song is echoed later in the book, it seems to recall this wonderful "spot of time." Did you do this on purpose?

No. But that "spot" was on the bay behind our house and I loved it so much. Some of what went into my feeling about that passage is that my son John wrote that song. He wrote it when he was a teenager (when I was writing the book; now he makes flutes). Here's another thing I didn't do on purpose. Will turned out to be a lot like John. Furthermore, in one drawing in the book, Will looks a lot like John.

Did the illustrator ever meet John?

No.

I would like to hear more about Mairi in Shadow in Hawthorn Bay, *about how she functions in the two geographies of Scotland and Canada.*

Mairi is terrified of the lowland. As you know, she won't go into the forest; she's not happy with it. That's why she likes the open. That's why she likes the little house in the clearing, partly because it had belonged to her aunt and uncle, but also because it is in a clearing. It's right near the water. She's drawn to the water. That's the ghost part of the story, but she was also drawn to it, in the beginning, because it was clear. It was open. She likes the water. She's a creature of the hills and the mists of Scotland.

Here's a passage about her connection to those hills; I wondered if you could comment on it. "Mairi loved the land fiercely. She felt as though she had been born out of its earth, that she was kin to the whin and broom and heather that grew so profusely over the hillsides, that were tiny unseen roots growing along her body, reaching out for the land, drawing nourishment from it. Once she had told Duncan, 'When I am old, I will lie myself on the hill and my roots will push themselves into the earth and I will sleep. The grass will come to cover me then, and I will be part of the hill forever'" (10–11).

Even hearing this passage read over, I want to cry. This is how I have always felt about hills. I always thought I would be buried in one, just as Mairi does. The hills of Vermont are my heart's home. Totally, completely. It's not the United States. It's not even the state of Vermont. It's those hills in that particular piece of Vermont that I love. So, for me, bringing a hill person into low land was something I could do because her feelings about it came out of my own deepest feelings. I was born in Dallas, Texas; I was only six months old when we moved. We lived, for a year and a half, or two years, with my mother's family in New Jersey (just across the Hudson River from Manhattan), because my father was out of work. That was in 1929. Then we moved to Vermont. So my earliest memories are there. We left when I was ten and my life has taken me to other places, to meet other people. And my life, my work, my husband, and my children and now their children and grandchildren are or have all been here in Canada. I'm happy here, I would not go back to those hills to live, but, like so many people who have left their first nests, my heart's home is in mine.

Mairi, like you, is finally happy in Canada. Shadow in Hawthorn Bay *ends with that and with her comment about "the old ones." Who are "the old ones"?*

I knew the end of Mairi's story before I started the book. It was what the book was about. As a sort of aside, I remember once hearing Timothy Findlay, in a lecture, saying, "I never know the end of my book." I can't imagine that, I always know the beginning and the end of my stories. The adventure is the journey.

As I said before, I had had this experience of reading Catherine Parr Traill's books and then later I read works by Douglas Le Pan and George Grant about the loss of mythology to the settlers in Canada. Because I was neither a poet nor a philosopher, but a story maker, that loss came to me as story. I could almost feel the pain of losing that mythology. For me, moving from one culture to another, even though there is little difference between the US and here, though not excruciatingly painful, was not easy. Even moving from one part of Canada to another is not easy. Friends of mine who grew up on the prairie and have lived their lives in Ontario feel dislocated. (Interestingly, there's a piece in *Canada's History* magazine that talks about this very thing.) And in the nineteenth century, the distance was so far because travel and communication was so slow and expensive, the change so big, and, sometimes, the customs and beliefs so different. Those very practical New Englanders who settled this province (I grew up with them; I know them well), would have no patience with Mairi's understanding of the world. And to her all those familiar mythical creatures were real, her second sight was certainly real and ghosts were real. (They are real to me, too.)

And they, Mairi and Duncan and Luke and their like, become the mythology, the old ones. But after a very long time. We European immigrants haven't been in the country that long. Only the native people have been here long enough to be the old ones. So far.

Old ones for Mairi in Scotland would have been those creatures, the fairies and other mythical creatures, but in Canada she and Luke would be the ones others would tell stories about. Those stories would become more mythical as the centuries went on. A lot of centuries. Think about Irish history. Think of the Tuatha Dé Danann. Were they real? Nobody really knows. Living in countries like Ireland or Scotland where people have been almost since people first migrated from Africa, the old ones are the mythology. Our old ones are still in Europe. They live in the shadows of time. We don't know who they were. Give this country time.

I am in my eighties now and life has come to seem a little surreal. Quite often, I find myself pulling back from the everyday world, thinking about the length of human history and how the parts might fit together, and then about us as a species. What a funny set of critters we are, what an exotic part of the animal kingdom!

How does that fit in with your giving Phoebe a bear and a cat in The Hollow Tree?

We're not separate from the other animals. We share DNA from the very first mammals in the time of the dinosaurs. And when people talk about loving nature as out there, separate from us, I think, as Desmond Morris wrote in *The Human Zoo*, that we have created a zoo to live in. But we are part of nature. We still bleed and breed, live and die. We are still part of everything. Not separate, though a great many of us seem not to care to know that. I do care to know it.

Let's go back to Phoebe in The Hollow Tree. *She too, like Mairi, faces the woods.*

She doesn't hate the woods the way Mairi does. It's her habitat.

How did you come to make her journey through the woods to Canada so real?

Well, I grew up in the 1930s on a hillside. Farms and villages in Vermont in the 1930s weren't little openings in the deep woods, but we had plenty of woods. We lived a mile and a half outside the village and the slope on the other side of our brook was covered with woods. I loved it. Now this was not deep forest, but it was easy to imagine that it was. And we read in school about the early settlers carving those little clearings in those deep forests. Little clearings full of stumps. I could always imagine that. But the forest that I described in *Shadow in Hawthorn Bay* wasn't what I had imagined as a child. It was the one my husband Richard's great-grandfather told Richard's father about. The primeval forest. Great-grandfather was the one who said that the trees were so wide apart you could drive an oxcart through them. That's not very long ago.

I stayed with a Scottish writer, Molly Hunter, a few kilometres outside Inverness, when I was researching for *Shadow in Hawthorn Bay*. I told her about the primeval forest the way Great-grandfather had described it. She could hardly believe me. She said, "And my people have been farming here for two thousand years." She lived where I set the opening of *Shadow in Hawthorn Bay*. That difference in landscape is central to the Canadian story. We are all immigrants here, all but the Aboriginal people—including the French who have been here longer, but by only a hundred years or so, not thousands.

Did you think of The Hollow Tree *as a more political book than* Shadow in Hawthorn Bay *or the others?*

I didn't think of it as political, but of course it is. It's an antiwar story. My focus, when I was writing it, living with it, was with Phoebe Olcott and how she was going to deal with what happened to her, which is what all my stories are about.

In a way, *Shadow in Hawthorn Bay* is political in that it deals with the roots of this country, of the New World and the Old. A Scottish girl coming to Canada is such a Canadian story. When I came to Canada I was seventeen. I came here to Ottawa and went to Notre Dame Convent School. I had a little radio. I was not Catholic (the Mother Superior made sure I went to my own church on Sunday). When there was a Mass or some other religious event going on, I would be in my room, often listening to my radio, and I would hear a lot of Scottish music. Even at seventeen, I knew I was in a very Scottish place! That isn't true anymore. But this was right after the war in 1946, so ethnically, it was still like pre-war Canada.

One of the remarkable things about your three novels is the interconnectedness of the families. Isn't a love of home at the heart of them?

Home. The most magic word in the English language. I'll bet it is in every language. I loved writing about my own home in *The Root Cellar*. Richard loved that book. He was very moved by it. It was about his county. He was a Prince Edward County boy through and through. I used to say of him and his friends that they were all on elastic bands. Get them more than a hundred miles from home and boing! back they would go! The only consolation I had when Richard died was that I could bury him in his county with his mother and his father and his grandfather. Home. It's a thing immigrants think about a lot.

Any closing thoughts about geography in your books?

Closest to my own heart and the most geographical of my books is *Shadow in Hawthorn Bay*, because of my own feelings about hill country and low country, hill people and low-country people, and the culture that grows from each. The Vermont refugees in the later book, *The Hollow Tree*, came from hills, too (as I have certainly mentioned). They also end up in low country, but that book doesn't deal with it. It's story is a refugee trek. The geography is just whatever geography that they had to deal with on the way. Dealing with their new Lake Ontario country was to come later for them.

On a side note, did you have environmental issues in mind when you had Uncle Bob attend a conference on the health of the fresh water of Ontario and New York in The Root Cellar?

I wanted them there, but I didn't want to make any noise about it. In the same way I referred to the doctor who attended Aunt Nan only as "she." I like slipping things into stories without flagging them, things that not everyone will notice but are sometimes important to me—sometimes just fun. The name of one of Aunt Nan's books is *Shadowbrook Farm*, the name my grandmother gave to our farm in Vermont. I did that for my sisters.

Connections are so important in your books. For example, Rose finally connects with Aunt Nan when she writes her stories down for her.

I have always written about connections. I have always had strong feelings about connections. The connectedness of life, of the world—of the universe—has been at the back (and often the front) of my thinking all my life. With E.M. Forster, I treasure the words "Only connect." I sometimes think that the history of the world is like one great big long conversation. I've pictured it like a huge skipping rope. We all get into it one way or another. Some of us get in only a passing sigh and some of us get in lots of words— political figures, military big shots, a few artists. Some of us get in a word or two. But it's all part of the human conversation. I teach the writing of family stories in my local seniors centre and I really want all those stories to get into the public archives, so that historians and writers of historical fiction will have them. These family stories, every bit as much as the great events, are what knit us together as a people. This is why I write historical stories; I want to find—and pass along—the stories of those ordinary people behind the big events of the past, to get them into that conversation.

Thank you for bringing me into your conversation.

Two

GARDENS AND GREEN PLACES

8) How Does Your Garden Grow?

The Eco-Imaginative Space of the Garden in Contemporary Children's Picture Books

❦ MELISSA LI SHEUNG YING

> Gardens, like the wild places of nature, are the premises of transcendence.
>
> — Des Kennedy

═════════

As one of the greenest and most recognized landscapes present in the field of children's literature, the garden has always held a fine balance between delight and death. From the healing, protective space of Frances Hodgson Burnett's *The Secret Garden* (1911) to the sad place of parting for Lyra and Will in Philip Pullman's *The Amber Spyglass* (2000), the garden has become the ultimate site for the "renewal and regeneration [that] should follow a green death" (Carroll, "Green," 54). As such, the cyclical nature of the garden not only "establishes the link between green space and death" (54)—it also highlights the important place the child may occupy within that very system of growth and decay.

Although the foundation of death and regeneration sprung from the Edenic myth still figures greatly in children's literature today, there has been a notable shift towards the garden as a positive landscape when depicted in the format of contemporary children's picture books. Driven at times by "our current clamour to be 'green'" (Ricou 3), contemporary picture book authors and illustrators have chosen to display the garden with an increasingly optimistic view that both carries and teaches the undertones of an environmental awareness hidden—some more than others—just below the surface.

Educating children into a deeper consciousness and appreciation of the natural world around them is one powerful way in which we can, as Sidney I. Dobrin and Kenneth B. Kidd state, "develop the positive attitudes towards the environment that are so crucial to its preservation" (7). Thus, as we tie the task to educate with ecocriticism's "earth-centred approach to literary studies" (Glotfelty xviii–xix), the child's environmental imagination serves as a crucial link that comments upon the relationships between family history, love, memory, and the sense of self that is cultivated in the imaginative space of the garden. It also accentuates how that environment truly nurtures the intellectual, psychological, and physical growth of the child.

Each of the three picture books I have chosen to bring together—Lane Smith's *Grandpa Green* (2011), Peter Brown's *The Curious Garden* (2009), and Andrew Larsen and Irene Luxbacher's *The Imaginary Garden* (2009)— strongly value the environmental experiences their child protagonists undergo in order to reach a deeper understanding of themselves and of the garden spaces around them. But before we turn to the books themselves, we must first look at how the fields of ecocriticism and children's literature come together in ways that explore "the real and imagined wild places of childhood [through] the interplay of children's texts and children's environmental experience" (Dobrin and Kidd 14, 1).

~ *Ecocriticism and Children's Literature*

ECOCRITICISM in its best-known and perhaps most quoted form is "the study of the relationship between literature and the physical environment" (Glotfelty xviii). The term was said to have been coined by William Rueckert in his 1978 essay "Literature and Ecology: An Experiment in Ecocriticism"; however, it was with Cheryll Glotfelty and Harold Fromm's

1996 collection *The Ecocriticism Reader: Landmarks in Literary Ecology* that the field started to become truly defined in its academic scope and collective movement.

Glotfelty and Fromm's anthology demonstrates the importance of studying how writing reflects and influences the ways we interact with the natural world. How nature is represented in a sonnet, what role the physical setting plays in the plot of a novel, how our metaphors of the land influence the way we treat it, and how the concept of wilderness has changed over time are only some of the questions that, according to Glotfelty, ecocritics and theorists should be asking (xviii–xix). These queries lead to a much higher level of earth awareness that also invites us to think about the ways in which environmental crisis is seeping into contemporary literature and popular culture, as well as what affect that has on our processes of literary analysis (xix).

The fundamental premise that there is a connection between how human culture affects and is affected by the physical world remains at the root of all ecological criticism and shares a common motivation: that we have arrived at the very moment of potential global catastrophe. Our awareness of the fact that we are reaching or have reached the age of environmental limits has impacted the ways in which the humanities are attempting to contribute to environmental restoration. Literary scholars, in particular, are contributing to environmental thinking through their questioning of values, meaning, tradition, point of view, and language, while reaching out for more interdisciplinary methods or approaches to expand how we may think about the canon in light of environmental problems (xxii).

The future vision of environmental dialogue expanding across multiple disciplines is also supported by another influential name in the field: Lawrence Buell. Ecocriticism for Buell has shown itself to be a multidisciplinary field in its ability to take on all literary discourses and histories. However, its emergence as "an increasingly heterogeneous movement" has prevented it from being on par to the standings achieved by other fields such as gender, postcolonial, or critical race studies (*Future* 1). Buell believes that ecocriticism will find its footing, but only after overcoming "a path bestrewn by obstacles both external and self-imposed" (1).

One reason why environmental discourse appears to have become more crucial in the last third of the twentieth century instead of at an earlier time is that the environment has become front-page news:

As the prospect of a sooner-or-later apocalypse by unintended environmental disaster came to seem likelier than apocalypse by deliberate nuclear *machismo*, public concern about the state and fate of "the environment" took increasing hold, initially in the West but now worldwide. (4)

The cross-disciplinary alliances arising from environmental issues "have had the positive and permanent advantages of stretching [ecocriticism's] horizons beyond the academy and of provoking a self-examination" that has encouraged the movement to evolve beyond its "initial concentration on nature-oriented literature and on traditional forms of environmental education" (7).

In his more recent article "Ecocriticism: Some Emerging Trends" (2011), Buell continues his support of the cross-pollination of disciplines from ongoing debates over how we can now rethink the frameworks and vocabularies used for currently existing ecocritical categories. Ecocriticism, according to Buell, can be broken into two waves of thought: the first typically privileges rural and wild spaces over urban ones; the second works against this by arguing that metropolitan landscapes and industrial environments should also be "considered as at least equally fruitful ground for ecocritical work" (93). Regardless of which wave one chooses to pursue, the environmental imagination remains crucial in linking how literature and the environment come together to discuss humans and—as well as within—nature.[1]

For both Buell and Glotfelty, ecocriticism still faces "a broad scope of inquiry and disparate levels of sophistication" (Glotfelty xix). But as the two prominent figures considered by many to have initiated the environmental movement into a field with their publications in the late 1990s, Buell and Glotfelty have rightly anticipated the "greening" of the humanities in a fitting and productive way.

That said, much of the earlier ecocriticism available has focused mainly on environmental literature for adults—a point Greta Gaard notes in her article "Children's Environmental Literature: From Ecocriticism to Ecopedagogy" (2009).[2] The field of ecocriticism slowly began its shift towards children's literature when a number of publications began to notice and address this missing piece of scholarship. Special issues such as *The Lion and The Unicorn*'s "Green Worlds: Nature and Ecology" (1995) and the *Children's Literature Association Quarterly*'s "Ecology and the Child"

(Winter 1994–95) raised significant "critical questions about environmental rhetoric in children's books" (Gaard 325).

Carolyn Sigler's article "Wonderland to Wasteland: Toward Historicizing Environmental Activism in Children's Literature" (1994) has also been invaluable in tracing the development of "the traditionally pastoral ethos of children's literature" (148) into ecocritical studies. In her overview of three hundred years of nature representation in children's literature, Sigler uses Glotfelty as a key reference to explain how ecocriticism as an interdisciplinary field of inquiry "decenters humanity's importance in nonhuman nature and nature writing … and instead explores the complex interrelationships between the human and the nonhuman" (148). By the time Sigler's article was published, ecologically aware books had been flourishing for just over twenty years, and arguably, this pace has yet to slow as we fast-forward to today, nearly twenty-five years later. Contemporary articles by scholars such as Jenny Bavidge ("Stories in Space: The Geographies of Children's Literature" [2006] and "Vital Victims: Senses of Children in the Urban" [2011]), and collections like *Experiencing Environment and Place through Children's Literature* (edited by Amy Cutter-Mackenzie, Phillip G. Payne, and Alan Reid [2011]), continue to demonstrate the interest in and investigation of the child's relationship with diverse environments and places.

In the process of "greening" children's literature, Sigler cautions that "concern for the nonhuman environment and a questioning of humanity's place within that environment" are concepts that are not as new as we might think, for they "can be traced back to the literary and cultural transformations that accompany the development of children's literature from the eighteenth century" (148). This return to a more biocentric setting—to the interrelations between the human and non-human—in contemporary literature for children indicates that "even isolated wonderlands and secret gardens [can be] endangered" (150) in the attempt to take on more activist measures.[3]

But just as how Glotfelty and Fromm's *Ecocriticism Reader* and Buell's *The Environmental Imagination* was said to have initiated and organized the field itself, environmental children's literature did not become formalized until Sidney I. Dobrin and Kenneth B. Kidd's 2004 publication of *Wild Things: Children's Culture and Ecocriticism*. As the first collection of its kind to join children's literature, children's culture, and ecocriticism together, *Wild Things* promotes educating children into developing a positive understanding of the natural world in order to ultimately preserve it. "Children are still presumed to have a privileged relationship with nature, thanks largely to the legacy of romantic and Victorian literature, which emphasized—often to the point of absurdity—the child's proximity to the natural world and consequent purity" (Dobrin and Kidd 6); and the complexity of applying and maintaining this concept in the twenty-first century is highlighted in the fact that children today participate in activities that isolate them from the natural world and, as a result, "may never achieve the familiarity with nature that is vital to environmental planning and activism" (7). What constitutes necessary and suitable nature education, its representation, and the balance between instruction and entertainment are at constant pull with one another; and although Dobrin and Kidd and the authors of the essays within the collection all acknowledge that "the child may still have a privileged relationship with nature, he or she must be educated into a deeper—or at least different—[environmental] awareness" (7).

∽ Ecocritically Reading the Child's Garden: Greenly, Curiously, Imaginatively

LOOKING AT CHILDREN'S LITERATURE with an ecocritical eye may be a fairly recent scholarly concept, but the influence of the natural environment on children, and on their literature, is not. In exploring the "interplay of children's texts … and children's environmental experience" (Dobrin and Kidd 1), reading children's literature through the lens of ecocriticism allows us to see how various urban and rural landscapes can be used to display the effects of global warming, air pollution, and the disappearance of animal and plant ecosystems through what publishers, parents, and teachers deem as appropriate representations for educating a young audience. As a result, the choices authors and illustrators make have a strong impact not only on the picture book as an object and product, but also on how we read and interpret the resulting connections between nature and childhood.[4]

In our first picture book, Lane Smith's *Grandpa Green*, the lasting impact of garden nature on a growing child is poignantly highlighted through a deceptively simple narrative about a boy, his great-grandfather, and the topiary garden they tend together. As the nameless little boy follows the trail of gardening tools left behind by Grandpa Green in his forgetfulness, a symbolic gesture towards how the environment may take shape through lively acts of the imagination brings the topiary shapes to life. One example of this is when the little boy comes across topiary in the shape of a little girl with braided pigtails, who is balancing on one foot with arms outstretched and reaching for a green, ball-shaped bush. The page itself is split horizontally in two so as to show a before-and-after effect, which compliments the strategic splitting of the sentence describing how Grandpa Green stole his first kiss in junior high school (Smith, *Grandpa* 12). As the boy looks coyly behind him before stepping on the ball and stretching up on his toes to successfully kiss the topiary girl, this lighter moment creates a doubling effect: the garden not only records the ordinary life of the boy's extraordinary great-grandfather, but also allows the boy to playfully use his imagination to re-create and interact with the garden itself.

As an alternative to a photo album or other device to show a life, author-illustrator Smith "thought it would be [visually] more interesting to show it all through trees and plants" (Zenz; *TeachingBooks*). From the opening illustration of a large bush shaped like a wailing baby—whose fountain of tears is revealed on the reverse page to be from the garden hose the great-grandson is holding—to a frowning little boy-bush sprouting stray leaves on top of his head, while red berries signifying the chicken pox have him stuck in a bed made of tree trunk bedposts, Smith's allusive artistry conveys a sense of eternal growth and displays the young great-grandson in his role as a momentarily carefree caregiver to the garden.

As the topiary shapes begin to display more mature moments that reflect Grandpa Green's transition from young boy to young man, so too does the great-grandson's imaginative perceptions of the greenery. Puffing gently on the fluffy remains of a dandelion, the great-grandson's peaceful moment alludes to both the wishing of a bright future and the last bit of youthful innocence that comes with momentous change. This image, accompanied by half a sentence detailing Grandpa Green's dream to become a horticulturalist after high school (Smith 13) carries similarities to the aforementioned split sentence that recounts the experience of a first kiss. However, the noticeable difference of one sentence on one page that is contained over two panels versus one sentence split over a page turn emphasizes that the change to come is a life altering and sober one for the boy, Grandpa Green, and the reader. According to Judith Graham, the page turn or "turning the page" can "reveal surprising information or effects" (213)—and we discover this when the last part of the sentence reveals that Grandpa Green enlists to fight in a world war on a double-page spread featuring a cannon.

The strategic placing of the great-grandson in both cannon-related illustrations (from a central position with a height that occupies three-quarters of the page to a much smaller figure bending down to pick up Grandpa Green's glasses, located just under a branch extending from the cannon itself and holding a cannonball suspended in mid-air) momentarily departs from the lighter side of previous illustrations and takes on a seriousness that mirrors the life-and-death metaphor mentioned earlier in regard to the space of the garden. Additionally, the scene is beautifully horrific as the colour green—a "metaphoric representation of nature and environment [and of] growth, to be new" (Dobrin 265)—merges with yellow, orange, and red flowers to indicate the firing of the cannon, while red spider-like plants scattered across the green grass indicate bloodshed and loss. This moment, in "the dark[er] side of topiary" (K. Graham 95), marks the coming-of-age of Grandpa Green, as a man who has lived through and wants to remember the experience of war, and of the great-grandson, who has learned about that very experience through the memory preserved in the garden's living shapes.

Age truly begins to affect the topiary garden, as well as alter the great-grandson's attitude from playful to pensive, when the leaves begin to change from vibrant greens to rust reds near the end of the book. In a double-page spread that features an ancient-looking tree with a trunk as wide as it is tall, the great-grandson comes to realize that Grandpa Green's memory is beginning to fade. As the gnarled tree limbs draw our eyes upwards to the top edges of the pages, the green vibrancy and denseness of the leaves on the left-hand side provide a stark contrast to those on the opposing page as the limbs of the same tree are laid almost bare by the wind blowing its fall-coloured leaves away. This colour transition marks the natural processes of aging—for both humans and nature alike—but also reveals a formative moment in the internal growth of the great-grandson, who, as he swings from the lower branch on the right-hand side of the tree, watches a single red leaf float silently to the ground before him.

Although the gnarled tree is a sign of the natural cycles of life and aging, the memory and link between generations of family found within the tree forms a parallel to the landscape's ability to reflect, as well as exude, an emotional state on those who interact with the garden. "In looking at how children's literature deals with the environment, we are looking at how children's literature deals with the world" (Greenway 147); and in the case of the great-grandson and his Grandpa Green, it does not matter that the latter is becoming forgetful as the garden commemorates all of these meaningful events for him. In the final pages of the book, the great-grandson has finally caught up to Grandpa Green, whom he finds working away on a new topiary shape. The outline, featuring one arm and one leg, is strategically cut off by the edge of the page, which provides an intriguing lead into the book's gatefold which presents all of the topiaries in context. Like a map that acts "as a real instrument for the traveller [and allows] for the discovery of new spaces[,] new territories" (Meunier 34), and—I might add—new views of one man's entire life, Grandpa Green's topiaries together become a literal and metaphorical journey of growth and development in the symbolic green space of the garden. The newest topiary addition to the garden is revealed in the gatefold to be that of the great-grandson, wielding a sword as he stands on guard against a sea monster. A playfulness momentarily returns when the great-grandson mimics the *en garde* stance of the topiary boy by using a pair of garden shears, while Grandpa Green energetically cheers him on. Nevertheless, the great-grandson's positioning beneath the sword of the much larger topiary boy echoes that of the cannon illustrations mentioned earlier. In the book's final illustration, the image of the great-grandson picking up his own shears in an attempt to sculpt his own bush into the likeness of Grandpa Green indicates that a new level of maturity has been reached within this role reversal—the boy has now become the very embodiment and continuation of everything Grandpa Green has taught him.[5]

Although *Grandpa Green* is not an explicit example of an environmentally oriented activist tale, it nonetheless harnesses the power of the garden in the processes of a child's self-cultivation and the development of an imagination that comes alive through the garden itself. This "imaginative geography," as Aïda Hudson touches on in her introduction to this collection, is a perspective that also mirrors what is seen in the (child's) mind's eye. Moving away from the enclosed space of the topiary garden, our next picture book—Peter Brown's *The Curious Garden*—takes the concepts of *Grandpa Green* and places them in an urban setting that ultimately highlights the challenges one boy must face when he comes across a struggling garden along an abandoned elevated train track on Manhattan's west side.

Inspired by "the idea that nature could spontaneously grow in unexpected places, like in the middle of a 'concrete jungle'" (*Wardrobe*), Brown begins his narrative of little boy Liam's love of the outdoors not with the expected green spaces of the garden, but instead with a drab, monochromatic city puffing black smoke into the air and empty of any people on the streets. This particular scene will provide a striking contrast to the city seen at the end of the book, which displays how Liam's efforts with the initial little garden he learns to help grow expands into a citywide project and allows children and adults to come together to make their urban dwelling the greenest—literally and metaphorically—it can be. According to Clare Bradford and her colleagues, "the moral and political orientation of personal development becomes intensified when linked to actions informed by ecocritical perceptions, that is, by perceptions that nature, the environment, earth itself, are endangered and in need of appropriate management" (91). Liam intuitively picks up on this sense of responsibility at the beginning of the narrative when he first spots the lonely patch of dying plants in his dreary city; and while this continues the trend started with *Grandpa Green* with regard to the child's special relationship with nature, it also emphasizes that "close contact with nature can be dangerous, but so, too, can our evasion and denial of it" (Dobrin and Kidd 2).

Not a natural gardener but knowing he can help, Liam starts to work on making the little garden stronger. Although he nearly drowns the flowers and prunes the plants a bit too much, his efforts are slowly rewarded by the vibrant colours that begin to make their way back into the illustrations. An interesting parallel now develops between Liam and the garden that reflects how the child's "perception of the self as part of a larger, unified natural world" (Sigler 148) becomes mirror-imaged: "As the weeks rolled by, Liam began *to feel like a real gardener,* and the plants began *to feel like a real garden*" (Brown 9; emphasis added).

As the garden expands to all corners of the railway over the next few months, Liam's impact on the environment and his growing knowledge of how to care for his garden are made especially clear on the first double-page spread to reflect the transformation the city is slowly undergoing. With a glorious blue sky above and lush green grass that has grown like a carpet over the abandoned railway tracks below, Brown creates a meeting point between the two by placing red-haired Liam in the middle, smiling and waist-deep in a sea of wild white daisies. Significantly, the daisy is symbolic of loyalty to love, commitment, and innocence (Kirkby, *Victorian*, 167)—all of which Liam embodies at this stage in the narrative, and is reflected in the garden's untamed development across the tracks in response to Liam's attentions. This otium, nevertheless, is interrupted by the presence of a single brick smokestack puffing its black exhaust into the air on the far right-hand side of the page.[6] The small but noteworthy contrast between the whiteness of the daisies and the blackness of the smokestack alludes to the impending challenges that await Liam and his garden as summer turns into autumn, the season of hibernation.

But "rather than waste his winter worrying about the garden, Liam [spends] it preparing for spring" (Brown 18). We are shown Liam sitting at his snow-covered square window, which frames him in a picturesque way as he reads a book about gardening. The child and nature here are unified again, and it can be said that both are—in a sense—in a natural cycle of hibernation where Liam and the garden take the time to prepare themselves, whether it be intellectually or innately. There is, nonetheless, a "contradistinction between cyclical vegetative regeneration and the linear progression of the individual human life [that] throws into relief the transience and impermanence of the human condition" (Carroll 50)—and perhaps even a potential change of heart that could threaten the garden's future. In the illustration showing the toll that winter has taken on the previously green garden, Brown places Liam in opposition to a little brown tree that has survived the cold but is in dire need of Liam's help. The identical miniature size of both Liam and the tree is yet another mirror image—the child sees himself in nature—that marks a pivotal point in Liam's ability to come to terms with the dystopian destruction that now greets him with questions about whether he can truly make a lasting difference in his city.

Liam's decision to continue nurturing the garden—and, in turn, be nurtured by the green space it provides—is reflected in the garden's desire to explore parts of the city it had not previously reached. The garden expands rapidly from the railway with its colourful blossoms and vines, and Liam encourages this while prudently keeping tabs on the places where it does not belong, such as on the face of a stop sign or wrapped around a fire hydrant.[7] But it is not just greenery that is popping up in surprising ways—so are new gardeners, who now realize how their city can be and join Liam on his quest. In six separate images spread over two opposing pages, we move forward in time to see the parallels between the growth of the garden—as it now reaches the dock where little children play on gigantic lily pads; sprouts topiary moose, elephants, and giraffes that march along the sidewalk; and grows a slender tree that makes its way up between two apartment complexes while people lean out their windows to pick the fruit or trim the leaves—and the growing up of Liam, who now walks barefoot down grass-carpeted stairs and joins his special someone for a picnic outside. In the words of Sherie Posesorski from her review of the book in *The New York Times*, "[*The Curious Garden*] is an ecological fable, a whimsical tale celebrating perseverance and creativity, and a rousing paean, encouraging every small person and every big person that they too can nurture their patch of earth into their very own vision of Eden" (n.pag.).

"Emphasis on the responsibility and agency of individuals" (Bradford 93) to make a difference in their urban environment is expressed in this cluster of images, which accentuate the citizens' rediscovery of the garden as a "site for 'the purest of human pleasures'" (Bacon, qtd. in Carroll 52). Those same images foreground the extent to which "they [all] share [in Liam's] strong desire to bring about a new world order in which nature and culture are rewoven and the world is made green again" (Bradford 104). These acts of environmentalism remind us of how we may interpret garden spaces as part of imaginative geography. After all (and to borrow from Hudson's introductory chapter), if imaginative geography allows us a way to see the world, then environmentalism can be thought of as a positive (re)action that does something to save it.

The focus on Liam's lifelong impact on his city culminates in the two double-page spreads that close the book. The first, featuring an adult Liam pruning a large tree while his wife and two children look on, returns to "where it all began" (Brown 30): the place where Liam originally found the dying little garden as a child. The little brown sapling that barely made it through its first winter after Liam found it is now a tall, strong tree, able to hold both grown-up Liam and his little son on its two branches. This harmonious image of a family unit is symbolically balanced on both sides of each page—Liam's wife and daughter on the left looking towards Liam and his son on the right—to accentuate the garden's own balanced relationship between itself and the city. "More than anything," states Brown in an interview, "I want kids to feel a sense of wonder and amazement with nature. I want them to imagine a world where people and nature lived more harmoniously, and to realize that such a world is possible" (*Wardrobe*). And indeed it is, as Liam's own little Eden opens up and the tight viewpoint widens to encompass a sweeping view of the entire city—smog-free and bustling with people tending their own pieces of imaginatively curious green space.

The challenge the urban environment poses for a garden, and a child's nurturing of that garden, continues in our final picture book, *The Imaginary Garden* by Andrew Larsen and illustrator Irene Luxbacher. Continuing with the themes of family history, the expression of memory within the physical space of the garden, and the development of environmental awareness inspired by the imagination, *The Imaginary Garden* highlights how one girl brings nature back into her grandfather's life after he moves to a high-rise apartment in the city.

The double-page spread that opens the narrative shows us the wide-open space of Poppa's backyard garden—a green space symbolic "for human renewal and spiritual regeneration through pleasure and relaxation" (Carroll 50). Thus, it is fitting that we find little Theo and her Poppa sitting across from each other in matching lawn chairs, surrounded by a vibrant and peaceful garden in full bloom. To make the flowers look strikingly real—they appear to almost spring off the page—Luxbacher uses a collage of bits of patterned paper—the same method she uses for Theo's and Poppa's clothes, and a contrast to the softer-edged paint strokes that will re-create their new apartment garden on canvas used in the rest of the book. As "Poppa [tells] Theo all about the different flowers while they [sit] together under the maple tree" (Larsen, *Imaginary*, 3), emphasis is placed on the fact that this particular outdoor garden is a key instigator for Theo's intellectual, psychological, and physical growth.

However, just as with *The Curious Garden*'s fall into the winter season, Poppa's real garden disappears on the following page, which indicates he has moved to an apartment and left the old garden behind. The illustration seems almost sterile with its white brick buildings outlined in thin ink-drawn lines, and the change of season into a windy fall offers a bleak new beginning for any growth. Nevertheless, Theo's suggestion that they "have an imaginary garden" (Larsen 4) fill the entire balcony gives Poppa an idea. As fall turns into winter, the next illustration shows the two of them sitting side-by-side practising how to paint flowers on white sheets of paper. The "hibernation mode" expressed here is parallel to that of *The Curious Garden*'s Liam and becomes a moment where the child's own growth can be seen within the hidden nature that is to come back again with the first sign of spring.

Signified by the appearance of a "great big blank canvas" (Larsen 7) propped against the brick wall of Poppa's balcony, spring arrives. Under Poppa's tutelage, Theo learns to mix paints and turn those colours into essential garden elements. They begin with the basics; and as Poppa paints the garden's wall "stroke by stroke, stone by stone" (9), his controlled method—as indicated through Larsen's sibilance—provides a contrast to that of Theo, whose soft blue sky is created above the stone wall through rapid, zigzagging lines that signify an excitable sense of innocence and inexperience. But as the day ends and the canvas expands to the very edges of the page, the smooth, complete strokes of the sky indicate that Theo has indeed begun to master her Poppa's teachings and sense of patience. This particular illustration—featuring a smoothed blue sky, a long stone wall, and earth-toned soil—marks the beginning of several single- and double-page spreads where Theo and Poppa physically appear within the growing canvas garden itself. According to Bette Goldstone, "children's picturebooks have always been magical in their creative offerings, but postmodern picturebooks present startling new ways to read and view a page [that] create[s] a tableau for ingenious and out of the ordinary fictional and artistic representations" (322–23). As such, Poppa's and Theo's "physical" appearance within the canvas garden draws not just upon their shared eco-imaginative state, but also in Dobrin and Kidd's aforementioned link between the child, nature, and "the wild places of childhood—both real and imagined" (14).

The vibrant yellows, purples, and whites used by Poppa and Theo as the crocuses begin to bloom among the now abundantly green scilla is a reflection of how the canvas garden continually achieves the stages of a real one through Poppa's and Theo's nurturing sense of imagination-turned-art. But when Poppa decides to leave for a holiday at the height of the canvas garden's growth, a new sense of self takes hold within Theo and ultimately becomes reflected in how the garden continues to develop while Poppa is away. With Poppa's wise words—"'When you see the garden, you'll know just what to do'" (Larsen 18)—Theo returns to the apartment and the canvas garden with her mother to see that the garden has magically transformed from the crocuses and scilla of before to beautiful leafy stems and vines that have made their way up the stone wall. Theo, in this illustration, is once again physically standing in the middle of the garden as the viewpoint narrows and the picture bleeds to the edges of the page. As she begins to surround herself with tulips and daffodils, Theo realizes that something is missing as she holds her thumb in front of her eye in a painterly fashion and pink paint drips from her brush: "Blue. Blue. Blue. Forget-me-nots, Poppa's favourite flower" (24).

In an echo of the relationship between Grandpa Green and his great-grandson, Theo's association of forget-me-nots with Poppa is an important comment on the special bond of grandchild with grandparent cultivated within the imaginative space the garden offers. In an interview with *TeachingBooks*, Luxbacher reveals that the forget-me-nots' double-page spread "is important to the story because it would have to illustrate the important relationship Theo has with Poppa and it would also have to show the amazing and powerful feelings art can stir in us when we use it to express ourselves" (n.pag.). The extreme close-up of Theo's smiling face and the crisp lines of her gardener's hat contrast with the bright, bold colours and soft edges of the forget-me-nots; but it is her closed eyes and rosy cheeks that remind us of the "subtle warmth she feels at that particular moment" (Luxbacher, "Imaginary") and brings us to the end of the book, where Theo—after completing the garden with a pair of lawn chairs—steps out of the canvas and beams happily at the reader as she signs hers and Poppa's names at the bottom of the colourful masterpiece.

Whether it exists in a secluded place, an urban setting, or simply in the brushstrokes made on canvas, the powerful and colourful space of the garden as portrayed in *Grandpa Green*, *The Curious Garden*, and *The Imaginary Garden* ultimately participates in "the thriving new crop of 'green' literature for children [that] tap[s] into children's energy and optimism, as well as their sense of wonder" (Sigler 152). In tracing how family history, memory, and the process of self-cultivation comes to be reflected within the greenness of nature, each child protagonist invites us not only into an awareness of the larger environmental concerns that currently exist for our earth, but also into a truly engaging "appreciation of natural beauty [and our own] long ecological coming-of-age" (152). Undoubtedly, the garden as a positively lush and rejuvenating green space is an important part of fostering a child's environmental imagination, curiosity, and awareness. But whether the role of natural beauty—be it wild or cultivated—in these three picture books is to ultimately support humans or to simply be itself, one thing is clear: we are still negotiating that delicate balance in children's literature.

Notes

1) The "environmental imagination" equates the environment with acts of the imagination. For more, see Buell's *The Environmental Imagination*.

2) This includes Buell and Glotfelty, but also others who joined the ecocritical field later on (such as Michael P. Cohen, Laurence Coupe, and Glen Love).

3) Anthropocentric views also existed and are, arguably, the first turn to ecocriticism in children's literature.

4) Picture books, according to Judith Graham, "are almost invariably the first books that children encounter" (209)—and at a time when green spaces are limited, they may also be their first introduction to a garden. Because picture books have the power to shape aesthetic tastes as well as introduce the principles and conventions of narrative, written text is usually kept to a minimum so as to not impede the processes of inspection and enjoyment that are intentionally built into the illustrations (209, 211). "No approach to the picturebook can overlook the importance of medium and design as part of the reader's experience. Nor can we pretend to be unaffected by pictures we encounter in picturebooks" (Moebius 312).

5) Not noted in my descriptions of the illustrations is the fact that Smith drew both the boy and great-grandfather in the matching outfits of a striped shirt, overalls, and green rubber boots. The closing illustration for *Grandpa Green* and the role reversal that occurs is, as such, echoed within the very likeness of the great-grandson as the embodiment of continuity and the carrier of heritage.

6) According to Carroll, "the pleasure associated with green space is an *otium*, a drowsy reinvigoration of the spirit, a return to a quieter, more considered lifestyle which repairs the damage of the urban environment" (50).

7) Just as with the topiaries of *Grandpa Green*, the desire to shape nature into "unnatural" shapes is not always seen as a positive thing. According to Kathryn V. Graham, "from its beginnings, topiary has been a craft associated with patriarchy, leisure-class privilege, and willful human domination of nature … But at all times the metaphor of control—whether its particular representation emphasizes domesticating the wild, disciplining the unruly, or gentrifying the common—holds great intuitive or intelligible power; for the child, in Rousseau's famous formulation, is 'a young plant'" (108).

9) Into the (Not So) Wild

Nature Without and Within in Kenneth Grahame's
The Wind in the Willows

❦ ALAN WEST

I n her introduction to this present volume Aïda Hudson considers the notion of "imaginative geography" at length, particularly with respect to Edward Said's *Orientalism* and his "dualistic vision" (xiv) by which an imagined division of landscape as "a familiar space which is 'ours' and an unfamiliar space beyond ours which is 'theirs,'" an "our land– barbarian land" split, supports a self-justifying division between "'us'" and "'the[m]'" (Said 54). This division and its ramifications are certainly visible in Kenneth Grahame's *The Wind in the Willows* (1908), in which the imaginative geography encompasses the River Bank, the Wild Wood, and the Wide World. Of these three areas, the one that represents safety and security is the River Bank; both of the others represent potential danger, with the book's "barbarian land," being the "menacing" Wild Wood (Grahame 47). This symbolic tripartite division is similar in some ways to that which Susan R. Brooker-Gross has identified as present in the Nancy

Drew series of children's novels. In Brooker-Gross's observed model one of the three landscapes is urban, rather than Grahame's Wide World, the largely taboo human sphere with which Toad alone engages, mostly to his detriment, and concerning which the Rat tells Mole, once the latter has mentioned it, "'Don't ever refer to it again, please'" (Grahame 10). However, Brooker-Gross notes that in the Nancy Drew stories "unfettered nature denotes an unpredictable situation, perfect for criminal activities," while "romantic and pastoral landscapes" are accorded "moral superiority" (62). This is very like the symbolism in *The Wind in the Willows*. Rat says to Mole that the Wild Wood's inhabitants are unpredictable and undependable, that "they break out sometimes," and that, with the notable exception of Badger, "you can't really trust them, and that's the fact" (9–10). The idea that wildness is something disturbing, something to be wary of, and that "break[ing] out" is alarming is true of the text as a whole. There is an emotional boundary between the civilized, mannered, constrained, safe world of the River Bank and the "wild" forces (within as well as without)—animality, unpredictability, and unconventionality—that threaten it. That these forces lie within is visible in the temptations to which most of the central characters are subjected, with Toad—the most impulsive, unpredictable, and hedonistic of them—the most frequent offender. Like Henry Jekyll in Robert Louis Stevenson's *Strange Case of Dr. Jekyll and Mr. Hyde* (1886), Toad, Mole, and Rat all seek to unshackle themselves from the conventional, safe lives they live or that are expected of them; luckily for them and unlike for Jekyll, the consequences are not catastrophic. None of the attempts to enjoy freedom by travelling to an unknown landscape ultimately succeed.[1] Each returns, or is helped to return, to his ordered and comfortable existence; inevitably, life on the River Bank remains the best choice. Wild impulses, the text seems to say, need to be negated or at least contained. The River Bank trumps the Wild Wood, as it were, and indeed, in the last chapter of the novel, the River Bankers, aided by Badger, defeat the Wild Wood's inhabitants in order to return Toad's home— occupied by the Wild Wood folk—to him so that the Wild Wood becomes "successfully tamed" (144). This struggle between the wild and spontaneous, and the restrained and conventional, is central to the text.

The River Bank and the landscape that surrounds it are frequently celebrated. For much of the novel, the natural world is presented as generally benign and often idyllic, with a benevolent Pan (renamed the Friend and Helper) at its heart. Yet at the heart of the text itself is a dichotomy. The (aptly half-human, half-animal) Friend and Helper provides a manifestation of this dichotomy; there is an unmistakeable sensuality to his physical description, yet the more disturbing qualities and wild, anarchic aspects of the original Greek demigod are apparently absent. Similarly, while the text's central characters (Mole, Rat, Toad, and Badger) are all, in a zoological sense, wild animals, Grahame has removed most of what makes wild animals *wild* from them. They are so not-wild that they, a few lapses notwithstanding, live the lives of perfect gentlemen, which is just as well since, in the real world, the three smaller characters might well have formed part of the (omnivorous) badger's diet. Their relationship is decidedly not a manifestation of "Nature, red in tooth and claw." This de-wilding of the characters is paralleled by Grahame's treatment of topography in the text: those elements of the landscape that are benign are depicted positively, but while the environment's harsher qualities are also explored, those rougher aspects are associated with threats and fears. The reason for this is simple. Grahame's own wild side was something he kept suppressed. Although the text begins with Mole's escape from "the seclusion of the cellarage" into "sunshine … [and] soft breezes" (5), Grahame kept a part of himself confined.

Rat and the other central characters can be read as an idealized blend of the human and the animal. In this they are not quite furry gentlemen, not beasts in form only. When the Mole unknowingly passes close to the home he abandoned to live with Rat his nose automatically picks up the "mysterious fairy calls" emanating from his erstwhile domicile; it is powerful, a "telegraphic current" (50). However, animal though this ability of Mole's is, it is a benign animality. The home that is appealing to Mole through his senses is not a rough earthen tunnel, complete with worm larders, but "Mole End" with its "plaster statuary [of] Garibaldi [and] … Queen Victoria." In fact, Grahame in effect winks at his readers by having Mole have a lawn roller because he hates "having his ground kicked up by other animals into little runs that [end] in earth heaps" (53-54). That is, he does not like his yard disturbed by moles. So we have a Mole who finds what wild moles really do to be annoying, to be flattened out and thus suppressed. Elsewhere the novel's most animal animals, the threatening denizens of the Wild Wood, are rendered harmless—are defanged, in a sense—by the end of the novel. Grahame's text pushes animals to the forefront yet paradoxically ensures that much of what would make them animal is absent or eventually negated. Clearly Grahame believed that nature, without or within, should be made anything but wild.

For much of the book Rat and Mole stick to one of the three main locations in the novel, the landscape of the river bank, one that is far removed from nature in the raw; as David Sandner has noted, they "picnic and loaf … not in innocent and untrammeled wilderness but in a [domesticated] place of compromise between Nature and Comfort" (*Fantastic* 68). The river bank represents the known, the safe, the predictable: it is the novel's comfort zone. The other two regions that make up the landscape of *The Wind in the Willows* are suspect. Rat warns Mole to avoid them early in the book: the Wide World that partly (though not wholly) equates to the human world where Mole should not go if he has "any sense at all," and the (initially) dangerous Wild Wood, where Mole, having become lost, experiences "the Terror of the Wild Wood" (31). However, the wildness even of that Wild Wood is relative, as it would have to be were the fantasy to represent any sort of reality, since there is no truly wild England (in the sense of being unaffected and unshaped by the activities of people). This would be particularly true of the central England in which *The Wind in the Willows* is apparently set. The English landscape, shaped by the hands of many generations of people over the centuries, for agricultural or more arcane reasons, is not even particularly natural anymore. While love of nature has been rooted in English culture, and fed by its literature, since at least the time of the Romantics, it is a relatively benign nature that is the subject of that infatuation. The British countryside has its wild animals and birds, but none of them are particularly dangerous to humans; when Coleridge, in "Frost at Midnight," projected baby Hartley into a Wordsworthian boyhood, wandering "like a breeze … beneath the crags," there was no ominous spectre of bear or wolf to intrude on the imagined idyll. The notion of any genuinely Wild Wood in England, or of any dangerously wild aspects of nature resident there in any form, seems moot. The wood's wildness becomes a sort of trope, an imagined space upon which Grahame can project his fears of a sinister aspect to nature, conjuring a place capable of inducing a sense of panic in anyone who might unwittingly experience nature in the raw, and risk losing themselves in it. It is true, as Robert Macfarlane observes in *The Wild Places*, that the wild cannot simply be written off, since, although "time and again, wildness has been declared dead in Britain and Ireland" (8), vital nature is still there; "the wild prefaced us, and it will outlive us" (316). This idea of the persistence of the wild, of it outlasting humanity, is depicted in the "Mr. Badger" chapter *The Wind in the Willows*. However, it is still a relatively tame manifestation of the wild that replaces the human. Grahame flirts with the idea of nature in the raw but opts for a safer, gentled nature. The wood may revert to a kind of wildness, but it has its guardian Badger to stop things getting too far out of hand. "'Any friend of *mine* walks where he likes in this country, or I'll know the reason why'" (46).

Where the Wild Wood now is, according to Badger, the landscape once looked radically different because in ancient, presumably Roman, times, it was the site of a city. But the people left, to be succeeded by "ruin, and levelling, and disappearance," and then the forest grew. Badger remarks, "there were badgers here" before the city, and now "there are badgers here again" (45). Nature swallowed up the remains of the city; the animal replaced the human; the wild returned:

> The strong winds and persistent rains took the matter in hand, patiently, ceaselessly, year after year. Perhaps we badgers too, in our small way, helped a little—who knows? It was all down, down, down, gradually—ruin and levelling and disappearance. Then it was all up, up, up, gradually, as seeds grew to saplings, and saplings to forest trees, and bramble and fern came creeping in to help. Leaf-mould rose and obliterated, streams in their winter freshets brought sand and soil to clog and to cover, and in course of time our home was ready for us again, and we moved in. Up above us, on the surface, the same thing happened. Animals arrived, liked the look of the place, took up their quarters, settled down, spread, and flourished. (45)

This is reminiscent of the first, apocalyptic, part of Richard Jefferies' 1885 novel *After London*.[2] In that text civilization is overwhelmed by some unspecified cataclysmic event, London collapses into ruin, nature takes over, and there is a decline into "barbarism." One might be tempted, therefore, to see the collapse of the city and the return of nature and animal life in *The Wind in the Willows* as a metaphor for social degeneration or even atavism. However, that is not how it works in the text. Instead, a comfortable balance between animal and human is achieved in Badger's subterranean home; when fellow earth-dweller Mole "is staggered" by the extent, and impressive architecture, of Badger's digs, the latter explains that he simply made use of the sunken city (45). He happily lives among what was, at least once, human, and, with his kitchen where friends could "smoke and talk in comfort" (38) and eat eggs and bacon (41), lives like a human. Moreover, the Wild Wood underneath which Badger's home is situated is wild only in name by the time the book ends, its inhabitants having been rendered harmless by Badger and his friends. Grahame believed that nature, without, and more especially within, should be mild rather than wild.

The Wind in the Willows revolves around notions of the natural and the wild. Grahame has animals be his protagonists yet paradoxically ensures that all taint of "the beast" is removed. This is not coincidental. Peter Green has remarked that although Grahame anticipated D.H. Lawrence in believing "that man has fatally neglected the instinctive, animal side of his nature," he ultimately "swerved away from its implications ... [and] refused the challenge" (qtd. in Haining, 88). Certainly *The Wind in the Willows* itself expresses an overt concern with the balancing of animal and human qualities. According to the text, human evolution has not been a process of unqualified progress because we have lost too many of our instinctual powers. The narrator deplores that loss; in this regard, the appeal to the senses emanating from Mole's home (above) is impossible for modern humans to detect. Indeed, we do not even have a suitably subtle vocabulary to express an animal's sensory interaction with its environment:

> We others, who have long lost the more subtle of the physical senses, have not even proper terms to express an animal's inter-communications with his surroundings, living or otherwise, and have only the word "smell," for instance, to include the whole range of delicate thrills which murmur in the nose of the animal night and day, summoning, warning, inciting, repelling. (50)

In an earlier essay, "The Lost Centaur" (1893), Grahame commented on this loss more strongly. In that piece he laments the snapping of the threads that make up the cord that "somewhere joins us to the Brute," and he complains of a "forlorn sense of a vanished heritage"; however, it is not the more savage side of our animal heritage that he thinks of in terms of loss, nor is it our connection to "tigers and apes"; rather, it is our kinship with more "pleasant cousins in hide and fur and feather" (92). Remarking upon what he saw as a "lamentable cleavage" between the animal and the human, Grahame remarks that we should think of a "kinship once (possibly) closer" with animals and wonders if a "wrong turning" during the course of human evolution has lost to us the potential for a "perfect embodiment of the dual nature: as who should say a being with the nobilities of both of us, the basenesses of neither" (93–94). He seems to have tried to create such a "perfect embodiment," with no disturbing "baseness," in his novel, in the generally genteel blend of animal and human that the central characters represent. In a sense, Grahame is a Dr. Moreau, humanizing the animals, but without the pain and failure that accompanies Moreau's attempts in Wells's novel, though perhaps one can see in the Wild Wooders of Mole's experience "the old animal hate" of Moreau's Beast People, held in check by "the Law" (Wells 93) provided by Badger.

Meanwhile, the human world itself, by and large, impacts negatively on the animal characters in the book. Toad's splendid new caravan ends up "an irredeemable wreck" when a reckless driver scares the horse (23). In Chapters 6, 8, and 10, Toad's own recklessness, in his excursions into the Wide World, result in a twenty-year prison sentence and a desperate escape; as he makes his way home, an encounter with a barge woman results in her flinging him into a canal; later, he falls in a river while fleeing the police and has to be rescued by Rat. Elsewhere in the text, field mice and harvest mice move home as harvest approaches to avoid the "horrid [reaping] machines" (93). In a sense, the animal world is constantly at threat from humanity but manages to avoid being overwhelmed by it. To this end, the novel's deity, the Friend and Helper, promises to *spring the trap that is set* and *"loose the snare"* to protect animals from humans (80). Thus there is a strong sympathy for animals with respect to their dealings with people, and a constant struggle to preserve the animal from the human threat. Yet there is also a sense in which the animal—the instinctive self—is also a threat, resulting in an underlying uneasiness in the text about exactly what should constitute a comfortable relationship with not only external nature but also internal nature.

In the short pieces he wrote for publications like the *National Observer* in the late 1880s and 1890s (including "The Lost Centaur," above), Kenneth Grahame frequently extolled the pleasures of experiencing the English countryside first-hand. He wrote of rowing the upper reaches of the Thames, of walking its banks, and of hiking from dawn to dusk along ancient tracks like the Ridgeway. Woven into some of these pieces are reminders of the human history of the landscape, of the Roman soldiers who "passed silently to Hades" ("Romance" 46), and of the "Saxon levies" who fought "Danish invaders" on a local ridge ("Northern" 75). But Grahame often situates a figure in this landscape that is mythological rather than historical, and neither English nor wholly human. That figure is Pan, the deity who, in Grahame's 1891 essay "The Rural Pan," pipes "his low sweet strain" to the "chosen few" (32) who visit his "remote … haunts" (32). Pan is to be found in a number of literary forms in Late Victorian and Edwardian England. One of Grahame's biographers, Peter Green, remarks that "'Pan,' in fact, was a convenient short-hand term for their [writers and other artists'] own anarchic emotional urges; he had become a literary symbol, without any highest common factor except a vague full-blooded animalism" (94). Certainly his appearance in the fiction of the period is often associated with liberation from various constrictions. Ronald Hutton notes that it is "beyond doubt that the shocking, menacing, and liberating aspects of the god's image were exciting many writers by the opening of the twentieth century" (46). He observes also that Pan "could function as a liberator of those types of sexuality that were either repressed or forbidden by convention," and that sexuality that was forbidden was "in the Edwardian context … gay sexuality" (48). Sexual liberation through the intervention of Pan, though his presence is signified to the reader concretely only by his hoofprints, is, for example, central to E.M. Forster's short tale "The Story of a Panic" (1902): a repressed youth, Eustace, on holiday in Italy with his generally disapproving aunts, experiences a sexual awakening after an apparent encounter with Pan and flees off into the countryside, never to be seen by his family again. Rebranded as the Friend and Helper, Pan is also central to "The Piper at the Gates of Dawn" chapter in *The Wind in the Willows* and is, on one level, integral to the book as a whole, but overtly liberating he is not, other than in the sense of freeing animals from traps (see above). In fact, what disturbance he does cause in the text (albeit in a beneficial way) is dissolved with the "last best gift … [he] is careful to bestow on those to whom he has revealed himself … the gift of forgetfulness" (77). Thus the continuing lives of those who have a vision of him are not diminished by being contrasted with their epiphanic experience.

In "The Lost Centaur" Grahame talks of Pan as the "Goat-foot" who has little in common with humans other than a "love of melody" and whose "sympathies are first for the Beast" (93). Half-human and half-animal, god of the wild, Pan is a particularly fitting spiritual focus for a text in which the central characters (a mole, a rat, a toad, a badger, an otter, and sundry creatures associated with woods and farmland) are anthropomorphized wild animals. The fact of his being neither human nor animal but a blend of the two is symbolized by where he appears in the text, on a "small island" (*Willows* 76) that is neither quite land nor river, but associated with both. What Pan himself symbolizes in the text, however, is more complex. Grahame obliquely acknowledges the more alarming qualities of the original god by affirming that the "great Awe" that Mole feels in the Friend and Helper's presence is "no panic terror" (76), using the adjective derived from that fear felt by people in remote wild places that was supposedly engendered by Pan. Grahame's version of Pan does not cause that fear; in fact, he is a remarkably benign version of the deity, a god of the wild devoid of anything much that actually *is* wild. There being a Wild Wood in the text, one might reasonably expect Pan to dwell there, but this is not the landscape in which he is discovered. Rather the Friend and Helper is found reclining on a "little lawn" surrounded by "Nature's own orchard trees" on a "reserved, shy" island with a scented "flowery margin" (76). Flowers, a lawn, an orchard: it all sounds more mildly domestic than ruggedly natural. One quality traditionally associated with Pan is apparently missing in the Friend and Helper: the ruttish tendencies of the faun-like, half-goat figure who was so fond of nymphs. While Grahame's equivalent is "holding … pan pipes" and is sensually described, there's no suggestion that those pipes used to be the nymph Syrinx who had become a reed to escape Pan's attentions; although Nature looks on, "flushed with fullness" and seemingly "hold[ing] her breath," one does not get a sense that this Pan is all that interested in female company.

A lack of interest in females (female animals being almost completely absent from the text), and absence of genuinely wild qualities, are true also of the book's protagonists, who are eminently civilized. They *are* animal in shape and *do* exhibit a few suitable traits; for example, with a nod to hibernation, the narrator tells us, concerning the animals, that in winter "all are sleepy—some actually asleep" (*Willows* 40). But they wear clothes, eat human food, speak English, have furniture in their dwellings, are rarely inclined towards violence, and so on. The balance between human and animal qualities is tilted quite definitely away from the animal. In a novel in which a significant amount of the story takes place beneath the landscape (the homes of Mole, Badger, and Rat are subterranean, and an assault required to reclaim Toad Hall from squatters in the final chapter is via an underground tunnel), the animal qualities of the animals are largely buried. In fact, the book as a whole reflects the need to keep things buried, to keep jack in the box and put him back firmly if he escapes, to recognize where one belongs, and generally to maintain stability. There is no room for the wild in *The Wind in the Willows*, no room for anything animal in the animals, and any disturbance that does occur in the text has been successfully negated or crushed by the time it ends. The one real explosion of violence, of a relapse into savagery, by the four central characters occurs when they burst into Toad Hall from a secret subterranean passage in order to attack and then evict the Wild Wood folk who have claimed it:

> What a squealing and a squeaking and a screeching filled the air! Well might the terrified weasels dive under the tables and spring madly up at the windows! Well might the ferrets rush wildly for the fireplace and get hopelessly jammed in the chimney! Well might tables and chairs be upset, and glass and china be sent crashing on the floor, in the panic of that terrible moment when the four Heroes strode wrathfully into the room! … They were but four in all, but to the panic-stricken weasels the hall seemed full of monstrous animals, grey, black, brown and yellow, whooping and flourishing enormous cudgels; and they broke and fled with squeals of terror and dismay, this way and that, through the windows, up the chimney, anywhere to get out of reach of those terrible sticks … Up and down, the whole length of the hall, strode the four Friends, whacking with their sticks at every head that showed itself. (136–37)

However, this explosion of violence is justified in the context of the novel in the sense of the end justifying the means, or a sort of wild justice. It is not a giving-in to an urge, not a burst of impulsivity or lack of restraint, but the climax of a planned expedition. It does not introduce anarchy or uncertainty but rather dispels it, restoring the status quo and returning the community to a state of stability.

In fact, the whole thrust of the book is towards maintaining stability. Impulsiveness, not listening to reason, can soon put one in danger and deep water, as when the naive and inexperienced Mole refuses to listen to Rat's advice when they are in the latter's boat, grabs the oars, and overturns the vessel: "Over went the boat, and he found himself struggling in the river … He went down, down, down … How black was his despair." He needs Rat to rescue him (13). Upsets are also caused by several of the breakings out and escapes in the novel, most of which have to be thwarted or cause someone grief because they upset the status quo in some way. There are two escapes in the novel, however, both by Mole, which, while they are driven by a sort of animal instinct, are benign and therefore no threat to the even tenor of life. The first of these is seen as positive in the text because of where Mole ends up: the river bank. The river bank is the book's geographical and social centre, and the lyrical descriptions afforded it reflect its importance; at one point Grahame devotes a whole page to the scenes, flowers, and activities of summer on the river bank without any furthering of the plot (28). So when, at the beginning of the novel, Mole responds to the call of Spring and its "spirit of divine discontent and longing" and is lured away from whitewashing his home (*Willows* 5), he is drawn towards something positive: the river bank, a place where he can experience "long waking dreams" (8), appreciate aesthetically scenes like the "so very beautiful" mill stream (10), and, of course, establish lifelong friendships. That his choice in shifting locales represents progress is signalled by his new status as an "emancipated mole" (15). Later in the text he has the reverse urge; he picks up the scent of his old home, its "telegraphic current" (50) draws him back briefly, and he appreciates having "Mole End" as an "anchorage" but has no desire to return to it permanently, preferring his "new life and its splendid spaces" (59). So his visit home actually ends up underscoring the pleasures of his new rural lifestyle on the river bank.

Between these two urges to follow his nose, as it were, Mole makes a misguided attempt to find Badger's home in the Wild Wood, despite Rat's warnings. In doing so he believes (illusorily) that this experimentation, this excursion into the wild, is safe. It is winter, and Mole, wanting to seek out Badger, convinces himself that he likes the "bare bones" of the "undecorated" landscape; so he heads away from the security of river bank, imagining that the lack of foliage has enabled him to see "intimately into the insides of things" now that the countryside and its flora have "exposed themselves and their secrets"; he continues despite the fact that the Wood appears "low and threatening, like a black reef" (29). In venturing into what is, for him, uncharted territory, he endangers both himself and his friend Rat.

Of course, it is not so much the wood itself that turns out to be threatening, but the mean-spirited woodland creatures of the weasel family, with their "hard eyes" and "evil wedge-shaped face[s]" (29), who cause him to feel "the Terror of the Wild Wood" (31). Once Rat comes to rescue him he feels more secure, and snow, "a gleaming carpet of faery," makes the wood seem more benign (31). However, the transformation wrought by snow on the woodland landscape is itself problematic, and both Rat and Mole are punished for Mole's foolhardiness in striking out into the unknown; they are soon lost, "dispirited, weary, and hopelessly at sea" (32). They seem in danger of being lost for good: "there seemed to be no end to this wood, and no beginning, and no difference in it, and, worst of all, no way out" (32–33). Luckily they stumble upon Badger's home, and Mole gains a new friend, so it all ends positively. However, Mole has ignored good advice and given in to temptation. The sequence illustrates the dangers of turning from the safe path along the river bank and indulging dangerous desires. Better to stay, content, on safe territory. When the friends leave Badger's for home, they look back; the Wild Wood looks "dense, menacing, compact, grimly set in vast white surroundings," and they hurry towards "the river that they knew and trusted in all its moods, that never made them afraid." Mole recognizes that he is an "animal of the tilled field and hedgerow" and "the cultivated garden plot"; he must keep to the "pleasant places" and avoid the conflicts of "Nature in the rough" (47). The wild is too dangerous, whatever its illusory temptations.

Later in the text it is Rat's turn to be tempted to wander away from the river bank, this time into the Wide World and away to the Mediterranean. It is the Ancient Mariner–like Sea Rat that provides the final motivation, transfixing Rat with his tale, "holding him fast with his sea-grey eyes" (102) to such an extent that Rat's own eyes turn "a streaked and shifting grey" while his movements become those of "a sleep-walker" (103). However, the Sea Rat is only a catalyst; Rat is already unsettled because it is the end of summer, birds are migrating, and other animals are moving to winter quarters. Despite being "rooted to the land" (92), Rat himself feels this seasonal urge; as swallows chatter about "the call of the South," he feels, "vibrating at last, that chord hitherto dormant and unsuspected," an awakening of a "*wild* new sensation" (95; my italics). In a text in which being wild is not to be admired, Rat must suffer having this urge to leave quashed, and Mole pushes him back over his own threshold and pins him down until he collapses into hysterics and then torpor (103–4). Rat's return to normalcy is begun by Mole casually relating what he would have missed by affirming the picturesque quality of the local landscape, combined with the benign rural activities that would take place on it, with the harvest wagons and teams of horses, and then the "large moon rising over bare acres covered with sheaves, "and of the produce—the "reddening apples … nuts … jams and preserves and … cordials"—that would be derived from it. And, of course, the "snug home life" yet to be enjoyed (104). So the unknown, with its potentially new vistas, is trumped by the old familiar and comforting landscape.

Toad is another character who breaks out, and with disastrous consequences. When Badger, Mole, and Rat intervene to try to cure him of his motor car obsession, he struggles with the confinement to which he is subjected, compensating for his inability to drive by creating a make-believe car and temporarily satiating his urges through imaginary drives. On one level, this is reminiscent of children's play. On another, it has more than a hint of auto-eroticism:

> When his violent paroxysms possessed him he would arrange bedroom chairs in rude resemblance of a motor-car and would crouch on the foremost of them, bent forward and staring fixedly ahead, making uncouth and ghastly noises, till the climax was reached, when, turning a complete somersault, he would lie prostrate amidst the ruins of the chairs, apparently completely satisfied for the moment. (65)

After escaping from this house arrest through a ruse, Toad cannot resist stealing the first motor car he comes across. If we look at Grahame's choice of words when Toad is tempted, we see it is all to do with succumbing to desire:

> As the familiar sound [of the engine] broke forth, the old passion seized on Toad and completely mastered him, body and soul. As if in a dream he found himself, somehow, seated in the driver's seat; as if in a dream, he pulled the lever and swung the car round the yard and out through the archway; and, as if in a dream, all sense of right and wrong, all fear of obvious consequences, seemed temporarily suspended. (69)

As he drives, indulging his wants, we see that he has given in to a deep-seated urge, heedless of the consequences: "the miles were eaten up under him as he sped he knew not whither, fulfilling his *instincts*, living his hour, reckless of what might come to him" (69; my italics). Later, after he has been convicted of theft (among other things) and has then escaped from prison, when a combination of coincidence and cunning puts him at the controls of the same vehicle that he stole earlier, he becomes victim to the same compulsions as before: "[He] tried to beat down the tremors, the yearnings, the old cravings that rose up and beset him and took possession of him entirely. 'It is fate!' he said to himself. 'Why strive? why struggle?'" (116). He finds out why he should have struggled when he crashes, is chased, and escapes only by falling into the river. So, again, we see how giving in to a strong impulse is dangerous and can result in serious consequences.

Moreover, Toad's absence from the river bank enables another disturbance. Early in the text Rat had said that the Wild Wood inhabitants "break out sometimes" (9). Their actual breaking out later in the text ironically involves a breaking in, and the weasels, ferrets, and stoats burst into Toad Hall, thrash the caretakers in the form of Badger and Mole, and make themselves at home (123). Toad makes it back to the river bank to find that his return has not established any normalcy. The aggressive behaviour of the Wild Wooders has upset the status quo, and, as related above, it takes the combined efforts of Badger, Rat, Mole, and Toad, accessing Toad Hall by means of an underground passage, to restore it by defeating and ousting the squatters. The quelling of this "rising and invasion" results in the reform of the Wild Wood's inhabitants, and we're told that the Wood has been "successfully tamed" and rendered completely safe for the four victorious friends, who are greeted "respectfully" on their visits (144). So the only relatively wild part of the geographical and social landscape has been effectively gentled.

To conclude: *The Wind in the Willows* has an author who lamented the loss to humans of the animal self. It seems somewhat paradoxical, then, that when he created animal characters, he made them animal in shape but human in habits, and gentlemanly in demeanour, with almost all trace of animal behaviour removed. He set his story in a rural landscape with one relatively wild feature, then cancelled out that wildness by the end of the book. The River Bank and the "Arcadian vision" (Carpenter 157) that it represents offer not only a leisured male lifestyle of all play and no work but also stability, and any threats to that stability are swiftly nullified. The book seems to suggest that a pleasant and sociable life demands that one avoid responding to desires or urges and instead maintain a united front with one's (male) chums, and not rock the boat. Why might this be?

I think the answer can be found in the notion that Grahame expressed in "The Lost Centaur" (above)—the idea that a union of human and animal might once have been possible, with "the basenesses of neither." "Baseness" would seem to suggest something coarse, something sordid, something sexual, perhaps. His motivation for employing animal characters was to avoid anything sordid; he wrote that he intended "by simply using the animal, to get away ... from weary sex-problems," and he proclaimed *The Wind in the Willows* "clean of the clash of sex" (qtd. in Green, 117). In 1900, Grahame read the manuscript of a work by gay writer and fellow *Yellow Book* contributor Frederick William Rolfe—otherwise known as Baron Corvo—

for the publisher John Lane. In his report to Lane he comments that, while Corvo had included "a good deal of *boy* in it ... lithe limbs ... subtle curves ... tawny skin of boy + so on," there was "not a jot of uncleanness or suggestiveness" and that, anyway, "all artists [know] how much finer an animal the male is than the female" (qtd. in Gauger, n26, pp. 175–76). This does throw some light on a reading of Grahame's description of the "Friend and Helper" in *The Wind in the Willows*:

> [They] saw ... the bearded mouth [that] broke into a half-smile at the corners; saw the rippling muscles on the arm that lay across the broad chest, the long supple hand still holding the pan-pipes only just fallen away from the parted lips; saw the splendid curves of the shaggy limbs disposed in majestic ease on the sward. (77)

It is almost impossible not to read that as homoerotic. So perhaps that explains the need to maintain stability, to not break out, to not respond to the wild, animal self and its urges, to discourage your friends from doing so, and to be the perfect gentleman at all times. For Grahame, like Mole, it was best to keep to the "pleasant places" and avoid the conflicts of "Nature in the rough." An ill-advised foray into the wild wood was dangerous, a walk on the wild side that could unleash who knows what. A walk on the mild side was much safer.

———

Notes

1) A partial exception to the negativity associated with giving in to temptation and kicking over the traces is Mole's exodus at the very beginning of the book, when he succumbs to spring's "spirit of divine discontent and longing" and leaves home (6). However, Mole travels to the ideal landscape, the River Bank, not away from it.

2) Peter Green remarks that "Jefferies' books had a considerable influence on the shaping of *The Wind in the Willows*" (143).

10) Earth, Sea, and Sky Writing in *Becca at Sea*

❧ DEIRDRE F. BAKER

═══════

Deirdre Baker teaches children's literature in the Department of English at the University of Toronto and since 1998 has been children's book reviewer for *The Toronto Star*. She also reviews and writes regularly for *The Horn Book Magazine* and has reviewed for *Quill & Quire* and other publications. She is co-author, with Ken Setterington, of *A Guide to Canadian Children's Books* (McClelland and Stewart). She is the author of several scholarly articles: her favourites to date are her articles on children's literature for *The Oxford Companion to Sugar and Sweets* (2015) and *The Oxford Companion to Cheese* (2016). She was fortunate in being able to organize two formative conferences on children's literature at Trinity College, University of Toronto, "The Particles of Narrative" (2009) and "The Environmental Imagination and Children's Literature" (2011). She earned her graduate degrees in Medieval Studies at the Pontifical Institute for Mediaeval Studies and the Centre for Medieval Studies at the University of Toronto. She is the author of two middle-grade novels, *Becca at Sea* (Groundwood, 2007) and *Becca Fair and Foul* (Groundwood, 2018). Deirdre grew up on the north bank of the Fraser River and at the edge of the coastal mountains in Maple Ridge, BC. Since early childhood, she has been lucky enough to spend summers on Hornby Island, where the natural physical surroundings—sea and mountains—and the flora and fauna that inhabit them are very much in the foreground, vibrant and vigorous reminders that humans are just one species among a vast realm of lives on earth. *Becca at Sea* and *Becca Fair and Foul* are born of wonder, observation, and appreciation for the ways in which lives, art and natural surroundings entangle.

⁓

It seems to be a role of fantasy to invent geography, if by geography we understand world, landscape, and earth-writing at its most literally figurative; thus also the cultural geography that arises from the invented world—or for which it exists. I say "for which it exists" because quite often, imagined geographies are at the service of—of what? ideas? theology? psychology? philosophy? dogma?—in short, any number of intellectual activities and abstract notions. Certainly this is true of some of the earlier fantasies, including the one that is probably the most immediately influential in children's literature, *The Pilgrim's Progress*. Would there be a Slough of Despond if there was no sad human capacity for despair? No. That feature was created by some spiritual process, and not by the workings of earth, air, fire, and water. The landscape Bunyan constructs reflects his own sense of the map of the human soul and body.

For Bunyan, "real world" or concrete geography is a useful metaphor for the condition of the human soul and its relationship to God. But let me turn to an inversion of that relationship between spiritual and concrete geography: that is, a reading of an extant, rather than a fictitious or fantastical, geography in terms of what it shows us of the human condition: a twelfth-century Latin poem by Alanus ab Insulis, "Alan from the islands," or Alan of Lille as he is also known. His poem "Omnis mundi creatura" begins "every creature of the world is like a book or a picture or a mirror for us, a faithful little sign of our life, our death, our status, our lot …"[1] For Alan of Lille, all of the "real world"—everything that geography affects—has one divine purpose: to tell us about ourselves, to reflect us back to ourselves so that we understand ourselves truly and he implies, act well.

Both of these imaginative representations and readings of geography and its features establish the human as the raison d'être of the physical features of the earth, whether of the imagined land in Bunyan's case or of the one familiar to us, in Alan of Lille's case. Both appropriate geography and all that "geography" implies to the service of humans and ideas pertaining to the human. Now, I am a regional writer: by this I mean I write of a particular, "real" place with its own noteworthy, idiosyncratic physical features of earth, atmosphere, and ocean. I write out of and about a small island in the Salish Sea (also known as the Strait of Georgia), BC, an "unimaginary" geography, one you can go visit without resorting to Platform 9¾ (Rowling, 68 inter al.). Nevertheless this "real," "unimaginary," or "unimaginative" geographical area inspires its own claims to fantasy on the part of Tourism BC, with advertisements for "Super, Natural British Columbia."

The geographical area in which *Becca at Sea* takes place is not "supernatural", whatever Tourism BC may claim. It may be "super"; it may be "natural"—but most important to me is that it is a real place in the physical world, and if there's anything its geographical elements tell us, it's that geography's raison d'être is not the human. In this essay, as a writer of this "real" geography, and a writer for the young, I will ponder some of the challenges of representing this unimaginary geographical area in such a way that it approaches non-humanocentricity, or at least, a balance between human and non-human geographical elements. This challenge is both literary and, ultimately, political. I'm going to focus on three questions.

One, how does one convey adequately the fullness and variety of the effects and features of such a geography? How to evoke the experience of superabundance and distraction that is endemic to this (real) place with its always active, always geographically induced features?

Two, how does one convey, effectively, inexplicitly, and continuously, that humans are just one small species in a multitude of geographically influenced features? For this is a matter that seems essential to writing the environment—the non-humancentric question.

Three, how to convey that all elements of geography—landforms, atmosphere, water, and all species of plant and animal sustained and affected by these—are gloriously, essentially alien to us as humans?—yet at the same time, a source of meaning, illumination, and imaginative sustenance? For it is my purpose not just to tell a good story about humans, but also to try to allow geographical features to be themselves. To celebrate the natural world's Otherness, and humans' capacity to apprehend and be moved by that Otherness.

So: my first question: How does one convey adequately, to the young reader, the fullness and variety of the effects and features of such a geography? The absolute in-your-faceness of geography in this island area?

Let me tell you something about the situation: first, there's its geology, a varied history of sedimentary deposits, compression, and upheavals that have created a landscape in which rock is always rather startling, causing even permanent residents to exclaim with wonder. Its sandstone beaches are eroded by rain, wind, and sea into lumps, bumps, pockmarks, pillars, protuberances, criss-crossed lines, splits, crevices, and shelves grainy to the sole of the foot. There's conglomerate, too—stretches of dark, uneven rubble, cobbles in a matrix of prehistoric silt, evidence of some ancient riverbed. Now the conglomerate is the home of—but that's a different geographical element: the life it sustains.

An exceptionally rich variety of marine life thrives here. Babies of *Strongylocentrotus franciscanus*, the red sea urchin, flourish in the protection of pockets created by cobbles that have loosened and dropped from their stony matrix. Tiny orange cup corals stipple the undersides of sandstone shelves; anemones red, green, brown, white, orange, and pink wave fronds in rocky crevices; purple-hinged pectins, jingle shells, nudibranchs, brittle stars, red stars, leather stars, bat stars, purple stars, sunstars ... a veritable cosmos of invertebrates inhabits the intertidal shore. As for the pelagic: six-gill sharks, the ling cod, rock cod, snapper, and seasonal herring and salmon busily do their thing under the skin of the sea in among the kelp beds, sea lettuce, eel grass, rockweed, evocatively named Turkish towel, and other abundant marine plants.

Plankton floats, swims ubiquitously in the strait, in tide pools, and in the very drops our bodies shed when we emerge from a swim—living organisms in all stages of growth, invisible to the naked eye, bioluminescent at night, a constant presence of a multitude of species zoological and botanical. Among the plankton swim the celebrity mammals: orcas, harbour seals, California and Steller sea lions, river otters, a lone elephant seal, a gray whale and its baby.

The physical features of earth form this river of sea that runs from Queen Charlotte Sound through Johnstone Strait and on through the

Salish Sea with its various straits and sounds. It's a body of water that fosters the life upon which all this other life I've mentioned depends. Hundreds of creeks and rivers race from the snowy peaks of Vancouver Island's mountains and the mainland's Coastal Range and pour into the strait's salty flow, changing its speed, temperature, and salinity and thus welcoming some, and discouraging other, forms of life.

The rivers pour from mountains, which are the sort of features many of us think of when we think of "geography." From anywhere on this island you're surrounded by mountains, mountains behind those mountains, and mountains behind those mountains. This is celebrity geography: The Scenery. Nice backdrop, you might say. The inspiration for the slogan "Super, Natural British Columbia."

Not so much a backdrop: more an immanent presence. Air flows over the earth's topography the way water flows over a riverbed, responsive to contours, rushing down valleys, climbing slopes, fanning out over plains. And so this mountain, sea, and island area has its own ornate invisible writing, a daily tide of cooling and warming air flowing one way in the morning and the other in the afternoon; a rush of airy tributaries that hurtle down fjords and inlets, eddy, rise, fall, lapse into languid pools, or blast in from the west with the breath of the open Pacific in their wake.

Air brings me to birds: oystercatchers, kingfishers, bald eagles, ospreys, turnstones, mergansers, hummingbirds, kinglets, pine siskins, bats ... and land creatures: mink, deer, possums, raccoons, mice of all sorts ... and vegetation: the west coast's very own orchid, lady's tresses, and salal, Indian pipe, arbutus, balsam, alder, fir, gumweed, cactus, camas, chocolate lilies, erythronium, sea blush ...

Okay, I'll stop. It's clear now, perhaps, why Alan of Lille's and John Bunyan's allegorical readings might seem inadequate to me here; why writing geography as a metaphor for the human condition isn't what I want to do; why my instinct is to try to make this geographical superabundance, distraction, and Presence-with-a-capital-P available to a reader, even while telling a story of human relationships.

I notice that I do it in several ways. One is to make geographical inflections a pervasive presence, to enfold them even in seemingly ungeographical exchanges and relationships. Here is Aunt Fifi on the first evening of a summer visit to Becca and Gran: "Aunt Fifi stood up to eat, gazing out to sea and pacing the sandstone in between bouts of picking sea asparagus fibres from her teeth" (Baker 52). The human point of this sentence, which stands in the middle of a conversation, is about character (Aunt Fifi's restlessness) and relationship (Aunt Fifi's annoyance with her mother's weird cooking, and, implied in the narrative point of view, Becca's wariness about both her aunt's and her Gran's strong personalities). But the geographical elements are equally present, if you choose to notice them: the sea flowing past in front of the three humans, the sedimentary rock on which they stand and sit (as well as its composition, sandstone, and all that implies to the senses), and sea asparagus, a tough, fleshy plant with a skeleton of silicon (which is why Aunt Fifi has to pick her teeth)—incidentally one of the few plants in which silicon appears naturally. The sea asparagus also tells you, should you have ears to hear, of a nearby habitat just below the high tide line—but you can look in your compendium of intertidal marine botany for that information.

That is one way to convey geographical abundance. It involves recognizing that there is no "general" geography: every habitat is specific. It doesn't take a lot of fuss or long descriptive passages. It can be enfolded into a single sentence, one whose primary momentum is towards character and plot.

Another way to convey the immanent presence of geography and geographical features is to have a plot that depends on them. There is a well-worn way to do this, and Canadian writers have often embraced it: survival. Or, as we were conditioned to categorize it when I was in high school: *man versus nature*.

I myself eschew *man versus nature*, a conflictual paradigm. I would rather consider how human change happens in tandem with geographically inflected events. Becca and Gran collect oysters in the middle of the night because that is when the tide's low on that day, by that moon, in that year; they take their first run in the inflatable Zodiac the day the herring begin to spawn; Merlin's brother-in-law's boat is almost wrecked because of a stiff northwesterly wind and a bay open to that wind. In fact many of the episodes in *Becca at Sea* involve humans explicitly engaging geography and its features, and the action is brought about by natural forces.

But human interaction is important too; for a story to work, most human readers must somehow see themselves there (like Alan of Lille). I want both elements, the human and the geographical, to be noticed equally: how can I invite readers to see this super, natural geography and its features to be as much the story as the peccadilloes of the human characters? How can I invite readers to perceive at once that place is not "a beautiful backdrop" for the "real" story, the one involving humans, but au contraire *is* the story, every bit as much as the human characters?

One way to do this is by a kind of antiphony, two or more alternating voices making comments that are both related and unrelated—in this case, pertaining to whatever is going on in the non-human world (animals, plants, weather, topography) and whatever is going on between the characters. That sort of antiphony happens in this passage, which occurs when Becca, Gran, and Aunt Fifi are enjoying after-dinner coffee on the deck with Merlin, the Shakespearean-actor-turned-plumber, and Mac, a neighbour. Aunt Fifi is arguing with Merlin about Shakespeare:

> "So you feel the young lovers are drippy," [Aunt Fifi] said.
>
> "Saps, to a man," Merlin agreed. "Lovely coffee."
>
> "It could be that that's not the most important thing about them," said Aunt Fifi sternly.
>
> "Of course it is! Think of Orlando's awful poetry!"
>
> "Poetry!" cried Aunt Fifi. "What do you know about poetry?"
>
> "Look at that boat," Becca said. "Isn't it kind of windy to be out in a boat?"
>
> "It is," said Mac. "There'll be a gale force warning in effect by tonight and gusts from the northwest up to fifty knots. That boat won't last a minute."
>
> "What boat?" Aunt Fifi asked, but she hurried on. "Merlin, I don't know why you have this anger, this unreasonableness, about poetry! A man who used to act Shakespeare!"
>
> "What's unreasonable about it?" Merlin demanded.
>
> "Nobody would be silly enough to go out in a small boat in this weather," Gran assured Becca. (83–84)

Here, the macaronic or antiphonal structure sustains the energy and passion of Aunt Fifi and Merlin's argument (an argument that some readers hope is a courting ritual), and at the same time, through Becca's, Gran's, and Mac's comments about a boat in the bay, sustains awareness of the ferocious wind and the great breakers foaming onto the beach. The passage demands that the reader read inferences: Becca's diffident question, "Isn't it kind of windy to be out in a boat?," tells us that the wind is strong and continuous (something that's only been hinted at up to this point), and presents the vulnerable boat bobbing about on the waves, thus beginning a new thread in the story. Mac's observation is a mode of decoding inferences itself—weather forecasting being a way of reading the invisible air writing. Eventually, the passion of the Shakespeare argument and the forcefulness of the weather become a confluence ending in the literal immersion of the human characters in the sea, the action of which geography has caused.

I like the way this antiphonal structure works, but sometimes the natural world demands a longer passage, one that conveys (and not just by contrast of length) that at this moment, geography overwhelms; that here, it commands the attention of all our senses.

When Becca and Gran find themselves in the middle of spawning herring, the comic plot line involves their deflating inflatable boat, the Zodiac, and Mac, who helps them with it. But that's a little story in the context of the great, seething event of the herring spawn, which is clearly even more not-to-be-missed by the wildlife than it is by humans, whether Becca and Gran or those in the fishing industry:

> With all the seals and sea lions breeching and diving, the Zodiac pitched and bucked. Sea lions' urgent barking, gulls' skirling, the weird high laughter of eagles and the helicopters' clatter filled Becca's ears. Tails slapped. Beaks ripped and tore and sliced. Eagles stretched out knife-sharp talons and raked up their prey. Water streamed as seals and sea lions dove and surfaced, exploding out of the water with their mouths bristling with fish.
>
> "The fishermen aren't allowed to start fishing until the herring are actually laying their eggs—when they're ripe," Gran hollered, her binoculars trained on ospreys and eagles. "That's why everyone's hanging about here. They're waiting for the Fisheries people to tell them they can start!"
>
> "The sea lions aren't waiting!" Becca yelled, pushing the tiller hard over to dodge a fat seal.
>
> Gran didn't seem to hear her. Gulls flapped madly, splashed down and lifted themselves out, gullets stopped with herring. Wheeling and crying, beaks open, tongues stiff with noise-making—wings beat, throats vibrated. (24)

This passage of activity and a multitude of species—including the humans waiting for the go-ahead from the Fisheries department—shows one of the ways I try to allow geographical elements their own space. It is an extended passage of description that privileges what is happening outside the human interaction. I use a lot of nouns pertaining to non-human species—seals, sea lions, gulls, eagles, tails, beaks, talons, water, fish. A multitude of verbs pertain to these species—breeching, diving, barking, skirling, slapped, ripped, tore, slice, stretched, raked, streamed, dove, surfaced, exploding, bristling, flapped, splashed, lifted, stopped, wheeling, crying, noise-making, beat, vibrated. Even what Gran says puts the "natural" world first—humans must wait until there are enough herring to ensure a successful spawning before they begin assuaging their own economic need.

Balance, then, in the straightforward measurable way of space, words, and time, is one of the ways a writer allows geographical elements a presence that can hint more truly at humanity's position on earth. We are contingent on geography—as geography's broadened definition acknowledges. Geography won't speak for itself, but we can draw attention to the need to notice and respond to it partly through giving it presence in writing—a presence that can't be relegated to backdrop for a human story.

This brings me to my third question: how to convey that paradox quintessential to a human encounter with geographical features, especially at their most overwhelming and alien? The longing that comes from an intimate interaction with the Other—another species, a breathtaking but terrifying environment—and the powerful recognition that one is excluded from that Other and has no place there at all? Becca experiences a kind of intimacy with the Other when she looks after a baby seal for the morning, but although she gives it a name and talks to it as if it's a selkie, she also marvels at its strangeness and is agitated about its incomprehensibility to her. When she swims through the seal channel near Camas Island, she's both enchanted and alarmed because she knows she's out of her element: "the water was the seals' home, and she was a visitor here" (157). It makes her think of the sea lions during the herring run: "They didn't care about people" (158). Must imaginative geography be in the service of the human? To me, in my writing right now, it must not be: that energetic dialectic between privileged intimacy and exclusion, identification and alienation, satisfaction and longing, is critical to the deepest kind of wonder—and to a sense of responsibility to and about environment. It is the way, also, to avoid a kind of mental colonization or appropriation of place—the sentimentalizing or romanticizing we can impose on place once we've left it, unconsciously smoothing out discomforts—bitter southeast winds, chilly water, lion's mane jellyfish that leave you in agonies, the pappy, brownish soup that is an August algae bloom and feels like grease on the skin.

No matter how much I might succeed at interleaving the geographical with the human, the sea, mountains, winds, plants, and animals with Becca's family and relations, I know that every story is a human construction. But writing a "real" geography, so to speak, allows for a possibility that goes beyond the human—real or imagined. When the book is shut and the fictitious characters have done all they're going to do in this volume, the life of the Salish Sea goes on. Its meaning for us comes from our capacity to apprehend it; but its life is not contingent on human observation. It goes on, even when we're not there:

> The thermal updraft is drafting upward; the flounder is flapping out of the sand; the heron has fixed its gaze on a movement at the edge of the water. Thousands of sand dollars wiggle their tiny black spines two fathoms deep in the mouth of the bay, the tide is coming in, and the winds are from the northwest, 5–15 knots.

———

Note

1) "Omnis mundi creatura / quasi liber et pictura / nobis est, in speculum; / Nostrae vitae, nostrae mortis, / nostri status, nostrae sortis, / fidele signaculum." Alanus ab Insulis, *Omnis Mundi Creatura*, ll. 1–6; translation my own.

Three

━━

FANTASY WORLDS AND
RE-ENCHANTMENT

11) The Imaginary North in Eileen Kernaghan's *The Snow Queen*

❦ JOANNE FINDON

———

Although Hans Christian Andersen's tale *The Snow Queen* begins and ends in the sunlit safety of a Victorian rose garden, Kay's departure with the Snow Queen launches Gerda on a journey north into unknown lands where the heartless Snow Queen rules. As critic Roger Sale has commented, by the time Andersen sat down to write his original fairy tale, he had travelled extensively and had come to ask "what it means to be captured by the queen of the snow." At this point in his writing career, Sale argues, "for once, trusting what he [knew]," Andersen was able to "release his story from the personal bondage that ties up so much of his other work" (Sale 73). In other words, instead of once again rehashing his own tale of rags-to-riches success in the face of unrequited love, Andersen turned to contemplating what it really means to be a northern writer. Andersen's North is as much an *idea* as a geographical space.

Canadians are familiar with this concept. Glenn Gould's famous radio documentary *The Idea of North* explores that region not only as a place but as a powerful concept, an imaginative geography that has formed our sense of Canadian identity (Gould 1967). This "idea of north" is connected with the exotic, the mysterious, the strange, and continues to inflect the ways in which we see ourselves in the larger world. The tragedy of the lost Franklin expedition, which continues to garner news coverage even now that both ships have been found, still shapes our view of the North as seductive, implacable, and haunting.[1] This North is icy and vast, a place where we can lose ourselves and go mad, or undergo extreme tests and trials and return triumphant. Notably, those who pit themselves against its power are seen as heroes. This North is, in a word, epic.

Eileen Kernaghan's Young Adult novel *The Snow Queen* takes this concept of the North further in a profound reimagining of Andersen's tale.[2] Kernaghan's North suggests the limits of human imagination and endeavour, but it also becomes a metaphor for the extremes of adolescence: heroic love, rebellion against parents, and the single-minded questing for new identities and a place in the world. Kernaghan's Kai and Gerda are teenagers, not young children like Andersen's, and she enlarges the emotional canvas of the tale through her development of the Little Robber Girl as the Saami teen, Ritva.[3] Ritva is angry, the unwilling heir to her mother's shamanic powers, but her links to the land and its mysterious forces prove to be key in the quest to rescue Kai. Most important, Kernaghan infuses Andersen's North with the power of the Finnish epic the *Kalevala*, whose heroes become role models for Ritva, and whose Hag of the North is vividly brought to life in Kernaghan's Snow Queen. This Snow Queen is much more dangerous than Andersen's: as Kernaghan has noted, she is the *Kalevala*'s Drowner of Heroes and Devourer of Souls (Kernaghan 91; Wolf, "Revisiting"). Kernaghan reshapes the northern landscape as a fantastic realm of myth and shamanic power to tell a riveting tale of female friendship and heroism. In the process, she plays with standard gender binaries and blurs the conventional boundaries between masculine and feminine, crafting a true female *Bildungsroman*.[4] It is the female figures who inevitably preside over the most powerful spaces in the novel, and even those who are apparently tied to domestic spaces wield profound influence.[5]

In Kernaghan's version, Gerda's trek north maps her inner journey and transformation; it is not simply a quest to rescue Kai. As soon as she leaves behind the southern cities of Denmark with their pleasant but stifling domestic spaces, she traverses landscapes that increasingly mirror her emotional journey. She must pass through a series of spaces that become more and more liminal, spaces that partake of the "potentiality" that anthropologists have argued is inherent in "the state of being in between separate categories of space, time or identity" (Nagy 135). Liminality is "frequently likened to death, to being in the womb, to invisibility, to darkness, to bisexuality, to the wilderness" (Turner 95), and tends to challenge traditional binary categories of all kinds. In myth and folk tale such "in-betweenness" is often reflected in the mixing of Nature and Culture, human and animal, and in landscapes situated between wilderness and civilization (Read 16; Turner, *Ritual*). Since other conventional binary oppositions between masculine and feminine, active and passive, public and domestic space, and speech and silence, are also often at play in traditional tales, including folk tales and fairy tales, the ambiguous spaces between these categories can similarly disrupt conventional order. Liminality can be dangerous, but it can also be extremely powerful as a "'realm of possibility' where new combinations of cultural givens" can be tested and tried out (Ashley xviii). It is worth noting that liminality is also the natural state of the adolescent, who stands on that borderland between childhood and adulthood that is in a way pure potentiality. In Kernaghan's *The Snow Queen*, the progressively less familiar spaces through which Gerda travels mirror her restlessness and confusion as she struggles to carve out an adult female identity of her own.

Although Kernaghan's Kai is not abducted by the Snow Queen like Andersen's Kay is, in both versions Gerda is the heroic female who sets out alone on a quest to find him and win him back from the Snow Queen. Andersen's Gerda is astonishingly brave and active for a Victorian girl, and indeed one wonders if he might have been influenced by variants of the folk tale "East of the Sun and West of the Moon," collected in Norway by Asbjornsen and Moe, whose similarly intrepid female hero travels to the edge of the world to rescue her beloved (Lang, *Blue Fairy Book*, 19–29). Certainly, Kernaghan's rewriting of Andersen's tale develops Gerda's active nature even further, within the realistic social and historical context of late-nineteeth-century Scandinavia. Gerda's agency is restricted by her Victorian society and its expectations for women and girls, and she must at times use cunning and lies to achieve her goals. Indeed, she must increasingly act the part of the trickster—a figure who is most often male in traditional tales. Gender roles are subverted almost from the beginning when

Gerda realizes that Kai's parents, although sick with worry at hearing nothing from their son, cannot and will not travel north to seek him. Kai's mother and Gerda's, conventional Victorian women, have no power to act on their own; Kai's father is not physically well enough to undertake such a journey and thus cannot fulfill the role of protector to his own son. Only a determined girl who breaks the rules, lies, and takes decisive action will be able to remedy the situation.

Gerda's voyage north in fact begins with a neat bit of trickery. On the pretence of visiting her friend in Copenhagen, Gerda leaves her comfortable home with its warm smells of coffee and ginger cake, that familial nest of Victorian domesticity which also constrains her. She travels alone to Madame Aurore's mansion in Sweden, only to find that the woman has left for her summer palace in the North. Gerda is taken in first by the coachman's aunt in her thatched cottage, and then, through the introduction of the female adventurer Ingeborg Eriksson, by a princess who equips her with a coach, warm clothing, and provisions for her dangerous journey north to Vappa-Varra. Ingeborg Eriksson is an especially important figure; appearing at a threshold moment in Gerda's journey, at the point when she is about to leave "civilization" behind with no maps to guide her, Eriksson provides Gerda with the role model of a woman travelling alone in the North, along with practical advice and the voice of experience. She even sounds a prophetic note when she tells Gerda, "I've never yet met a man worth going to the ends of the earth for," then adds, "Well, you're young, you're entitled to your illusions" (Kernaghan 53).

The princess's coach and provisions provide no protection from other northern dangers, however: Saami robbers soon ambush the coach, kill the driver, and capture the terrified Gerda. The chief robber's daughter, Ritva, claims Gerda for her own and brings her to her father's ruined hall, where Gerda spends the winter as a virtual prisoner under Ritva's strict control. However, spring brings change, and Ritva decides to help Gerda. Once Gerda and Ritva depart on their journey north together, the girls are sheltered by two women—the old woman who writes a message on a codfish to her friend, and the woman in Finnmark, who binds the winds—both of whom give Gerda and Ritva wise advice on finding the palace of the Snow Queen. Notably, all of these stations on the way contain women who are powerful in one way or another—in age, in wisdom, and in magic. The Snow Queen is the one figure who seems invulnerable, but she can ultimately be defeated through female cooperation and the force of love.

It is worth noting that despite their strong associations with enclosed, domestic (even womb-like) spaces, all of these older women exert considerable power in their own ways. The most visibly enclosed of all—the woman who binds the winds—is in fact the most powerful, the one who is key in providing the bag of winds that is instrumental in defeating the Dark Enchantress.

In Andersen's tale, the Little Robber Girl acts as a helper to Gerda but does not travel north with her. By contrast, in Kernaghan's retelling, Ritva is a major character in the story, and until the midpoint of the novel, alternating chapters are told from her point of view. Her father's ruined hall functions as the first real threshold between Nature and Culture that Gerda encounters, being in itself a threshold: the ruined stone building, cracked and pitted with holes, home to both humans and animals, is *both* Nature and Culture, truly a space in-between. The men leave to hunt, but when they are home, they behave like wild beasts in their drunkenness and uncontrolled appetites. Moreover, Ritva's home is also a threshold of a different sort: it is the site of her mother's *boasso*—the sacred space where the shaman chants and drums and from which she makes her "soul-journeys." Most important, this liminal space with its forest surroundings engenders Ritva's own visionary dreams, dreams that increasingly link her to the heroes of the *Kalevala*. These heroes are traditionally male, but Ritva, a female, is encouraged to take on their roles and identities, thus crossing the boundaries of gender as well. Ritva's mentors are the spirit of her dead grandmother, also a shaman, and her animal guardian the elk, who grants her a vision of the way north to Madame Aurore's palace. Thus Ritva's home is a space between "civilization" and the northern unknown inhabited by the dangerous Snow Queen.

Once Gerda is captured by the robber band of Ritva's father and claimed as a "pet" by Ritva herself, she enters a liminal period of suspended time, spending the winter in increasing despair and losing all hope of deliverance from her captivity. But this interlude, this "realm of possibility" (Ashley xviii), is critical to the growth of both girls. As the daughter of a shaman and the heir to her powers, Ritva occupies a powerful space in the novel—that of a mediator not only between Nature and Culture but also between this world and the otherworld. As a human who has one foot in the spirit world, whether she likes it or not (and at the beginning, she detests it), she is the perfect companion for Gerda, whose quest is as much spiritual as physical. Gerda needs to figure out who she is, as *Gerda*, not just as "the girl who is in love with Kai." At the same time, Ritva needs to pay attention to the spirits haunting her dreams who are trying to help her understand who *she* is. When spring comes and Ritva realizes that Gerda will die if she remains her captive, she has a terrifying vision of a bear who rips her own human flesh off. Reduced to skeleton form, in a state between life and death, Ritva is "flooded with calm, and lightness, and power ... pared down to the hard imperishable bone" (Kernaghan 82); in that purified form, she hears the spirits of the forest, the voices of the river and the trees, urging her to take Gerda and journey north to help her in her search for Kai. At this point, Ritva's inner life seems to snap into focus. Before this moment she has been sullen and angry, refusing the dreams of fire and death that torment her, fighting her mother and the destiny that must inexorably claim her. But having accepted the call to adventure, she is energized.

On the girls' journey north together, the huts of the two old women, in conjunction with the wilderness landscape, function symbolically in mapping the inner growth of both these female adventurers. As Gaston Bachelard notes, huts by nature embody safety and a kind of power rooted in the earth, since "the hut appears to be the tap-root of the function of inhabiting" (Bachelard 31). In Kernaghan's novel the huts of the wise women are also liminal spaces, barely distinguishable from the landscape: the first is "a solitary turf-covered, dome-shaped dwelling ... at the edge of the forest" that looks "like a small grassy hill with a thread of white smoke curling out of the top" (Kernaghan 89), while the Finnmark woman lives in a similar hut with no door but only a smoke hole, which is huddled in the shelter of a cliff (92). Despite their humble dwellings, these are women of power who know the perils of seeking the woman they know as the Dark Enchantress, and who give the girls helpful advice and directions; the Finnmark woman also provides the magic bag of winds that will ultimately propel them to and from the Snow Queen's palace. Significantly, meeting these wise women in their earth-like huts seems to connect Ritva with the primal power of the land, power she needs to access as a shaman. But the women also affirm Gerda's worth. As the Finnmark woman says, the quest must be accomplished by Gerda herself: she tells Ritva, "You will have your part to play, my girl. But if the boy's soul is to be set free, it is your little rabbit who must do it" (95). Gerda has her own brand of inner power, which Ritva cannot yet see or understand: the power of goodness and love.

Gerda's apparent softness and innocence masks an iron will and cleverness that the Finnmark woman seems instinctively to recognize.

As Gerda and Ritva travel ever farther north to the Snow Queen's palace beyond Spitzbergen, the terrain becomes more otherworldly—a fusion of the real northern landscape and the mythical geography of the *Kalevala*. In Andersen's original, the landscape does not play an important role at this point in the story; the focus is instead on Gerda's speedy rescue of Kay. In Kernaghan's retelling the quest is expanded and enriched to suggest that the land is a powerful force that both challenges Gerda and Ritva and mirrors their stubborn perseverance. As the girls leave solid rock and soil behind and begin travelling across ice, the ground itself becomes unstable. Near the Cave of the North Wind, we read this:

> There was something dreamlike, hallucinatory, about this northward journey. Always before there had been lakes and rivers, hills and forests to help them chart their way. Now there were no more landmarks, and the thin shell of ice upon which they walked was like a vast unfinished puzzle, the pieces endlessly lifted and turned and shuffled by a giant hand. (111)

Of course this last image foreshadows the ice puzzle that Kai is trying to piece together in the cruel and futile task given to him by Madame Aurore. This puzzle also recalls the evil mirror that features so prominently in Andersen's tale as the cause of all the trouble. Kernaghan's allusion will remind readers familiar with Andersen's tale that his explanation for Kay's change of heart is something external, something emanating from an outside demonic source; in contrast, Kernaghan suggests that human hearts can harden of their own accord and don't necessarily need external forces to become selfish and cold.

Kernaghan's addition of the *Kalevala* material transforms the story into a meditation on "the idea of North." In one of Ritva's dream visions, her dead grandmother tells her a fable about a shaman who sends a boy to the northern edge of the world. The third time he goes, he does not return, because "he learned that ice is death" (49). This is what the Snow Queen is: death. Kernaghan is perhaps echoing Andersen's depiction of his Snow Queen's abduction of Kay, in which her kiss has an effect that mimics freezing to death:

She kissed him on the forehead. Ah-h-h! Her kiss was colder than ice; it went straight to his heart, which was already half way to being a lump of ice. He felt as if he were dying, but only for a moment. Then he felt perfectly well, and he no longer noticed the cold. (Andersen 111)

Kernaghan's name for her, Madame Aurore, evokes the chill splendour of the Northern Lights, suggesting the seductive power and beauty of the North. Kernaghan harnesses the mystery and allure of the *aurora borealis* in her elaboration of Kai's seduction by the dazzling Madame Aurore. Although she is aligned with evil, she is more than a moral force: she is also a force of Nature, the embodiment of what Andersen and other writers understood the North to be—an implacable force that is hostile to humans. She presides over a physical landscape that is harsh and merciless, but also over an inner landscape of emotional stasis and death. Her palace is located beyond the Cave of the North Wind, and when Gerda and Ritva pass through the cave the experience is figured as a death and rebirth: they are sucked into its womblike space, and

> Then, as suddenly as they had entered, they were through, like a cork exploding from a bottle.
>
> On the other side of the mountain, beyond a narrow channel of dark water, lay a world of profound night. The moon hung like a great pewter dish in a cobalt sky. Trackless snowfields, stained with violet shadows, stretched away to the dark line of the horizon, where they vanished into a silvery mist ...
>
> Gerda said, "We've done it, Ritva. We've passed through the Cave of the North Wind. We've come to the Snow Queen's country, where earth and day end." (115-16)

The Snow Queen's realm is described as "a place outside of time, beyond geography: where snowfields flowed on and on under the frozen stars to the world's rim, where earth and heaven met" (117). When Gerda and Ritva finally see the Snow Queen's palace, it is a place of intense beauty:

> And all at once the sky was ablaze with arrows and archways and rippling curtains of flame. In the northern distance, across the shadowy snowfields, stood towers and turrets and parapets of crystal, glimmering rose-pink and gold and apple-green. (118)

The palace is a perfect example of the type of fairy tale "tower" discussed by Thomas in connection with fairy tales—a place of the past, where time stands still (Thomas, "Woods," 125-26). The castle with its towers, Thomas argues, embodies the supernatural, even the fairy realm. The fairy tale castle may suggest happy endings, yet despite its exquisite beauty, *this* palace is no place for humans:

> Nothing had prepared Gerda for a palace so magnificent—and so utterly devoid of warmth and comfort. *No one human could live in this place*, she thought. And she shuddered at a sudden chilling intimation: living here, what might Kai have become? (Kernaghan 118)

The mirror, which was the demonic source of Kay's problems in Andersen's tale, is in Kernaghan's retelling a symbol of the quest for all knowledge in which Kai is doomed to fail. His nightmarish attempts to put the ice puzzle together highlight the Snow Queen's power over him. Thirsty for scientific knowledge and the enlightenment that he thinks it will bring, he has been lured by the promise of reward on completion of an impossible task:

> He was like a figure carved from ice: blind, deaf, unfeeling. He stooped, and began to shuffle some small jagged pieces of ice from one place to another, as though trying to see a pattern in them. There was a terrible sense of futility, of defeat, in his stiff, painful movements. (120)

In Andersen's tale, Gerda is able to rescue Kay with relative ease: the Snow Queen has simply gone away to the south, leaving her palace unguarded, and Kay is free to leave. However, this is not the case in Kernaghan's version. The Snow Queen is determined to keep Kai for herself and sets the girls three impossible tasks before she will release him—*if* he wishes to depart. And in setting these tasks she explicitly invokes the fairy tale tradition:

> "Well," said the Woman of the North, "if you would play at being heroes, then I will join you in your game. Three impossible tasks to be performed—three challenges. Isn't that how the rules go? You know the fairy tales as well as I." (123)

These tasks involve both ingenuity and great risk. It is worth noting that in traditional tales from myth and folklore, both the figure who demands such impossible tasks and the figures who perform them are almost always male; here again, Kernaghan violates gender expectations and depicts a contest in which the combatants are all female.[6] The Snow Queen's demands are completed only through the cooperation of the two girls, aided by Ritva's shamanic magic. Accomplishing the second task—catching a great silver pike for the Snow Queen's dinner—involves Ritva's interweaving of materials from the natural world on the one hand—in the form of discarded fishbones and strands of her own hair—and a magical song in which she invokes and identifies herself with the male shaman Vaïnö on the other. She tells Gerda the tale of how Vaïnö made a magic harp from the jawbones of a giant pike that he had killed:

"Am I not a shaman, and the daughter of a shaman? What old Vaïnö could do, I can do." As she worked, Ritva sang cheerfully to herself,

> *As he played upon the pike-teeth*
> *and he lifted up the fish tail*
> *the horsehair sounded sweetly*
> *and the horsehair sounded clearly ...* (127)

The enchanted music of Ritva's harp lures the pike to her, and he is killed and delivered to Madame Aurore's table. For the third task, the Snow Queen demands a cover for her jewel chest that is embroidered with yarn dyed with the "purple of the lichen that grows on the stones beside the River of the Dead; a white swan's feather floating on that river; and the blood-red of the berries that grow at the entrance to the Dead Land" (131). Significantly, this third task that the Snow Queen demands for the release of Kai takes Gerda and Ritva to the very gates of the Dead Land. The lichen is found by Ritva's reindeer Ba, and to obtain the feather Ritva lures the swan by playing her fishbone harp. Then to make the dye that Gerda needs, Ritva risks her life swimming in the icy water to bring back the berries, whose blood-red colour speaks of sacrifice and life itself. Gerda and Ritva work together to produce the dye and the yarn, made from the fluffy hair of the palace's dog and twisted by Ritva's nimble hands into yarn. Gerda does the embroidery—a typical domestic activity, a "woman's art," but one that is transformed in this episode into a painful, heroic test for Gerda, whose fingers have been numbed by the punishing cold:

> Over and over, with fierce determination, she forced the blunt bone needle through the hide. For my sake, she told herself, Ritva risked her life in the icy currents of the river. Though my fingers freeze, though my blood turns to ice in my veins, I will not complain about this simple seamstress's task. (137)

Thus, this third and most difficult task is completed through a combination of shamanic magic and female cooperation that makes use of the specific landscape and the materials it provides.

Yet the Snow Queen reneges on her promise to release Kai and demands a sacrifice. Will it be Ritva's aged reindeer Ba's life in return for Kai's? Gerda refuses to allow the sacrifice of Ritva's faithful animal friend:

> She took Ritva's cold, trembling hands in her own.
> "Don't you see what she wants, Ritva? How cruel and calculating she is, to give us a choice that must tear the two of us apart?"
> "But you love Kai ..."
> "And you love Ba. Surely, Ritva, the two of us together can outwit her. She does not deserve to win." (140)

This betrayal—the refusal of the Snow Queen to play "by the rules" of fairy tales—marks a turning point for Gerda. Abandoning the last shreds of her conventional Victorian identity as an obedient girl, she takes matters into her own hands. At this point the two girls become tricksters in earnest. In a scene that echoes an important episode in the *Kalevala*, Ritva uses her magic fishbone harp to play a lullaby that puts the whole court to sleep, and it is then that the girls steal Kai.[7] Even this is not easy: Kai is so cold and weak and so enchanted by the Snow Queen's promise of attaining all knowledge that at first he will not come, unable to believe that Ritva's sleeping spell could have worked:

> "No one leaves here without her permission," Kai said in a dreary monotone, as though he were repeating a formula by rote. For answer, Gerda seized his hands, and dragged him to his feet.

"Nonsense," she said. "We need no one's permission. Show him what you've done, Ritva." She took firm hold of Kai's arm and led him, weakly protesting, to the edge of the frozen lake." (143)

But Kai insists that they steal the Snow Queen's rune chest, which is the source of her power. In a telling gender reversal, he is so weak that he can barely keep up with them, dragging the heavy chest behind him. In the final confrontation with the Snow Queen it is Ritva's shamanic power that saves them all. She asks Gerda for the flint and tinder they brought, saying:

"Give it to me. Am I not the daughter of shamans? Have I not accomplished every task the Snow Queen gave me? Have I not stolen her treasure chest from under her nose? I am a hero, like Väinö. What Väinö could do, I can do also." (148)

It seems that in her battle with the Dark Enchantress of the *Kalevala*, Ritva comes to truly accept her shamanic identity and the power available to her. She becomes not Ritva but one of the ancient heroes like Väinö or Ilmarinen:

Then she drew back her arm and with all her remaining strength hurled the fragment of flint over the stern. At the instant it struck the water, the flint began to grow ... At that moment, for the first time, she [Gerda] glimpsed the true nature of Ritva's power. This was no illusion, no conjurer's trick, but real stone, solid and impenetrable, created out of a bit of flint, and air, and sea-spume. *Truly,* thought Gerda, *Ritva is the heir of the magical smith Ilmarinen in the old tales, who forged a new sun and a new moon for the heavens, and welded the arch of air.* (149)

Notably, Gerda expresses no surprise that Ritva should step into this *male* role and be able to wield magic like the heroes of old. Here at the edge of the world, conventional gender roles have melted away. Kai is the passive victim, the damsel in distress, dependent on both of his female rescuers to defend him against the cruel Snow Queen.

Like most quest stories for young adults, Kernaghan's *The Snow Queen* maps the inner growth of its characters through their progress on the journey as they discover their own power by overcoming tests and trials.[8] As Trites points out, most young adult novels focus on the ways in which the characters' growth is linked to power and how they learn to "negoti-ate the levels of power" that reside in the social structures around them (Trites 20). Thus, Gerda finds ways to resist the conventions of Victorian womanhood that would hold her back. But Kernaghan also uses the Snow Queen and her power of enchantment to explore the perils of pursuing intellectual curiosity above human relationships. Kernaghan has said that in writing the novel, she researched the history of Arctic exploration, and was particularly interested in the context of the first great flourishing of scientific inquiry in the late nineteenth century (Wolf, "Revisiting"). For Gerda, as for some of those Victorian explorers, alongside the excitement of new discovery there is a lingering dread that emotional certainties have now become as unstable as the sea ice. Indeed, looking at her old friend Kai, the young man she had hoped to marry, Gerda wonders:

> But where, in that thin, worn stranger's face, was the friend of her childhood, the kind, clever boy for whom she had dared so much?
> Had the Snow Queen stolen Kai's true self—or had he simply lost it somewhere, laid it aside and forgotten it like a cap or a half-read book?
> And the thought came to her, like cold fingers clutching her heart—if you lose your self, can you ever find it again? (Kernaghan 156)

In light of the litter of shipwrecks and shattered dreams in our own northern regions (of which Franklin's *Erebus* and *Terror* are but two of many), Kernaghan's Snow Queen also seems emblematic of the seductive power that an obsession with scientific exploration can wield over human hearts. The novel suggests that the thirst for knowledge must be coupled with common sense and compassion. Kai appears to turn away from human connection and feeling when he rejects Gerda at the beginning and sets out to study with Madame Aurore, whom he idolizes as a "woman of great learning—a Doctor of Philosophy" (26). Great learning can indeed coexist with emotional bonds between humans, but in Kai's case it seems to distance him not only from Gerda but from his whole family. Anders-en's demonic mirror has been replaced by a more complex and unsettling psychology. This is brought home when Gerda accuses the Snow Queen of turning Kai into a monster and she replies, "No, my dear child. I think he did that to himself" (122).

Kernaghan also complicates the standard quest plot by having the female characters achieve their goal only to find that the quest has yielded unexpected results. Gerda's beloved friend and potential mate Kai has indeed been found and saved, but he has been transformed into someone she barely recognizes. For Ritva, the journey has given her a way to accept—even to welcome—her inheritance of shamanic power; she is now free and ready to move forward, accepting of her gifts and destiny, although not necessarily ready to return to her mother's community. For Kai, whatever and whoever he was before this novel began has been lost: perhaps his pursuit of knowledge to the exclusion of all else has robbed him of emotional warmth and would have done so even without the Snow Queen's enchantment. It is significant that Kernaghan's Prologue begins with the moment to which Gerda traces the tragic change in him, when he dismisses her poem about roses as childish. Unlike Andersen's opening, with the two dear friends sitting together in their garden, here the fond memories of such moments are consigned to the past even before the novel begins. We as readers never see the kind and clever Kai that Gerda loves so much, and as the novel closes we wonder if he ever even existed outside of her memory.

Andersen ends his tale with "little" Gerda and Kay safely back in their sunny rose garden, with the memories of the awful Snow Queen rapidly fading away. They seem to have completed the circle of their journey without undergoing any profound change or inner growth. Kernaghan's ending is radically different. The novel concludes with Gerda, Kai, and Ritva still in transit on a ship sailing southward to Denmark, in an evocation of their still-fluid and developing identities. Curiously, considering how much Kai has been through, he seems to have learned very little, like Andersen's child Kay. Indeed, his behaviour calls into question the very meaning of Gerda's quest, as Ritva notes when she pointedly asks, "And Kai? Did he love you as much as that? If the Snow Queen had stolen you, little rabbit, would your Kai have set out across the frozen seas to save you?" (156). Gerda's inability to answer, and her horror at the thought that Ritva will put that question to Kai herself, consolidates the impression that as a "damsel in distress" figure in need of rescue, Kai isn't really worth the trouble.

The final conversation between Gerda and Ritva takes place in that liminal geographical space between the North and its adventures on the one hand, and the South with its conventional gender roles and expectations for young women on the other. The future is anything but clear, especially for Gerda, who is forced by Ritva to reassess the fairy tale ending of domestic bliss that she has been dreaming of. Instead, Ritva and the whole adventure have shown her that there are other ways of living. Gerda asks her,

"Don't you ever mean to marry?"

"Me? Not a chance. Can you see me darning trousers, and stirring the stewpot? I am a shaman, little rabbit. I am a woman of power. I have travelled to the spirit kingdom. I have defeated the Dark Enchantress, and brought you safely back from beyond the world's edge."

"A woman of power," repeated Gerda, liking the sound of those words.

"As you are too, little rabbit," said Ritva, surprisingly. "It was you who saved Kai's life. And see how grateful he is, how he gets down on his knees to thank the hero who rescued him." (155)

Although Gerda jumps to Kai's defence, she knows that something inside her has shifted. Instead of the hardships and suffering that her quest has entailed, what she remembers most of all is the brilliant beauty of the northern landscape and the thrill of vanquishing the Snow Queen:

> She had been to the farthermost edge of the world, where earth and day end. There was no road, now, that she would be afraid to travel. How could she be content to dream away her life in a southern rose garden? (157)

Kernaghan's novel thus ends with Gerda turning away from the sentimental comfort of Andersen's tale. Instead of gazing at her beloved Kai, she is, in the end, focused on her heroic companion Ritva. Kernaghan concludes her version with a tender moment between Gerda and Ritva and the suggestion of possible future travel for the two female heroes whose journey into this magical North has changed them forever.

> Gerda leaned forward, put her arms around Ritva's shoulders, pressed her own chapped, windburned cheek against the robber-maid's.
>
> "Come soon, dear friend," she whispered. "While the roads to the north are clear. I will be waiting." (158)

In resituating Andersen's tale within the imaginary North of the *Kalevala*, Kernaghan provides a new vision of female heroism that is rooted in both the ancient traditions of shamanism and the passionate liminality of adolescence. At the same time, she anchors her story in the turbulent intellectual world of Victorian scientific discovery and the clash between old and new beliefs—not only between religious faith and science, but also between competing visions of what young women could and should be in the world (Wolf, "Revisiting"). In the process, the mythical North becomes a landscape on which Gerda and Ritva map their own personal journeys through fear, determination, self-discovery, and love.

──────

Notes

1) In recent years, Parks Canada has funded a series of expeditions to search for the lost Franklin ships, and as of September 2016, both have now been found: see Hong and Winter, "HMS Terror, second ship", https://www.thestar.com/news/canada/2016/09/12/hms-terror-from-doomed-franklin-expedition-found.html. DNA analysis is providing even more intriguing information, such as the fact that several of the bodies found were women: see Daley, "DNA Could Identify the Sailors (Including Women)," http://www.smithsonianmag.com/smart-news/dna-extracted-doomed-franklin-expedition-sailors-180963031. This latest news suggests that Gerda's and Ritva's strategy of disguising themselves as boys in Kernaghan's novel is not far-fetched but in fact reflects the behaviour of some historical women.

2) Kernaghan is the author of numerous works of fantasy, science fiction, and historical fantasy, several of which are written for young adults. She has always been interested in ancient civilizations, the power of folklore and myth, and the power of the past. For more on her writing, see her blog at eileen-kernaghan.blogspot.ca.

3) Kernaghan has said that this character had a long development: "Years earlier, I had written a poem called 'The Robber Maiden's Story'; later I expanded the poem into an adult short story, focussing on the relationship between Gerda and the robber-chieftain's daughter. Finally, using the Andersen story as a framework, I started work on a young adult novel." Wolf, "Revisiting."

4) In her study of archetypes, Annis Pratt considers the female *Bildungsroman* to be an impossible thing because of the very different and more restricted lives that women live; however, like the more active female heroes of some fairy tales, and like Psyche in the Cupid and Psyche story, Gerda and Ritva defy gender norms and act to change their world.

5) In this respect, the novel's two main characters challenge the ingrained gender binaries that most certainly governed Victorian society: binaries that saw women as closer to Nature than to Culture, as naturally passive rather than active, and as fitted for roles in domestic rather than public space. See Ortner's classic discussion, "Is Female to Male as Nature Is to Culture?" 67–87.

6) Kernaghan was likely inspired by the *Kalevala* episode in which Louhi, the Mistress of Pohjola (the Old Crone of the North, the equivalent of Kernaghan's Snow Queen) demands that the smith Ilmarinen perform three perilous tasks in order to win her daughter in marriage (Kirby, *Kalevala,* vol. 1, Runo 19, 211–24.) Although there are more recent translations from the Finnish available, it is clear that Kirby's is the translation that Kernaghan used.

7) In the *Kalevala*, Runo 42, "The Capture of the Sampo," tells of how Väinämöinen plays his *kantele* (harp) to put the whole court to sleep so that he can steal the *sampo*, the magical object that is clearly the model for the Snow Queen's rune chest in Kernaghan's novel. See Kirby, vol. 2, Runo 42, 168–82.

8) For more on the quest, and particularly the female quest, see Abel, Hirsch, and Langland, *The Voyage In: Fictions of Female Development*, and Annis Pratt, *Archetypal Patterns in Women's Fiction*. However, Kernaghan's characters do not really fit any of Pratt's archetypes, since they are acting in male roles; a better model is the Cupid and Psyche myth, as discussed by Ferguson (Abel, Hirsch, and Langland, 228–43), or perhaps more pertinent (and possibly known by Andersen himself), the folk tale "East of the Sun and West of the Moon," in which the young woman journeys across the world in search of her lost lover.

12) Camping Out on the Quest

The Landscape of Boredom in *Harry Potter and the Deathly Hallows*

✾ SARAH FIONA WINTERS

═══════

Popular reception of the novel *Harry Potter and the Deathly Hallows* (2007) and the film *Harry Potter and the Deathly Hallows Part I* (2010) included, among all the praise and delight, a significant amount of dissatisfaction over the long "camping" section that makes up nine chapters of the novel and approximately forty-five minutes of the film. Christopher Hitchens in the *New York Times*, for example, described this section of the novel as "an abysmally long period during which the threesome of Harry, Hermione and Ron are flung together, with weeks of time to spend camping invisibly and only a few inexplicable escapes from death to alleviate the narrative," while Rick Groen of the *Globe and Mail* said of the film that "surprisingly, [the director David] Yates repeatedly sticks the threesome into vast natural landscapes, posing them in aerial shots on rocky promontories overlooking sweeping plains or on windswept dunes by the rolling sea. It's almost as if, in the absence of Hogwarts, he's making the void literal, uncaring that his lead characters are getting swallowed up by his postcard vistas." Long, vast, uncaring: these are all ways of saying that the camping section is boring, both for the characters to live through and for readers and viewers to sit through.

According to Michael Raposa, "Boredom thwarts the spirit of play and signifies art's failure. Work may be boring but play is not supposed to be. Few comments are more damning of a work of art than 'it is boring'" (37), so such responses would indicate that the camping section in both novel and film is an artistic failure. I argue the opposite—that the camping section is deliberately dreary, and that both J.K. Rowling and David Yates intentionally created a geography of boredom to suggest that resistance to evil *is* often boring, and that the quest for a meaningful life involves large vistas of meaninglessness in time and space.

Indeed, Rowling and Yates turn time *into* space, replacing the rhythms of the school year—with its Quidditch matches, Hogsmeade excursions, and Hermione's exam revision timetables—into unstructured months during which Harry observes the changing of the seasons through the foliage of the woods and the dropping of the temperature. *Deathly Hallows* replaces the varied days and unchanging landscape provided by the school story in the first six books and films of the series with the varied landscapes and unchanging days of the quest-romance genre. While every volume of Rowling's series is constructed through a blend of different genres, the absence of the school story through most of the seventh volume allows not just the quest-romance but also religious allegory, the *Bildungsroman*, and the historical allegory of Second World War Resistance to Nazism to inflect the camping section with thematically rich treatments of boredom.

The quest-romance is certainly the most prominent of these shifts in genre: *Deathly Hallows* creates a new geography where Harry can nurture a new aspect of his role and destiny as hero, for in Britain's woods and wide-open spaces Harry moves beyond the schoolboy hero he has become in the safe enclosure of Hogwarts to become the hero of romance. As a result of this move to romance, the imaginary geography of the text evokes the imaginary geography of other romances, most prominently John Bunyan's *Pilgrim's Progress* and J.R.R. Tolkien's *The Hobbit* and *The Lord of the Rings*.

Bunyan begins his dream vision with a conventional account of the dreamer falling asleep. But this opening sentence of *The Pilgrim's Progress* could also be read as the experience of a man bored by the landscape who escapes his weariness by escaping into fantasy: "As I walk'd through the wilderness of this world, I lighted on a certain place, where was a Denn; And I laid me down in the place to sleep: And as I slept I dreamed a Dream" (10). Even the den, or the prison where the real Bunyan was confined, suggests the indebtedness of his text to traditional images of boredom, "often manifested as the experience of being trapped, confined, and afflicted (like the desert monk in his cell)" (Raposa 36). This den in the middle of the wilderness of the world corresponds to Harry's tent in the middle of the wilderness of the British woods: like Bunyan's dreamer, he is simultaneously lost in a wilderness and trapped in a prison. But it is Christian's experience with the Slough of Despond that most strongly resonates with Harry's experience in the camping section. The Slough turns up in this section both metaphorically and literally. In Bunyan's allegory it represents "the descent whither the scum and filth that attends conviction for sin doth continually run ... for still as the sinner is awakened about his lost condition, there ariseth in his soul many fears, and doubts, and discouraging apprehensions, which all of them get together, and settle in this place" (17). Sin aside, this sad condition perfectly describes the experience of camping for Harry and the experience of reading or watching the camping section for many readers and viewers; the whole section is a metaphorical slough. In this metaphorical landscape, Ron takes on the role of Pliable, at first eager and supportive of his neighbour's quest, but soon asking Harry, like Pliable asks Christian, "*where are you now?*" (16). Like Christian, Harry is unable to give a better answer than "Truly ... I do not know" (16); like Pliable, Ron is "offended" at this answer, and like Pliable "gave a desperate struggle or two, and got out of the Mire" (16), abandoning Harry (and Hermione) to struggle with despondency alone. But the Slough also shows up literally as the small pool of water in the Forest of Dean in which Harry nearly drowns while trying to retrieve the sword of Gryffindor. Like Christian, who "because of the burden that was on his back, began to sink in the Mire" (16), Harry, burdened with the locket Horcrux, begins to sink; at this point Ron, transformed from Pliable into Help, rescues him: "he gave him his hand, and ... drew him out, and set him upon sound ground" (16). And after this return and rescue, the tone of the camping sequence changes from despondency and boredom to hope and activity.

The camping section in the novel actually begins, however, in beauty rather than despondency, with Tolkien rather than Bunyan. The first line of the chapter that begins the sequence is "Harry opened his eyes and was dazzled by gold and green" (221), gold and green being, in the words of

Northrop Frye on romance, "traditionally the colors of vanishing youth" (200). Green and yellow are the colours most favoured by hobbits, J.R.R. Tolkien tells us at the start of *The Hobbit* (10), and like Harry, Bilbo Baggins will see them vanish in his endless journey through a forest once called Greenwood but renamed Mirkwood:

> It was not long before they grew to hate the forest as heartily as they had hated the tunnels of the goblins, and it seemed to offer even less hope of any ending. But they had to go on and on, long after they were sick for a sign of the sun and of the sky, and longed for the feel of wind on their faces. There was no movement of air down under the forest-roof, and it was everlastingly still and dark and stuffy. (132–33)

Unlike Bilbo, Harry does not remain for weeks in one literal forest, nor is he literally deprived of the sun, sky, and wind, but the paradoxical fact that, unlike Bilbo, he has no path to wander off *from* turns all his landscapes into metaphorical Mirkwoods: the camping section in *Deathly Hallows* has a metaphorical effect on its readers similar to the literal effect Mirkwood has on its characters.

Like Tolkien, Rowling is a self-declared Christian, and the Christian element in the texts of both writers emerges from the purpose of each quest. Although, like Bilbo in *The Hobbit*, Harry is on a quest to find treasure, the Horcruxes, he is also, like Frodo in *The Lord of the Rings*, on a quest to destroy the particular treasure he bears, the locket Horcrux. This action of carrying a burden that must be destroyed causes both heroes, Frodo and Harry, to resemble the scapegoat of the Old Testament. Although the word *scapegoat* has come to suggest one who is destroyed, the ritual origin of the scapegoat concerns not death but exile. In the words of Gary Anderson:

> Through this ritual act, Aaron symbolically puts the *weight* of Israel's sins upon the animal. Once the animal has assumed this burden, it can carry out its responsibility. As in much of the ancient Near East, the wilderness could serve as a portal to the underworld, the domain of the demonic ... Because this area was thought to be beyond the reach of God, the sins would fall outside the range of his supervisory powers ... The ritual of the scapegoat, therefore, is dependent on the imagery of sin as a heavy burden that requires a beast of burden to bear it away from the realm of human habitation. (22–23)

Both Frodo and Harry take on such a burden, in their cases an item of jewellery worn around the neck, and bear it away from other people into a wilderness beyond the supervisory powers of their godlike guides, Gandalf and Dumbledore. Rowling departs from Tolkien in having the trio in her text share the burden, but she echoes him in having that trio split, for a time, into the suffering hero (Harry), the faithful companion (Hermione), and the faithless, jealous companion, for Ron abandons Harry and Hermione just as Gollum abandons Frodo and Sam: indeed, the sequence that narrates Ron's departure begins "on a riverbank in Wales" (240), with Ron "prising a fishbone out from between his teeth" (241) in a startlingly Gollum-like moment.

The landscape of boredom in *Deathly Hallows* is, then, partly religious in nature. Specifically, the boredom experienced by both the characters and the readers/viewers resembles *acedia*, defined by Michael Raposa as "a medieval term for spiritual sluggishness, dullness in prayer, boredom with the rituals of devotion" (2). Raposa concentrates on the boredom of the religious life, including boredom in ritual and prayer, a process wherein the practitioner interprets the signs involved in religious devotion as empty, or pointing not to the presence but rather to the absence of God. In *Deathly Hallows* this process manifests itself in the following passage: "As Dumbledore had told Harry that he believed Voldemort had hidden the Horcruxes in places important to him, they kept reciting, in a sort of dreary litany, those locations they knew that Voldemort had lived or visited" (237). During one such scene, before his departure, we are told that "Ron yawned pointedly" (238), his interpretation of the litany being that it is empty and pointless. After his return, however, he recites the same litany in a very different spirit: "'Three Horcruxes left,' he kept saying. 'We need a plan of action, come on! Where haven't we looked? Let's go through it again. The orphanage ...'" At this point, the narrator picks up the refrain: "Diagon Alley, Hogwarts, the Riddle House, Borgin and Burkes, Albania" (354). Ron's *acedia* has been vanquished: he now interprets signs as meaningful, as rituals full of promise, leading somewhere. His story here echoes that of another classic text of religiously inflected quest-romance for children, C.S. Lewis's *The Silver Chair*. In that novel, Aslan gives Jill and Eustace four signs to remember in order to achieve their quest. But the journey soon becomes hard and boring: "The children ... were sick of wind and rain, and skinny fowl roasted over campfires, and hard cold earth to sleep on ... The road led through endless, narrow valleys ... the ground was all stony" (92). As a result, "Jill gave up her habit of repeating the signs over to herself every night and morning. She said to herself, at first, that she was too tired, but she soon forgot all about it" (92–93). The daily repetition of the signs is of course an allegorical representation of daily engagement with rituals of religious devotion, and when Jill gets bored with it, she fails to recognize meaning in the landscape itself: "'It's my fault,' she said in despairing tones. 'I—I'd given up repeating the signs every night. If I'd been thinking about them I could have seen it was the city, even *in* all that snow'" (118–19). Likewise, in *Deathly Hallows* it turns out that two of the Horcruxes are hidden in two of the places in Ron's repeated litany: Diagon Alley and Hogwarts.

But the boredom of the camping section is not just *acedia*, but existential boredom, a term described by Peter Toohey as

> in many ways ... a secular variant of the religious angst exhibited by Christians of many shades. What is God's plan? Where do I fit into God's plan and how do I live in it? What does God want me to do? Many God-fearing Christians will ask themselves questions like these. The more uncertain might respond to these perplexing questions with the same sort of passive, hopeless and depressive reactions that have been attributed to the sufferer of existential boredom. (26)

Toohey suggests that the three traditional ways to combat such boredom are drugs, sex, and travel. Only the last of these is generally considered suitable for children's literature, and this is the only one we see in *Deathly Hallows*. In his chapter on travel, Toohey discusses what he calls "the odd label of dromomania. That word means something like a mania for locomotion" (71), and it is certainly possible to read Harry's endless apparating into "postcard vistas," as Groen put it, as a form of dromomania, an effort to escape the lack of Dumbledore's plan, the absence of a path. Here the difference between Tolkien and Rowling is most clear: in *The Lord of the Rings*, Frodo walks, plots, trudges, staggers, and crawls, but he slowly progresses towards a point; in *Deathly Hallows*, Harry apparates and disapparates everywhere and anywhere in an instant, going nowhere. Frodo's quest of depression contrasts with Harry's quest of manic-depression. As Raposa puts it, "Boredom can result either from having nothing to do or from being overwhelmed with things to do" (39), and Harry deals with both these problems at once, and so suffers from the worst kind of boredom.

Another significant departure in Rowling's text from Tolkien's quest-romance is the absence of a map. While most fantasies set in a Secondary World post-Tolkien are published with maps, Rowling's, set in the Primary World of Great Britain, is not. In this lack of a map, the Harry Potter series resembles Susan Cooper's *The Dark is Rising* sequence (1965–77), in which the fantastic landscape arises from the myths, legends, and history of the local English and Welsh landscapes. One particular resonance between the two fantasies can be seen in the importance of the Forest of Dean in *Deathly Hallows*. In *The Dark is Rising*, the second and eponymous novel of Cooper's sequence, Merriman tells Will that he has travelled back in time to a forested England:

> That is where you were—in the time of the Royal Forests, that stretched over all the southern part of this land from Southampton Water up to the valley of the Thames here … Five hundred years ago … the kings of England chose deliberately to preserve those forests, swallowing up whole villages and hamlets inside them, so that the wild things, the deer and the boars and even the wolves, might breed there for the hunt. (54)

Merriman's history lesson is accurate:

> Medieval forests … varied in character and size; for example, the Peak Forest in the High Peak of Derbyshire was very different from Wychwood Forest (Oxfordshire) or from the Forest of Dean (Gloucestershire). The term "forest" had a legal meaning; it referred to an area that was under forest law … in which deer and other game could be killed only by the forest owner, usually the king … By 1500 there had been about 180 forests in existence at one time or another, about half of them royal. (Hey)

Will's adventure in this volume culminates in the remnants of one of these Royal Forests, Windsor, while the greatest turning point in Harry's quest occurs in another, the Forest of Dean—a most appropriate place, as one of the historical Royal Forests of Britain, for a white doe Patronus cast by a Half-Blood Prince to guide a young man called Harry to the winning of a medieval sword. However, up until Harry and Hermione apparate there (Hermione chooses it because she misses her parents who used to camp there with her), the places they camp in are *unnamed*, lost in a vagueness and emptiness that resemble a Dementors' mist of depression. The narrative gives us the characters' journey only in words, not maps:

> Every morning they made sure that they had removed all clues to their presence, then set off to find another lonely and secluded spot, travelling by Apparition to more woods, to the shadowy crevices of cliffs, to purple moors, gorse-covered mountainsides and, once, a sheltered and pebbly cove. (239)

The landscapes evoked here are also to be found in Frodo's journey, but—and here is the difference—in *The Lord of the Rings*, they all appear as part of named lands that the reader can look up on a map. Moreover, Frodo moves in one direction, even with the backtracking into Ithilien, for he knows his destination and what he is supposed to do when he gets there. Harry, by contrast, moves in circles or zig-zags, without a destination or path, a "pointless and rambling journey" (240) as he thinks of it. At his very lowest, his judgement on the disastrous trip to Godric's Hollow includes the following passage: "[They] had convinced … themselves that they were supposed to go back, that it was all part of some secret path laid out for them by Dumbledore; but there was no map, no plan" (287). Dumbledore, unlike Gandalf, did not leave his scapegoat a map through the wilderness; moreover, Rowling, unlike Tolkien, does not give her reader a map through her text. The reader of *The Lord of the Rings* can turn from Frodo in the landscape to the map of Frodo's progression through that same landscape, while the reader of *Deathly Hallows* has no equivalent map to turn to; that reader is every bit as lost as Harry, unable to take a God's-eye view of this pilgrim's *lack* of progress. And thus, even though Frodo moves through landscape much more desert-like and demonic, much more evocative of the scapegoat's fate to wander in the wilderness, Harry is, nevertheless, far more bewildered.

Both *The Lord of the Rings* and *Deathly Hallows* are quest-romances. But unlike *The Lord of the Rings*, *Deathly Hallows* can also be read as the final part of a *Bildungsroman* formed by the whole series. The imaginary geography of *Deathly Hallows* therefore can be read as a metaphorical portrayal of the boredom and depression felt by many young adults who have just left school and do not quite know what they are meant to be doing with their lives. This realism contrasts with the romance of middle-aged hobbits setting off on an adventure. Frodo is, in the novel if not the film, fifty years old and mostly content before he is thrust into his quest. Harry is seventeen, and leaving school should function for him as a plunge into adventure, but instead leaves him stranded in a premature geography of middle age for months. This contrast is amusingly highlighted in the detail of Harry's first meal in the woods. While a delicious meal of mushrooms serves as a reward for Frodo and his friends for their first days in the woods after leaving Hobbiton, the equivalent meal for Harry and his friends on *their* first day in the woods after leaving Grimmauld Place is completely miserable:

> They had nothing to eat except some wild mushrooms that Hermione had collected from amongst the nearest trees and stewed in a billycan. After a couple of mouthfuls, Ron had pushed his portion away, looking queasy ... [Harry's] insides, already uncomfortable due to their inadequate helping of rubbery mushrooms, tingled with unease. (228)

The hobbits in the woods of the Shire seem younger than the teenagers in the woods of Britain. As Raposa says, "We typically associate boredom with the landscape of the desert; with flatness, dryness, the color gray, midday, *midlife*; with tourists; with yawning" (37, my emphasis). Even though they are adolescents, Ron and Hermione do not act on their mutual attraction in this section, in spite of plenty of opportunity, partly because they are bored, prematurely bored with each other as if they'd been married for several decades already. And Harry is irritated with both of them. The theme of the stifling closeness of other people is perhaps best seen through the contrast between images of a barrier between eye and sky in Tolkien and Rowling. When Frodo spends his first night in the woods of the Shire, with Sam and Pippin, it is nature that partly hides the sky from him: "Thin-clad birches, swaying in a light wind above their heads, made a black net against the pale sky" (84). But when Harry spends his first night in the woods of England, it is one of his best friends who does so: "[Harry] looked up at Hermione, whose bushy hair obscured the tiny patch of sky visible through the dark branches high above them" (231). The beauty of the landscape is obscured in Rowling's text as it is not in Tolkien's by this vision of middle-aged domestic claustrophobia.

Another barrier between Harry and the sky is the tent itself, for unlike the hobbits when they camp in the woods, the trio do not sleep under the open skies, looking up at the stars, but beneath a "dark canvas ceiling" (231). Yet the tent is also inadequate, as shown in one of the most dreary paragraphs of summary, occurring after Ron's departure when Harry and Hermione zigzag wildly around Britain in the weeks leading up to Christmas:

> The weather grew colder and colder. They did not dare remain in any one area too long, so rather than staying in the south of England, where a hard ground frost was the worst of their worries, they continued to meander up and down the country, braving a mountainside, where sleet pounded the tent, a wide flat marsh where the tent was flooded with chill water, and [a] tiny island in the middle of a Scottish loch, where snow buried the tent in the night. (258)

Joe Morgenstern criticizes the tent in the film, claiming that "many of the production's deficits are baffling—the commonplace chases, the murky look, the indifferent effects—but none more so than the interior of the tent, which looks like a big and banal stage set before the enchantment sets in." But the banality is the point, as is the commonplace, the murkiness, the indifference: while the landscape of Harry's quest is somewhat vague and empty, his refuge is confining and inadequate because his quest is one of spiritual, existential, and age-related boredom and depression. There is no sex in Harry, Ron, and Hermione's three-way marriage, just as there is no magic in the mushrooms; as I have already mentioned, only one of three traditional ways to combat boredom—sex, drugs, and travel—is available to the heroes, readers, and viewers of the camping section in *Deathly Hallows*, and it proves itself inadequate.

The imaginative geography of the camping section therefore belongs to the genres of the quest-romance, the *Bildungsroman*, and religious allegory. It also, however, contributes to the text's status as historical allegory. While the parallels between the Death Eaters' persecution of Muggleborns and the Nazis' persecution of Jews are obvious, one particular resonance that has not been explored is that between Harry's time in the woods and Partisan Resistance during the Second World War. As with the Partisans of Europe, Harry's resistance to the rise of the fascist state after the fall of the Ministry begins with evading capture: he moves first to a safe house and then, as with many Partisans, to the woods. As with the Partisans of Europe, this move has both benefits and drawbacks: "The partisan forces could count upon their capacity for rapid movement into new, clear areas, and could return when the situation allowed [but] life was normally hard, marked by Spartan conditions, improvisation and constant danger" (Haestrup 465). And while not all readers of the novel or viewers of the film will be that familiar with the history of the Partisans, Ron's obsession with his radio would remind many readers and viewers of film scenes depicting members of the Resistance huddled around their radio, tuning into the BBC.

Daily life for the Partisans camping out in the forests is evoked vividly in George Millar's memoir *Maquis*, which describes his summer in 1944 helping forest-dwelling Resistance groups in France: the word *maquis* means "thicket" and the men who lived in the thickets were often, like Harry, young men. The first group of four organized by Millar consisted of Maurice "in the middle or late twenties," the Frisé, twenty-two, the Pointu, twenty, and Phillipe, who was "still younger" (152–53). Millar's account of this time is full of passages that resonate strongly with the camping scenes in *Deathly Hallows*, both novel and film. For example, Millar pays tribute to the pleasure of sleeping in a three-walled wooden cabin halfway up a hill in the forest, referring to "the height and peace and beauty of our life there ... In the mornings I could see the blue sky through the screen of pine-trees. Sometimes I would waken when the dawn was just coming up and the valley was still a misty, wonderful sea below us" (209–10). Elsewhere, he remembers "thinking, as I sat there in the sweetness of the dawn, that if the life were all like this, genuine boy scout stuff, I could live it for a year" (80). But his account also describes the squalor, discomfort, and claustrophobia of the life, including this account of a Maquis camp:

> It was a natural hole in the ground. Before the coming of man it had been a leafy bowl under the great trees. But the Maquis had changed all that. There were about twelve of them in that place, and to my mind they could scarcely have installed themselves worse. The bowl was about thirty feet in diameter. In this space the men cooked, disposed of the offal from the kitchen, lived, washed and slept ... Since the weather had recently been wet, the floor of the bowl was deep in mud, and a slight film of mud, now dry, now damp, coated every bed, every article of clothing, every cooking-pot, and every weapon. (133)

Millar also relates accounts of the Maquis executing "traitors" in these woods after forcing the people to dig their own graves, and recounts acts of sabotage and violence, concluding that "in places the valley smelled of death, and the woods were stained with a stain that would not wash out in years of wind and wetness, the stain of human rot and blood and pus" (347). The woods can shelter the Maquis temporarily from the Gestapo but not permanently from the sufferings of war.

But perhaps the following passage, describing the morning after an unexpected night in the woods on a bed of pine-tree branches, is the most evocative of the unsettling coexistence of beauty and anxiety, freedom and fear, freshness and privation in Millar's time with the Maquis:

> It was a noisy night, with a strong wind and much barking of dogs and ringing of cow-bells. We slept with the tension of men who imagine themselves in danger. And at six o'clock in the morning, chilled and hungry, we stole across the fields to the farm. The fields were full of beautiful mushrooms. I ate a handful of them as we walked. My trousers were soaked to above the knees by the copious dew. (174)

That particular memory of a tense and dangerous night in the woods dissolving in the dews of dawn to a farm of food and welcome through a field of delicious mushrooms is more likely to evoke *The Lord of the Rings* than *Deathly Hallows*, thus illuminating the different function the geography of forests and woods has in each text. As befits Rowling's allegory of the Second World War, the forests function as the scene of Harry's struggles; for Tolkien, however, although they can be dangerous places, they are seldom (with the exception of Mirkwood in *The Hobbit*) dreary ones, and for Frodo they function mostly as places of rest and relief, respite from the battle-scarred wide-open and treeless plains that resemble No Man's Land.

When Tolkien refuted the suggestions of readers in the 1950s that *The Lord of the Rings* was an allegory about the use of nuclear weapons to end the Second World War, he did so partly through his famous claim that "I cordially dislike allegory in all its manifestations, and always have done so since I grew old and wary enough to detect its presence" (Foreword, *Lord of the Rings*, 11), but also through his suggestion that his readers were focusing on the wrong war: "One has indeed personally to come under the shadow of war to feel fully its oppression; but as the years go by it seems now often forgotten that to be caught in youth by 1914 was no less hideous an experience than to be involved in 1939 and the following years. By 1918 all but one of my close friends were dead" (11). Thoughtful responses to representations of the Great War in Tolkien's epic include Tom Shippey's study of Tolkien's work as a response to twentieth-century traumas of evil; framings of the relationship between Sam and Frodo in the context of bonds between soldiers in the war by Anna Smol and Mark T. Hooker; and, especially, the full-length study of the effect of the war on Tolkien's entire mythology by John Garth. Garth includes the following judgment by Tolkien himself on the issue: "Personally I do not think that either war ... had any influence upon either the plot or the manner of its unfolding. Perhaps in landscape. The Dead Marshes and the approaches to Morannon owe something to Northern France after the Battle of the Somme" (310). It is the marshes and flatlands, treeless places, that Tolkien associates with his experience of war, not forests and woods. Indeed, Garth's history of the war experiences of Tolkien and his three friends, Rob Gilson, G.B. Smith, and Christopher Wiseman, suggests that for them, the woods meant respite, shelter, and even beauty, compared with the muddy battlegrounds on the fields and plains. For example, Garth describes Rob Gilson's experience of the Battle of the Somme thus:

> A thousand yards from the German line, Rob Gilson and his battalion spent the night in and around the small château in trench-riddled Bécourt Wood ... Even here, despite the unrelenting British bombardment, war seemed remote. Cuckoos called, nightingales sang, dogs barked at the guns; wild and garden flowers grew in profusion. A light rain pattered through the leaves for soldiers to catch in their hats to drink. (152)

Although the war encroached on the woods, and many soldiers died there, they were nevertheless not the main site of struggle in Tolkien's experience.

In *The Lord of the Rings*, the particular sequence of the hero's movement through the landscape that most resembles the same sequence in *Deathly Hallows* is Frodo's time in Ithilien. For example, the following passages from Tolkien could have come right out of *Deathly Hallows*:

> The day passed uneasily. They lay deep in the heather and counted out the slow hours, in which there seemed little change. (675)

> That day passed much as the day before had gone, except that the silence seemed deeper; the air grew heavy, and it began to be stifling under the trees. (723)

The sun rose and passed overhead unseen, and began to sink, and the light through the trees to the west grew golden; and always they walked in cool green shadow, and all about them was silence. The birds seemed all to have flown away or to have fallen dumb. (723)

But while empty and silent, the woods of Ithilien do not distress Frodo with boredom because he enters them in between two horrible journeys in landscapes that evoke the miseries of No Man's Land from the First World War: the Dead Marshes, and Mordor itself. For Tolkien, inspired by memories of soldiers fighting in that war, camping out in the woods is a relief from the landscape of Mordor and its surrounds; for Rowling, inspired (perhaps unconsciously) by images of Partisans escaping to and fighting from the woods in the Second World War, camping in the woods is not relief from the struggle, but the struggle itself.

Michael Raposa argues that the "failure to find some event or activity interesting is the result of an interpretation. *Boredom is the result of semiosis*" (36); it is a result of interpretation by readers and viewers of the camping section that so many find it boring. Such an interpretation misses the point that boredom is part of Harry's suffering, part of his growing up, part of his struggle against evil and for meaning, part of his resistance against fascism. Rowling writes the camping section of *Deathly Hallows* as a geography of boredom, and Yates incorporates her vision into his film, in order to explore and construct a heroism that embraces a contemporary understanding of boredom as part of the suffering and sacrifice inherent in the bewildering struggle against the meaninglessness of evil.

Fantasy Worlds and Re-enchantment

13) Sky Sailing

Steampunk's Re-enchantment of Flight

❧ CHRISTINE BOLUS-REICHERT

⸻

Flight goggles have become the nearly ubiquitous symbol of the literary and cultural movement called steampunk. More than a fashionable accessory, the oversized eyewear evokes an age when flying was synonymous with adventure. Steampunk, as a literary moniker, identifies science fiction (or science fantasy) set in the past, usually in the nineteenth or early twentieth century when important technologies, such as flight, might have turned out differently. All steampunk depends, therefore, on the imaginative geography of another time—to borrow only the first half of L.P. Hartley's pithy expression, "The past is a foreign country"; but steampunk for children particularly involves the extension of the child's imaginative geography upwards. Three recent novels for young readers—Kenneth Oppel's *Airborn* (2004), Philip Reeve's *Larklight* (2006), and Scott Westerfeld's *Leviathan* (2009)—take up what has become one of steampunk's most persistent fantasies: the re-enchantment of flight.

I borrow the idea of re-enchantment first of all from John McClure's important 1994 study, *Late Imperial Romance*. Writers of adventure romance like Joseph Conrad and Rudyard Kipling had to find a way to restore a "zone of magic and mystery" (essential in all romance) to a disenchanted world. The engines of disenchantment were, somewhat ironically, science and imperialism. Late imperial romances created the opportunity for "extreme experiences" that would "break the spell of secularizing discourses." Romances like *Heart of Darkness*, in McClure's analysis, must be regarded as "a kind of meta-romance: a quest for the very conditions that make romance possible, an effort to recover the right to dream romantically in a world that seems to be producing texts (the maps filled with words) that make romance impossible" (14).

The search for the spaces where romance can thrive did not end with "late imperialism"; on the contrary, all steampunk, which looks backward to a lost geography of romance, can be seen as meta-romance, as pastiche even, in Richard Dyer's definition, as imitation that knows it's an imitation, a combination of past works of art.

All three of these authors (Oppel, Reeve, and Westerfeld) re-enchant flight by taking it back to the age of sail. Oppel's work is most tethered to the possible, the what-might-really-have-been had a different technology for flight (dirigible airships) won the day, so he's most attuned to how a seemingly slight deviation alters our perception of the world. The novel's opening section plays with the terminology of sailing, no doubt fooling some readers (who might have skipped over the dirigible blueprint) into thinking the hero Matt Cruse was at sea. For Oppel, flying *is* sailing, and the upper world has its own geography, a mixture of old mysteries (the constellations as "a galaxy of adventures") and new mysteries ("things aloft in the sky, unseen by us") (Oppel 1, 20).

Westerfeld and Reeve take a slightly different approach, veering off from the probable or even possible but hewing closely to a kind of storytelling (and world view) popularized in the nineteenth century by writers like Marryat, Ballantyne, Stevenson, and Kipling. Reeve's *Larklight*, a self-conscious pastiche of Victorian adventure stories, proposes that Victorians are travelling through the far reaches of space, extending their already vast empire, by means of "aether-ships," a technology the British discovered during the great age of sail. Captain Cook in this alternate version of history becomes an "early spacefarer," and "Great White Hunters ... are forever setting off into the wilds of Africa or Mars in search of big game"

(Reeve 41, 59). Naturally the space pirates travel in ships that would not look out of place in a *Pirates of the Caribbean* film.

Westerfeld's *Leviathan* likewise combines core elements of the Victorian world view with a new mode of flight. An alternate version of pre–Great War Europe pits the Darwinists (Britain and its allies) against the Clankers (Germany and Austria-Hungary), with the Darwinists developing a technology based on gene-splicing, and the Clankers on steam engines. As in Oppel and Reeve, flight here is explicitly linked to the sea with the *Leviathan*, a British airship that's also a "huge beastie." The ship is a genetically modified whale that swims, or sails, in the sky. Westerfeld's Darwinist technology, like Reeve's alchemical engines and Oppel's blimps, allow the child characters (whose age is in keeping with that of the heroes of the great Victorian seafaring romances) to have an entirely different relationship to the realm of the upper world. It's more dangerous, more exposed, but the restoration of a lost imaginative geography—the "whole galaxy of adventure"—makes space for romance, transcendence, and entry into the "zone of magic and mystery."

McClure's *Late Imperial Romance* centred on questions of genre, but since its publication in 1994, re-enchantment has begun to be understood in broader terms as a deliberate strategy of modernity. In their 2009 intro-duction to *The Re-Enchantment of the World: Secular Magic in a Rational Age*, Joshua Landy and Michael Saler make the historical case for three kinds of enchantment: one associated with religion and spiritualism, which has gradually lost its power; a second, dangerous and self-delusory sort linked to global capitalism; and a third, "unjustly overlooked, which is the modern enchantment *par excellence*: one which simultaneously enchants and disenchants, which delights but does not delude" (Landy and Saler 4). Saler, Landy, and the contributors to the *Re-Enchantment* collection, along with political theorist Jane Bennett and literary historian Simon During, form a cadre of recent "antinomial theorists of modernity" aiming to "put on display a set of enchantments that are *voluntary*, being chosen ... by autonomous agents rather than insidiously imposed by power structures, *respectable*, compatible as they are ... with secular rationality, and *multiple*, being replacements, each one in its own way, for a polymorphous God" (Landy and Saler 7). The antinomial theorists regard modernity as messy, defined by "its contradictions, oppositions, and antinomies" (7). They admit loss, but don't accept it.

These antinomian approaches aren't really new and neither is the idea that enchantment can be voluntary, respectable, and multiple. Take George MacDonald in his 1893 essay "The Fantastic Imagination": "The best thing you can do for your fellow, next to rousing his conscience, is—not to give him things to think about, but to wake things up that are in him; or say, to make him think things for himself" (9). Or Tolkien in "On Fairy-Stories" from 1938: "Fantasy ... does not destroy or even insult Reason; and it does not either blunt the appetite for, nor obscure the perception of scientific verity. On the contrary. The keener and the clearer is the reason, the better fantasy will it make" (55). And he adds, irresistibly, that "if men really could not distinguish between frogs and men, fairy-stories about frog-kings would not have arisen" (55). Or Northrop Frye in his 1976 work, *The Secular Scripture: A Study of the Structure of Romance*: "The child should not 'believe' the story he is told; he should not disbelieve it either, but send out imaginative roots into that mysterious world between the 'is' and the 'is not' which is where his own ultimate freedom lies" (Frye 166). In Frye's analysis, romance is revolutionary and realism is not.

These earlier critics of romance—of fantasy—were antinomial in Saler and Landy's usage, regarding enchantment as indispensable to freedom, reason, and all creative endeavours. In his most recent book, *As If: Modern Enchantment and the Literary Prehistory of Virtual Reality* (2012), Saler argues that in the late nineteenth century, "Imaginary worlds became conceptual possibilities ... [because] they were also conceptual necessities, called into being as antidotes to disenchantment" (Saler 57). Because our world is that disenchanted one the late Victorians found themselves living in, the need to restore the "zone of magic and mystery" has become only more acute with the passage of time—indeed, Saler's purpose is to understand how we came to our present condition, wherein we are all involved in making—and perpetuating—imaginary worlds.

What to make, then, of this particular strand of our collective imaginary, the re-enchantment of flight in steampunk fantasies? I won't waste time arguing that flying, for most of us, has been thoroughly disenchanted. Barely a century after our species took to the sky, travelling by air involves more pain than pleasure. It's humiliating, confining, and dull—the opposite of what we imagined it could be. What would it mean to re-enchant flight? For Jane Bennett, in *The Enchantment of Modern Life* (2001), it boils down to this: "[enchantment is] a state of wonder ... the temporary suspension of chronological time and bodily movement ... It is to be transfixed, spellbound" (Bennett 5). Landy and Saler give us a longer list: for the world to be re-enchanted we must have mystery and wonder, but also order and purpose; there must be a hierarchy of significance, the possibility of redemption, and a locus for the infinite; and there must also be sacred spaces, everyday miracles, secular epiphanies (Landy and Saler 1)—in short, all that we once derived from belief in the supernatural can be generated by other means, such as the creation of imaginary worlds in fiction.

How do Oppel, Reeve, and Westerfeld give back to flying its mystery and wonder? How do they re-enchant the sky? In all three novels, as I have already described, the technologies of flight—Oppel's airships, Reeve's aetheric engines, and Westerfeld's hydrogen breathers—resemble sailing as much as they do flying. The process of defamiliarization compels readers to think all over again, or maybe for the first time, about what it means for human beings to fly and how we might do it differently. In all three novels, the technological shift results from a historical shift, backward and then sideways in time. Together they move the reader into a set of generic conventions that are less science fiction than scientific romance and less scientific romance than quest romance.

Reeve's example here is clearest. His ingenious mapping of the British Empire onto our solar system transposes the seafaring culture that enabled empire into outer space. But he keeps the sailing ships and the romances in which they were caught up. The narrator Art Mumby and his sister Myrtle escape their spaceship home Larklight in a "lifeboat" "shot out through the open doorway like a shell from a gun" (Reeve 38). Art narrates the boredom of being adrift and then shipwrecked in language that would not have seemed out of place in *Robinson Crusoe* or *Kidnapped*. But instead of a desert island, their lifeboat lands on the moon, a moon that was claimed in 1703 for Queen Anne and that has been "improved" with mines and a colony for convicts. "These villains, transported from England for sheep-stealing and machine-breaking, soon see the error of their ways after a few years' hard work in the thin air, and their descendants may one day populate the entire Moon" (45). The Moon takes the place of Virginia or Australia in Reeve's romance; but the Moon as a site for romance is already implicated in a very different kind of narrative, the scientific romance. The flora and fauna Myrtle and Art encounter, and their mode of locomotion across the lunar surface, are strikingly similar to those found in H.G. Wells's *First Men in the Moon* (only one of Reeve's many allusions to Wells's scientific romances). Throughout *Larklight*, Reeve crosses scientific exploration with the boy's own adventure. Not coincidentally, these are Art's favourite genres (science and adventure), and he continually references literature as a way to make sense of experience. The lunar romance leads directly to the pirate romance. A fifteen-year-old pirate, Jack Havock, and his alien crew (another blending of scientific wonder and adventure romance) save the Mumby children from certain death (hilariously using Jamaican rum to counter the venom of a lunar moth—Reeve gets a lot of jokes from the transposition of pirates to outer space). Jack turns out to be the "Terror of the Aetheric Main, a daring pirate chief who had been raiding ships of the Royal Interplanetary Company for the past three years" (68). His ship, *Sophronia*, resembles a sailing ship, right down to the barnacles:

> The metal bands which sheathed the timber hull were streaked with rust and speckled with space barnacles and clumps of hanging weed. She looked at least a hundred years old, and at her rear, above the bulbous exhaust-trumpets of her wedding chamber, the mullioned windows of a fine old stern gallery glinted with the light of distant Earth. There was gilded carving underneath the windows; bird and angels and a peeling, painted scroll that bore her name. (73–74)

Reeve mixes the new (and imaginary) with the old—for a reason. Space travel looks romantic when it's done in a sailing ship, because as readers, we possess a ready-made template for the kind of adventure in which this vessel belongs. Moreover, this sky sailing provokes "a state of wonder," to return to Bennett's definition of enchantment, because we don't expect to see an eighteenth-century ship anchored on the moon or sailing between the planets on "Sir Isaac's Golden Roads" (the exhaust of the "chemical wedding" that creates a "luminous bow-wave of alchemically-altered particles" [97]). The illustrations accompanying the text of *Larklight* let us see the marvellous ship, in a way the words can't, all at once—crucial, as Philip Fisher argues, to any experience of wonder.

Reeve's purpose is not only aesthetic, however. At the beginning of the story Art and Myrtle are firmly on the side of Empire and assume the worst about Jack. When the Royal Navy ship *Indefatigable* arrives to arrest Jack Havock and his crew, "the Union Jack [flying] out proud and bright in the lunar sunshine," Art is "filled with patriotic fervor at the sight of our brave tars" (87). Like the *Sophronia*, the *Indefatigable* looks and acts like a sailing ship, with "gold scrollwork at her bows" and "a white chequerboard of gunports stretching along her flank" (86–87). This familiar confrontation between the disciplined naval crew and the motley pirates, who load their "fat cannons with anything they could find" (87), quickly turns into an interrogation of the way empire is being conducted, when Jack and his crew escape with their hostages to the place where his story may be said to have begun, the abandoned colony on Venus. A mysterious plague turned all the settlers, except Jack, into trees. He is taken back to Earth, to the Royal Xenological Institute in London, where he is treated as a test subject, like the other aliens held there. Art and Myrtle, born at Larklight, see themselves as imperial subjects, though they dwell "in the back of the black" (17). Myrtle longs to "live in England, like a civilised family," complaining that "even in Bombay or Calcutta or one of the American colonies there would be visiting and so forth" (3). Their remoteness from the centre of the empire pushes them into more fervent displays of devotion than characters who are better placed. Jack, finding himself at the centre, can easily discern his marginalization. To drive home the point, Reeve gives Jack brown skin and parents of different (human) races. The *Sophronia* is Jack (and his crew's) escape from scientific torture and dehumanization in the broadest sense.

The idea for escape came from a book Jack read while in prison, *Sea Stories for Boys*. All of Reeve's characters are readers and understand their world through books. The barrage of textual allusions, to books real and invented, models for Reeve's child readers how different sorts of stories shape our values and influence our actions. Even though steampunk tries to re-create the lost geography of Victorian romance, as a way to re-enchant a disenchanted world, steampunk authors don't swallow the past whole, but instead use the geography of romance as a way to critique both past and present. Reeve, more than Oppel or Westerfeld, uses situations rather than characters to undercut the Victorian values of which he disapproves (racial prejudice, jingoism, and unethical science) and to promote those of which he does approve (curiosity above all, manifested in scientific exploration). *Airborn* and *Leviathan* are each structured around a male–female pairing, primarily aimed at undercutting the Victorian ideology of separate spheres. Kate De Vries and Deryn Sharp both take on traditionally male roles (scientist and sailor), and their stories get equal time with those of the boy heroes, something that can't be said of Reeve's *Larklight*—though to be fair, Myrtle is finally responsible for defeating the spiders who threaten Earth. Oppel and Westerfeld let their girls fly in a genre that used to be reserved for boys.

In *Romancing the Postmodern*, Diane Elam regards this "juxtaposition of differing historical periods" (our present and the Victorian past) as central to the effect of postmodern romance. It offers "the coexistence of multiple and mutually exclusive narrative possibilities without a point of abstraction from which we might survey them. Postmodern romance offers no perspectival view; it is an *ironic coexistence* of temporalities" (Elam 13; emphasis in original). This "coexistence of temporalities" is not ahistorical, as some critics of steampunk have suggested; in Elam's analysis, the lack of a perspectival view—the lack, I would argue, of teleology in steampunk, things don't have to be the way they are—"make[s] the past impossible to forget" (15). Alternate histories coexist with actual histories. The possibility of changing course, of re-enchanting the disenchanted present, underpins all steampunk. To take up again my most recent example, all three authors represent the historical discrimination women did experience; without it, what their female characters accomplish would be meaningless. But the coexistence of what was and was not throws into relief a present that still encompasses the past. Discrimination coexists with freedom.

The terrain of romance can be, therefore, both familiar and defamiliarizing, transporting the reader out of ordinary existence into a heightened reality. All three writers use what Bennett calls crossings to enchant their readers, to provoke in them a state of wonder. The magic of "metamorphing creatures," Bennett explains, "resides in their mobility ... in their capacity to travel, fly, or transform themselves ... Their presence carries with it the trace of dangerous but also exciting and exhilarating migrations" (Bennett 17). Oppel captures this pleasure and danger of mobility in his melding of two plots—the hero's coming of age and a girl's tracking of a mysterious flying creature. Matt Cruse was born in the air; his belief in his own lightness, in his capacity to stay aloft even if he falls, is connected to his wishful thinking about his father's death—his sailmaker father didn't die in a fall, but kept flying. Until the very end of *Airborn*, Matt can dream of his father only while he's airborne himself. He flees from the land that confines him, that restricts both his real and imaginary flight. The need to fly is physical for Matt—a point that's driven home when Matt and Kate, the balloonist explorer's granddaughter, find the cloud cat. It's not a bird or a bat, yet it can fly; it gives birth in the air; and it never lands. But there's one that did—one that through bad luck or some deformity could not unfold its wings. It fell to land, and learned to survive on the island where

Matt's ship crashed. Matt's identification with the strange hybrid creature is bodily. "Her legs were all wrong for walking ... She slouched, she slunk, as if revolted by the feel of the earth beneath her feet. I wished I could help her. I knew what it was like to have your wings clipped" (Oppel 221). Later, when he's cornered on the top of the airship, he tries to fly:

> I fell backwards and instinctively opened my arms, spread my legs. I felt the air pouring over me, felt how it parted for my head and over my shoulders and over my chest and down my torso to trail off my legs ... I was not frightened. This was how my father fell. It was the most natural thing in the world ... If my father could do it, I could do it. I was born in the air. (306)

But Matt doesn't fly, of course; he grabs hold of his ship—and he crashes. "All my life I'd told myself I was light and could soar free of things. I was light and I could outrun sadness. I could fly away and keep flying forever. But I could never catch up with my father. He had fallen like Gilgamesh, and I had not been there to save him with an all-powerful Enkidu hand" (307). The reference to myth here is crucial—all three writers re-enchant flight by remythologizing it. Oppel uses the Gilgamesh story; Westerfeld's hybrids all have mythological precursors: Leviathan, Medusa, Kraken; Reeve's heroes find a god, Thunderhead, in the swirling gases and cloud fairyland of Jupiter. Just when Matt concludes that the myth has no potency, an Enkidu hand comes to save him. The deformed cloud cat flies, for the first time, transfixing Matt's attention; she flies to her kin. Matt's enemy sees the "shine of amazement" in his eyes, and looks up. "A huge group of them wheeled over ... and I could feel the wind from their mighty wings as they passed. I couldn't help laughing aloud in delight at this glorious turbulence" (309). Matt is spellbound, forgetful of his imminent death, when a cloud cat clips the pirate who threatens him and knocks him off the ship. There's wonder in this scene, crossings between the human and the animal, the human and the machine, and a potent reimagining of a myth—of what civilization means and what friendship means.

For Bennett, this is the point of enchantment, or rather re-enchantment, to see, as Matt Cruse does, that the world is worthy of our love, that it can be wonder-full. Bennett looks for everyday marvels, such as cross-species encounters, because "it seems ... that presumptive generosity as well as the will to social justice, are sustained by periodic bouts of being enamored with existence, and that it is too hard to love a disenchanted world" (Bennett 12). Enchantment thus has ethical relevance: in Bennett's sharp formulation "you have to love life before you can care about anything" (4).

For Westerfeld's Deryn Sharp, loving life, being wholly herself, means flying, just as it does for Oppel's Matt Cruse: "She couldn't wait to get off the ground, the flightless years since Da's accident suddenly heavy in her chest" (Westerfeld 34). Her first flight as a boy recruit in the navy is on a medusa, a small hydrogen breather made mostly from the "life threads" of a jellyfish. When the medusa goes off course, she's picked up by *Leviathan*, a whale, or rather a whole ecosystem, that occupies the sky much as a blimp would: "The *Leviathan's* body was made from the life threads of a whale, but a hundred other species were tangled into its design, countless creatures fitting together like the gears of a stopwatch" (69–71). The novel seems at first to be setting up the genetic hybrids of the Darwinists as superior to the polluting, mindlessly destructive war machines of the Clankers; yet there are plenty of hints throughout the novel that the Clanker machines are just as enchanted, just as infused with imagination, as the fabricated beasties, which the Clankers regard as godless abominations. The steam-driven machines are modelled on organic bodies—the two-legged walkers, the eight-legged landships—and they're mythologized as often as the Darwinist creatures are. Alek, the Clanker protagonist, feels he is part of the walker when he's controlling it, much as Deryn senses how the human crew functions as part of *Leviathan's* body, just as the bats, hawks, bees, and sniffer hounds do. At the end of the novel, *Leviathan* can escape only by taking on the engines from Alek's walker, along with its five-man crew. The crossed flying whale has been crossed again and quickly adapts—develops a mind of her own, the crew says, enamoured as she is of her newfound power (418).

Crossing of a different kind is Reeve's aim in *Larklight*. There are interspecies crossings aplenty, the disease that turns Venusian colonists into trees, spiders piloting human automata, a four-billion-year-old Shaper that takes human form and becomes the mother of the book's two protagonists, and, most colourfully, the pirate ship crewed by a motley collection of aliens, most of whom escaped from Britain's Royal Xenological Institute. *Larklight's* villains are a very old race of spiders, the First Ones, who wove their webs throughout the solar system until Mrs. Mumby, then a Shaper, came along to set it free. "Left to themselves," she explains to her son Art:

Most stars stir up planets out of the clouds of dust and gas which gather around them in the aether. But the First Ones drift through space on their gossamer threads, and when they find a solar system in the early stages of formation they bind it, and tie it, and wrap it in their knots and cradles, and make sure that no worlds can ever form, and no sort of life but their own can ever thrive. (Reeve 325)

The Shapers "love life, in all its infinite variety," and their purpose was to help new solar systems bring forth life (324). By this point in the novel, readers will have had an object lesson in the wonders of infinite variety and will agree with Mrs. Mumby that it might be necessary to destroy in order to create, an ethical dilemma indeed, but one that the book has set out clearly for its young readers: Would it be better to have a universe in which only one form of life, the original form, is allowed to prevail, or a messy universe in which infinite variety truly is possible?

It might appear, from steampunk's interest in fantastic modes of flight, that it is uninterested in actual science. Yet an aesthetic engagement with science and technology is central to its purpose. Jake von Slatt, probably the world's best-known steampunk designer—or "contraptor," to use his word—regards the reaction against advanced technology as the beginning of the "Maker Movement." The Victorian era was, in his view, the last time "the average educated person was *expected* to have an understanding of how the machines that made modern life possible worked" (von Slatt 405). All the "components were visible and their functions obvious" and very often, they were also beautiful, intended "to glorify the technology that made them work" (405). One hundred years after the advent of electricity (the beginning of the end for makers and tinkerers), the guts of the machine are completely hidden, and technology is so advanced that most educated people cannot understand it; we can't do anything to or with it. Steampunk, for von Slatt, is "about falling in love again" with technology and taking control of the machines that we use every day (408). Above all, steampunk is about recapturing the sense of wonder that makes learning possible.

The antinomian critics talk a lot about wonder. Literary critic Philip Fisher, in his seminal 1998 work *Wonder, the Rainbow, and the Aesthetics of Rare Experiences*, calls wonder "the most neglected of primary aesthetic experiences within modernity," probably because it seems to contradict the dominant historical narrative of disenchantment (Fisher 2). Wonder "involves the aestheticization of delight, or of the pleasure principle" (2); wonder "has to do with a border between sensation and thought, between aesthetics and science" (6); "Wonder is the outcome of the fact that we see the world" (17)—it is always tied to the visual, to something we can take in spontaneously, all at once. Wonder is all these things, for Fisher, but it starts, significantly, in wondering: "I wonder." The rainbow is emblematic for Fisher because the "combination of pleasure and puzzlement" leads to "scientific explanation" (7). The two senses of wonder in English are both at work here: curiosity and admiration. Wonder is therefore also, always, tied to learning—we learn by wonder:

> Socrates' phrase "Philosophy begins in wonder" and Descartes's pro-posal that wonder be considered the first of the passions, because it is the very mechanism of an interest in differences upon which all learning depends, imply that human speculation and creativity, because of their link to wonder, are finite, declining as age increases, and linked to first experiences, and therefore to youth. (59)

While it would be difficult to dispute the prevalence of wondrous experi-ences in youth as compared to older age, the antinomian critics, past and present, regard efforts to re-enchant the world as continuous and at least intermittently efficacious. Re-enchantment entails renewal, waking up to the things that are in us, as MacDonald would have said or, in Jane Ben-nett's words, "be[ing] struck and shaken by the extraordinary that lives amid the familiar and the everyday" (Bennett 4). Tolkien called it "recov-ery," the "regaining of a clear view" (Tolkien, "On Fairy-Stories" 57). When fantasy is the tool of this recovery, we could be reading of "the centaur and the dragon, and then perhaps suddenly behold, like the ancient shep-herds, sheep, and dogs, and horses—and wolves" (57). Fantasy can stimu-late wonder, no matter the age of the reader, because it affects the way we think and the way we see, from a new angle, or through something that has been conjured in our minds by words on a page. This is why fantasy gets a central place in Saler's study of re-enchantment: it is one of the most important tools we have for renewing wonder, and, therefore, for learning, speculating, and creating.

Fisher would not allow that narrative itself could provoke the experi-ence of wonder; because of its unfolding in time, it cannot be taken in all at once. But he sprinkles his work with literary representations of wonder, in much the same way that Freud used literary case studies to build his theories of mind. Steampunk excels at such representations, straddling as it does the border between aesthetics and science, between "How wonderful!" and "I wonder." The works by Oppel, Westerfeld, and Reeve repeatedly invite the reader to take the position of a character caught up in the experience of wonder, the kind of wonder that, as Bennett puts it, suspends "chronological time and bodily movement," that leaves him or her "transfixed and spellbound" (Bennett 5).

Tolkien said: "Fairy stories were plainly not primarily concerned with possibility, but with desirability. If they awakened *desire*, satisfying it while often whetting it unbearably, they succeeded" (Tolkien, "On Fairy-Stories" 40–41). No doubt Oppel, Reeve, and Westerfeld have awakened desire in their readers—and for what? For adventures of the old kind. For the freedom that comes with mobility. For crossings that make us shiver, or laugh aloud in delight, as Matt Cruse did, when the cloud cat flew.

14) Mythic Re-enchantment

The Imaginative Geography of Madeleine L'Engle's Time Quintet

❦ MONIKA HILDER

T he prominent twentieth-century American author Madeleine L'Engle (1918–2007) asked the question, "What Is Real?" (1978). In her novel *A Wind in the Door,* the angelic Teacher Blajeny also asks the siblings Meg and Charles Wallace Murry, "What is real?" (57). In other words, is reality limited to provable fact as seen through the lens of scientific empiricism, or is there important truth to be explored in the realm of the mythic, seen through religious faith and grasped imaginatively?

In an era when dialogue in the academy was dominated by race, gender, and class, almost to the exclusion of religion, as Stanley Fish also points out (2005), L'Engle explored the existential questions of spirituality and related issues of moral virtue in her award-winning children's fantasy literature the Time Quintet, five novels that span two generations of the Murry and O'Keefe children's adventures: *A Wrinkle in Time* (1962), *A Wind in the Door* (1973), *A Swiftly Tilting Planet* (1978), *Many Waters* (1986), and *An Acceptable Time* (1989). While the academy tended to regard modern post-Enlightenment culture as one of "disenchantment"—a loss in which the premodern sense of mythic enchantment had been replaced by a narrowly conceived form of rationalism—L'Engle, like some of her peers, practised a mythic re-enchantment that challenged philosophical materialism. Through imaginative engagement with countercultural concepts rooted in a biblical vision and also informed by theoretical science, L'Engle invites readers to explore mythic reality through her imaginative geography: a geography of earth and fictional planets, biblical and invented characters and species, contemporary and other times, and, above all, the "inscape" or inner spiritual geography of place, event, and the individual soul. Unlike literary realism, which depicts social reality without the supernatural dimension, L'Engle explores human experience from the vantage point of a unified cosmos in which the natural and the supernatural are united. In her words, "The great mystics have always understood ... that the invisible world is as real as the visible"; citing J.B.S. Haldane, she emphasizes that "'the universe ... is stranger than anything we can imagine'" ("Kerlan" 5, 1).

∾ Mythic Re-Enchantment

Mythic re-enchantment might be described as a modern return to an earlier paradigm of the mystery of the supernatural cosmos. In *Many Waters,* the seraph Adnarel describes contemporary North America as "a time and place where [angels] like myself are either forgotten or denied" (134). In the words of the cherubim Proginoskes in *A Wind in the Door,* "I'm real, and most earthlings can bear very little reality" (84). Bishop Nason Colubra in *An Acceptable Time* explains, "We have lost a sense of the sacredness of space as we have settled for the literal and provable" (165). L'Engle spoke of this loss as leading to the evils of idolatry (*Penguins* 96), and she sought to recover what "we have lost as we have settled for Western, pragmatic thinking" ("Knowing Things" 7). In "Before Babel" she wrote, "I am so worried about the trend away from the unprovable realm of the limitless imagination to the terribly limited realm of the provable, which leads to the world of material things, which leads to the absurd" (667).

Rolland Hein describes the distinction between literary realism and literary myth:

Myths are, first of all, stories: stories which confront us with something transcendent and eternal. Myth exists because life has a duality to it. All events and circumstances exist on a physical level, which we apprehend sensuously, and on an anagogic level, apprehended with the intuitions of the spirit. Literary realism presents the former; literary myth attempts insights into the latter. The spiritual dimensions of any event are elusive, but it is only they that really matter. (3–4)

Similarly, both J.R.R. Tolkien and C.S. Lewis argued that imaginative fantasy literature might be thought of as a potential *"spell,"* which means, as Tolkien explained, "both a story told, and a formula of power over living men" ("On Fairy-Stories" 31). Tolkien spoke of its purpose to aid prisoners of materialism to escape into "more permanent and fundamental things," and suggested that its "true form" and "highest function" is its denial of "universal final defeat and in so far is *evangelium,* giving a fleeting glimpse of Joy, Joy beyond the walls of the world, poignant as grief" (62, 68–69). Lewis spoke of its potential "to wake us from the evil enchantment of worldliness" that much of modern education has inculcated as "the good of man is to be found on this earth" when "the truth [is] that our real goal is elsewhere" in "a transtemporal, transfinite good" ("Weight" 4–5). Corinne Buckland argues that "in postmodern culture the experience of the sacred tends to be unacknowledged or misunderstood," yet fantasy literature "which successfully evokes the full power of the numinous ... allows us to see the marvels of the natural world through a clean lens and to glimpse the sacred dimension of the Other that is both beyond and infused into the natural world" (17, 27).

L'Engle's imaginative geography is her creative response to desacralization: hers is a work of recovery in mythic imagination in which the material and the spiritual are shown as being one indivisible reality. Just as a wind blows open the Murry kitchen door with a bang on a windless night (*Wind* 219), L'Engle's mythic vision bursts rationalism. Her celebration of myth is rooted in her Christian vision—one of incarnation and sacrament in which the loving character of God reveals himself to humanity in the Bible, in nature, throughout history, and in people's lived experience. While the faith that such mystery calls forth is not verifiable by empirical means—as Mr. Murry insists in *A Wind in the Door,* "The naked intellect is an extraordinarily inaccurate instrument" (90)—neither is it, in L'Engle's view, so completely at odds with the pursuits of theoretical science.

In fact, the origin of Western science has its roots in Christian faith in a rational universe. And as Bernard Lightman, for instance, points out, "for Newton, science and religion were inevitably in harmony": "the conflict thesis" is a recent historical development perpetuated by agnostics and atheists, which "to the uninformed, [seemingly has] the complete backing of science" (167–68). Melody Briggs and Richard S. Briggs too warn that it is "too simplistic to equate 'spiritual' with 'non-empirical'": they insist, "Like fantasy, spirituality integrates the empirical and the metaphysical" (32). Indeed, L'Engle critiqued both the church and the scientific establishments for their slowness to recognize the compatibility of science and religion ("Kerlan" 3). In addition to her love of ancient biblical words and the writings of the early Christian theologians whom she considered more relevant to our contemporary world than the later ones—whom she moreover found "depressing" (*Walking* 88; "Allegorical Fantasy")—L'Engle also thought of cellular biologists and astrophysicists as her favourite twentieth-century mystics because "their questions are theological ones" (*Walking* 88–89).[1]

Like her primary mentor, whom she called the granddaddy of fantasy literature—nineteenth-century novelist George MacDonald,[2] a Congregationalist minister who also had training in chemistry—L'Engle responded to reductionist rationalism by exploring spirituality. She identified herself as a "universe disturber," one who is unafraid to challenge conventional paradigms because, in her words, "that truth, which would set us free,[3] usually runs counter to culture and so disturbs the universe" ("Do I Dare?" 675). Thus, L'Engle's characters challenge readers with the possibility of truth that is much larger than some contemporary paradigms. In *A Wrinkle in Time,* Mrs. Murry and Mrs. Whatsit declare that logical explanations might exist for things that limited human understanding cannot yet grasp (46, 75), and Mrs. Who cites Delille: *"How small is the earth to him who looks from heaven"* (86). In *Many Waters,* Dennys explains, "There's no such thing as an unbreakable scientific rule, because, sooner or later, they all seem to get broken. Or to change" (299). In *A Swiftly Tilting Planet,* physicist Mr. (Dr. Alex) Murry speaks of how the discipline of science encourages the possibility of faith beyond rational understanding ("My training in physics has taught me that there is no such thing as coincidence" [30]); the Puritan settler Ritchie Llawcae declares, "The ways of the Lord are mysterious, and we do not need to understand them. His ways are not our ways—though we would like them to be" (138);[4] and the unicorn Gaudior insists, "We do not always know what is important and what is not" (210).

Fantasy Worlds and Re-enchantment

Mythic re-enchantment might invite the question, "What do unicorns have to do with the maturation of young people?" Quite a bit, if we concur with Stephen Prickett's claim that "realism and fantasy are two sides of the same coin: that realism is as much an arbitrary and literary convention as fantasy, and that fantasy is as dependant on mundane experience as realism" (122). In L'Engle's mythic fiction, as, arguably, in her more realistic fiction also, the sharp division typically made between the world of provable fact and the realm of mystery blurs. In her fiction, dragons, or cherubim that look like dragons, interact with twentieth-century children. In fact, in these novels, the cherubim and the unicorns that burst into contemporary North America suggest a spiritual reality that is as real as so-called ordinary physical reality—or even more real, as Plato has it, "the eternally Real" (cf. Hein 3). In the words of the first-grader Charles Wallace, "There are dragons in the twins' vegetable garden" (*Wind* 1).

This study considers how in the Time Quintet L'Engle explores questions of spirituality through the imaginative possibilities of space–time relativity. In each of the novels L'Engle's characters are thrust into a battle between good and evil in which their moral choices are of utmost importance. In *A Wrinkle in Time* (1962), the children are initiated into spirituality through a fierce struggle against demonic possession. In *A Wind in the Door* (1973), the cosmic war occurs at the microbiological level. In *A Swiftly Tilting Planet* (1978), the young adults influence individual choices across the centuries, choices that could either prevent or escalate into planetary warfare in the present. In *Many Waters* (1986), teenage twins experience human society before Noah's Flood. And in *An Acceptable Time* (1989), young adults experience how their own choices interact with the practice of human sacrifice thousands of years ago.

L'Engle's imaginative geography extends from the microcosm of mitochondria to the macrocosm of the universe and heaven itself, and includes the span of human history. Her writing is highly charged with dramatic imagery, such as friends fighting evil while inhabiting the body of the child Charles Wallace at the cellular level (*Wind*), witnessing many creatures performing a cosmic paean of praise that fills them with indescribable joy (*Wrinkle*), and experiencing both the deep depravity and the surprising salvation of humanity at the time of Noah's Flood (*Many*). Her novels explore timeless questions of the human spirit: questions of identity, human value, ethics, and the transcendent. As C.S. Lewis has said, the creation of "plausible" secondary worlds means that the writer draws on "the only real 'other world' we know, that of the spirit" ("On Stories" 12). In *Renaissance of Wonder*, Marion Lochhead spoke of George MacDonald's achievement in mythic children's literature in terms of a joyous sense of tangible goodness (2). Similarly, L'Engle creates an attractive and credible landscape. From the particularities of family and village life in a rural northeastern state to the cosmos, L'Engle invites readers inside a tactile world of wonder. In this study, I will consider how L'Engle's mythopoeia—informed by a holistic, sacramental view of the world—creates an integrated vision of *matter, human choice,* and *spirituality.*

∽ Matter

Matter matters to Madeleine L'Engle because she believes that matter is the glorious stuff of creation that still reflects a loving Creator. Countering the existentialist view that the universe is meaningless, she declares that "there is structure and meaning to the universe: God is responsible to his creation" ("Before Babel" 668). She explains,

> It is an extraordinary and beautiful thing that God, in creation ... works with the beauty of matter; the reality of things; the discoveries of the senses, all five of them; so that we, in turn, may hear the grass growing; see a face springing to life in love and laughter; feel another human hand or the velvet of a puppy's ear; taste food prepared and offered in love ... Here, in the offerings of creation, the oblations of story and song, are our glimpses of truth. (*Circle* 206-7)

L'Engle's use of the religious word "oblations" for every feature of creation—that is, as a gift offered, the ultimate being that of Christ in the incarnation, God-made-flesh, celebrated in the Bread and Wine of the Eucharist, and therefore the context for all other gifts—provides an important clue to the extent to which she celebrates the physical world. L'Engle has described herself as "a particular incarnationalist" who "can understand God only through one specific particular, the incarnation of Jesus of Nazareth" ("Allegorical Fantasy").[5] Christ is the visible image of the invisible God, the Creator whose glory is reflected in His creation, and the One who holds it all together.[6] Thus all of nature is sacred substance. L'Engle cites Martin Buber: "'The world is not something which must be overcome. IT is created reality, but reality created to be hallowed'" (qtd. in

"Kerlan" 3). By contrast, evil is associated with virtual reality, such as the synthetic food on Camazotz (*Wrinkle*), and is defined by the nihilism of the dark angels, the Echthroi.

Readers relish the tactile world of wonder in L'Engle's novels, for example, Charles Wallace's midnight offering of hot cocoa and liverwurst-and-cream-cheese sandwiches (*Wrinkle* 10–11); the prize cucumber in the twins' vegetable garden (*Wind* 43); the preparations for a celebration feast when the battle for Charles Wallace's life has been won (*Wind* 216–17); the warmth around the Murry family Thanksgiving dinner where the loving prayer *Dona nobis pacem* (*Give us peace, give us peace, give us peace*) forms a protective circle in a world in crisis (*Swiftly* 19); their mother "making tea against the cold" (*Swiftly* 40); their farmhouse kitchen with its "familiar odor of fresh bread, apples baking in the oven, and warmth, and bright-ness, and all the reassurance of home" (*Many* 310); the glacial star-gazing rock in the North Pasture where mythic beings arrive (*Wind*; *Swiftly*); the blistering desert and the lush oases of prediluvian times where humans interact with angels (*Many*); the rich earthy smells and colours of a New England autumn, and the arrival of Anaral, the black-haired druid girl with honey-coloured skin from another circle of time pulling herself up through the window at the Murry family indoor swimming pool to talk to Polly O'Keefe (*Acceptable* 3–4, 28–29).

In L'Engle's incarnational view of reality, matter clearly has salvific value. Matter stirs memory and understanding: matter embodies hope. For instance, "the pungent scent of Gaudior's living flesh saved [Charles Wallace]": he inhaled unicorn sweat as the unicorn's "wings beat painfully against invisible wings of darkness beating at them" (*Swifly* 66–67)), later described as "an anti-unicorn" of "negative" energy (103). And various ani-mals, real and mythic, serve as comforters and guardians to the protago-nists on their dangerous adventures: the dog Fortinbras lies down across the threshold of Charles Wallace's bedroom "as though to bar death from enter-ing the room" (*Wind* 208); the snake and family friend Louise the Larger arrives to protect Meg when the Mr. Jenkins-projection appears in their gar-den (*Wind* 45–46); the dog Ananda, whose Sanskrit name means "that joy in existence without which the universe will fall apart and collapse" (*Swiftly* 40), rouses Meg to intercessory prayer that helps save her brother from another Echthroi attack (147); mini-mammoths nurture and also accompany the twins Sandy and Dennys into danger (*Many* 92, 292); a griffin defends the boys from nephil-inspired erotic assaults (*Many* 155–56, 175–77).

For L'Engle, matter has moral significance. The rootedness of things in the natural world is interconnected with well-being, with God. In her view, the universal is honoured in the particular, and only as we are willing to be so rooted are we able to mature and live, as the rebellious fara Sporos learns in *A Wind in the Door*. As G.K. Chesterton said, "The moment we are rooted in a place, the place vanishes. We live like a tree with the whole strength of the universe" (23).

∾ *Human Choice*

IN THE TIME QUINTET the celebration of the tangible goodness of created matter is just as apparent in the characters. Both human and mythic characters are memorable persons whose peculiar characteristics generate in readers strong responses of either affinity or dread. This illustrates L'Engle's vision that individual and collective *human choice* impacts our world for better or for worse, for good or for evil. So the wry humour of the angelic guardians in Halloween costume dress, Mrs. Whatsit, Mrs. Who, and Mrs. Which, and the gentle ministrations of Aunt Beast on the planet Ixchel, help the sometimes angry but essentially caring Meg to choose love over hate—and so save her brother Charles Wallace from demonic possession (*Wrinkle*). The sometimes grumpy but benevolent cherubim Proginoskes helps the teenagers Meg and Calvin O'Keefe and the stodgy, fearful school principal Mr. Jenkins to save Meg's dying brother, and then saves each of the humans, at last giving the ultimate self-sacrifice in his effort (*Wind*). The silver-horned unicorn Gaudior, whose name means "*more joyful*" (47), guides an intelligent adolescent out of dangerous pride into humble receptivity, thus saving the planet from nuclear destruction (*Swiftly*). The man with red eyes in *A Wrinkle in Time* embodies cold, calculating cruelty, as do the physically beautiful nephilim in *Many Waters*. When the young woman Mahlah gives up her liberty and very life to submit to the nephil Ugiel (*Many*), the loving tears of the seraph Alarid accentuate the magnitude of moral choice.

The issue of moral value is a contested one in postmodernity. By which standard do we make moral choices? Vigen Guroian describes our culture thus: "our society is embracing an antihuman trinity of pragmatism, subjectivism, and cultural relativism that denies the existence of a moral sense or a moral law" (4). L'Engle's moral vision, although shared by many, is therefore also countercultural in her call for a return to making choices that reflect intrinsic, objective moral truth—choices that are life-affirming, nourishing, and loving, instead of consumerist ("Heroic" 123) and destructive. Whether affirming the fundamental value of the disliked school principal Mr. Jenkins (*Wind*), forgiving the treacherous friend Zachary Gray (*Acceptable*), or entering a series of Might-Have-Beens in order to prevent nuclear holocaust in the present (*Swiftly*), every moral choice either heals or harms humanity. As Gaudior explains to Charles Wallace, "Everything that happens within the created Order, no matter how small, has its effect. If you are angry, that anger is added to all the hate with which the Echthroi would distort the melody and destroy the ancient harmonies. When you are loving, that lovingness joins the music of the spheres" (*Swiftly* 60). Indifference is not a viable option: Mrs. Who insists, "*To stake one's life for the truth*. That is what we must do" (*Wrinkle* 63). And L'Engle's heroic characters are "ordinary people who somehow or other manage to do extraordinary things"; they are fallible people who sin, repent, learn, and take risks for the sake of goodness, perfect only in the sense of "doing something thoroughly" ("Heroic" 122–25), and it is on their choices "that the balance of the universe depends" (*Wind* 148).

The arena of humanity's moral choice is the cosmic *spiritual battle-ground*. Heaven and hell are real in L'Engle's cosmos. In each of the novels, L'Engle illustrates how the powers of evil, the Echthroi or dark angels, seek to destroy humanity, while the powers of good seek to prevent evil, to restore and to heal. The Echthroi are identified as the fallen angels depicted in the Bible (*Wind* 101). Proginoskes describes them thus: "Sky tearers. Light snuffers. Planet darkeners. The dragons. The worms. Those who hate" (105). "They are the powers of nothingness, those who would un-Name. Their aim is total X—to extinguish all creation" (148). There is "war in heaven" because of the Echthroi (*Wind* 102), and if, for instance, the cherubim and Meg fail to Name Mr. Jenkins and therefore subsequently fail to save Charles Wallace from dying, "there would be rejoicing in hell" (106). In contrast, the glorious angels, like the seraphim, "have chosen to stay close to the Presence" (*Many* 59). They serve as comforters and guides, and they fight for the Good even to the point of self-sacrifice. Mrs. Whatsit describes the war in heaven as "a grand and exciting battle" fought throughout the universe by a great company: Jesus, artists, scientists, mystics, musicians, and many others (*Wrinkle* 88–89).[7] And Mrs. Who prophecies the outcome: "*And the light shineth in darkness; and the darkness comprehended it not*" (89).[8]

ꙮ Spirituality

Spirituality, then, is central to L'Engle's view of a created universe. In her words: "The spiritual world and the visible world are one, and so are we part of that Oneness. We are all kin. We have all been made by the same Creator, a God of love that is so unqualified that no one is excluded" ("Kerlan" 5). To affirm one's kinship with the divine and with one another is to join the creative energy of the universe. Thus, in *A Wrinkle in Time* Meg, Charles, and Calvin can tesser through space/time (take a wrinkle or shortcut) (78), and experience their connectedness, for instance, with an Edenic garden in which flowers moving about trees make music—a new song of praise that gives glory to the Lord [9]—and the "pulse of joy" flows through the children and Mrs. Whatsit (66–68). In *A Swiftly Tilting Planet*, the charge that all paganism is oppositional to biblical faith is countered in the People of the Wind's spirituality. L'Engle did not wish "to tell God ... where he can and cannot be seen!" (*Walking* 32). Thus, the native Indian wife Zylle explains to the hostile Puritans, Pastor Mortmain and his son Duthbert, "I do not know what pagan means. I only know that Jesus of Nazareth sings the true song. He knows the ancient harmonies" (126). In *An Acceptable Time*, Bishop Colubra calls on Christ to rescue them from final disaster (203, 237, 301–4). Healing comes not through "a vague kind of universalism" (*Walking* 32), but through Christ.

For L'Engle, a Christian incarnationist who therefore resists materialism, the stakes for recognizing the impact of good and evil spirituality cannot be higher. She cites the two words for power in earlier societies: "'mana'" for "benign power" and "'taboo'" for "malign power" ("Childlike Wonder" 107). She does not, for instance, conflate spirituality with psychology. In *A Wind in the Door*, the Teacher Blajeny prefers "the old idea of [demonic] possession" to diagnose the "phenomenon ... called schizophrenia" (131) in Mr. Jenkins. In Meg in *A Wrinkle in Time*, L'Engle depicts the seriousness of spiritual injury, showing how the inner disease of disappointment and hatred is as "dark and corrosive ... as the Black Thing" (171–72). The outcome of any spiritual battle is either annihilation or restored life: ultimate joy and hope.

Marek Oziewicz argues that the spirituality of L'Engle's mythopoeia "may be seen as stories instrumental to a radical reassessment of our civilization's assumptions, if we are to survive as a race" ("Envisioning" 78). In L'Engle's integrated vision of matter, human choice, and spirituality, several *countercultural spiritual principles* suggest themselves. Of these I will consider three: *listening, interdependence,* and *Naming.*

Listening

IN A CULTURE that tends to value decisiveness and sometimes aggression, the spiritual principle of *listening* as agency is undervalued. But in *A Swiftly Tilting Planet,* when a dictator threatens the world with nuclear war while the talkative and prayerful Murry family gathers around the Thanksgiving table, the noticeably silent youngest child Charles Wallace, now a young man of fifteen "with them big ancient eyes[,]" as Calvin O'Keefe's mother refers to him (27), answers his inquisitive brother with the cryptic words, "I'm listening" (22). Charles Wallace is listening to St. Patrick's Rune, which Mrs. O'Keefe gives him as a prayer to resist political evil. Although he is at this point in his life more sympathetic with the doubting rationalism of his older twin brothers than with the openness of his parents to runic power (34), and indeed nearly fails several times out of his arrogant desire for control, on this journey Charles Wallace learns to relinquish his intellect and his strength in order to listen to the Wind (200)—the divine force that guides him through interactive layers of time and space. He agrees to go "Within" various characters across the centuries to come alongside them as they make decisive choices. While on this quest Charles Wallace learns the importance of releasing pride and self-reliant action (150, 200), especially during an Echthroi Projection (a negative Might-Be), when the temptation to control things is at its height and could in fact bring about the intended evil (68). Charles Wallace acquires the "unhurried patience" of the unicorn (156), and with each act of humble listening he is able to make the moral choice that will help heal the planet and avert disaster. As the seraph Alarid asserts in *Many Waters,* "Part of doing something is listening" (279–80).

Throughout the Time novels, L'Engle emphasizes intuitive listening with one's whole being, "not just a fragment" (*Wind* 135), but with heart and mind in tune with cosmic love. In the original trilogy, the children learn to communicate with one another without words, and kything becomes a means to fight evil. L'Engle explains that the old Scottish word *kythe* "means to communicate with someone with love, beyond barriers of time and space. It is far more than ordinary ESP, which does not necessarily include love. Kything cannot happen without love" (*Trailing Clouds of Glory*

129). In *A Wrinkle in Time,* Meg saves her brother Charles Wallace from the disembodied brain IT by sending her love to the child caught inside the darkness. Love is the weapon that overcomes evil.[10] In *A Wind in the Door,* Calvin links such soundless communication with "communion" (177), and he, Meg, and Mr. Jenkins learn to kythe together to restore harmony to Charles Wallace's failing mitochondria. In *A Swiftly Tilting Planet,* Meg kythes with Charles Wallace as he travels Within each character, supporting him through her loving attentiveness.

In the subsequent two novels, intuitive listening is also on the personal and cosmic scale. In *Many Waters,* Sandy and Dennys discover that they know the "Old Language, the language of creation ... still in communion with the ancient harmonies" (101-2), because they, like Noah's son Japheth, can listen with "under-hearing" (21). Dennys, for instance, learns to hear the "gentle love song" of creation (280) beyond the violence of that world and his own worry that the young woman Yalith will drown in the coming Flood. He hears how the stars chime at him, *"Do not seek to comprehend. All shall be well. Wait. Patience. Wait. You do not always have to do something. Wait"* (280-81), and so learns to trust although "Goodness has never been a guarantee of safety" (277). He is "enfolded in a patient, waiting love" (280) and thus experiences the coming redemption from evil. In *An Acceptable Time* when lives are at stake, the druid Karralys tells Polly that their hearts are open to her as her own heart can be to them, and Bishop Colubra compares the transmission of such love to a "telegraph" (278). In Anaral's words, "the lines of love cross time and space" (340); and in Mr. Murry's language, he will try to have courage for his daughter (*Wrinkle* 200).

To listen well and therefore stay open to love is an exercise in discernment because, as the seraph Aariel instructs Yalith in *Many Waters,* "The nephilim are masters of mimicry" (48). The spiritual practice of *listening* requires active attention to the divine voice.

Interdependence

A SECOND SPIRITUAL PRINCIPLE, *interdependence,* is another major contribution to culture. Contrary to the modernist values of individualism and the law of the survival of the fittest, L'Engle dramatizes how each player is vital and connected to the whole—whether pregnant Meg at rest but intimately kything with her brother and thereby assisting in the spiritual battles (*Swiftly*); the brutalized young girl Beezie who becomes Mrs. O'Keefe, a mother and the heroic rune-bearer (*Swiftly*); the discerning snakes named Louise the Larger (*Wind*; *Acceptable*); Mrs. (Dr. Kate) Murry with her many Bunsen burner stews, which feed her family while she conducts work that leads to the Nobel Prize (*Wrinkle*; *Wind*; *STP*); the seemingly inept school principal Mr. Jenkins, who transforms into a robust saviour (*Wind*); Sandy and Dennys, who help bring about reconciliation between Noah and his father Lamech (*Many*); or Polly, who helps stop human sacrifice and initiates healing between two peoples at war (*Acceptable*).

Important conventional values are subverted in L'Engle's vision of cosmic harmony. As Matthew Maddox says in *A Swiftly Tilting Planet*, "Nothing, no one, is too small to matter" (236). Among the People of the Wind, "each person in the tribe knew what he was born to do, and no gift was considered greater or less than another" (*Swiftly* 62). Similarly, L'Engle cited actor and director Stanislavski: "'There are no small parts; there are only small actors'" (*Rock* 57). As Proginoskes admonishes the rebellious Sporos, "we all need each other. Every atom in the universe is dependent on every other" (*Wind* 194). Meg Murry O'Keefe speaks of symbiosis: "everything and everyone everywhere interreacting" (*Swiftly* 21). Elsewhere, L'Engle commented on "a world of total inter-relatedness, where everything affects everything else, and what happens in a quark can affect a galaxy." She added, "The great mystics have always understood the underneathness of things" ("Kerlan" 5).

In this universe where "everything is connected by the love of the Creator," it is "the dark angels who [a]re separators" (*Acceptable* 291). So in *A Swiftly Tilting Planet*, Pastor Mortmain tries "to separate the white people from the Indians" whom he has encouraged the English settlers to regard as "savage heathen" (109, 114). It becomes clear that his zeal for separation is linked to his vanity and lust for power (the belfry to his church was built more for his own glory than for that of God [137], and his witch hunt was an exercise in fear to coerce the settlers into submission). So in *Many Waters* in the time of Noah, individuals are often rejected: wounded Dennys is thrown away from Tiglah's family tent "like offal" into a "garbage dump" (54) (an instinct that Noah shares to some extent [76]); Anah's ailing grandfather is put out to the desert to die of exposure (81); the nephilim seek and use women with a cold lust (192–93, 204, 208). Separation is motivated by the greed for self-reliance, and in *A Wind in the Door* Mr. Jenkins recognizes in the fara Sporos this same insanity of pride, the desire to "control the world," that fuelled Hitler, Napoleon, and Tiberius (191). Likewise, in *An*

Acceptable Time, Zachary is so addicted to self-centredness that he is "out of [his] mind" (319, 330, 338), even willing to have Polly killed in the hope of extending his own supposedly independent life.

But those who seek love enter cosmic harmony, like Charles Wallace when riding the unicorn Gaudior on the divine wind: "Stars, galaxies, circled in cosmic pattern, and the joy of unity was greater than any disorder within" (*Swiftly* 58). As with Meg, Calvin, Mr. Jenkins, and later Sporos, their education takes place "Here, there, everywhere. In the schoolyard during first-grade recess. With the cherubim and seraphim. Among the farandolae … the entire cosmos" (*Wind* 61–62). They witness the birth of a single star as an image of the cosmic dance: "a dance ordered and graceful, and yet giving an impression of complete and utter freedom, of ineffable joy … wind, flame, dance, song, cohered in a great swirling, leaping, dancing, single sphere" (*Wind* 154). The company of friends joins the song of the Deepened, rooted farae in Charles Wallace's mitochondrion, Yadah, who sing, "We are the song of the universe. We sing with the angelic host … Our song orders the rhythm of creation" (*Wind* 150–52, 187). Instead of trusting in one's own intelligence or independent strength (*Wrinkle* 102, 165; *Wind* 193; *Swiftly* 200), the heroes act out of the "foolish weakness" of a community of love. Meg, Charles Wallace, and Calvin approach CENTRAL Central Intelligence on Camazotz holding hands (115). Meg returns to Camazotz seemingly alone, but supported by the "foolish weakness" of love that hatred cannot fathom or overcome (*Wrinkle* 201–2).[11] The seekers are not isolated, but part of a vast army of cosmic heroes, many of its best members hailing from planet Earth who, as Mrs. Which declares, "wwill cconnttinnue tto ffightt!" (*Wrinkle* 88–89). The seekers discover that they are called into a cosmic pattern that "will be worked out in beauty in the end" (*Many* 304). Life is not "chaos and chance" (299), nor is the pattern set, as in fortune-telling (*Acceptable* 275); rather, it is "fluid" (*Many* 304), interactive with the individual's choice to fulfill the pattern, like poets writing their own ideas in sonnet form (*Wrinkle* 198–99).

Naming

A THIRD SPIRITUAL PRINCIPLE, *Naming,* swims upstream against the current of rationalistic labelling (merely scientific and quasi-scientific prescriptions to human dilemmas) as well as outright disinterest, rejection, and hatred. While it may seem accurate to label a person's weakness with terms such as claustrophobia, kleptomania, or depression, to identify a person by a label is dangerous because labels are reductionist, limited, and therefore limiting (cf. *Wind* 127; Thomas 160). In short, labels are dangerous because they hinder healing and growth. Moreover, hateful labels are outright destructive. To hate is to un-Name, to destroy, so that people will not know who they are—and this is one of the main activities of the dark angels, the Un-Namers (*Wind* 102), who, like the nephilim, choose to surround people with their "circle of extinction" (*Many* 273). A judgmental labelling attitude, for instance, leads to "Xing" or annihilating someone, as Meg discovers in *A Wind in the Door* when she struggles with the task of Naming Mr. Jenkins; it is like murder, as with the witch-hunting Pastor Mortmain in *A Swiftly Tilting Planet* who condemns the loving woman Zylle to death. But to Name is to love so that people can become who they are meant to be because "if someone knows who he is, really knows, then he doesn't need to hate" (*Wind* 102). To properly Name another is to acknowledge intrinsic meaning and worth. Naming, unlike labelling (cf. *Rock* 241), identifies an individual as an invaluable being who is infinitely more than the sum of his problems, or her strengths, and therefore has purpose and hope in a meaningful cosmos. Words are powerful; they can bless or curse (*Acceptable* 142–43), release love or withhold, give life or kill.

In *A Wind in the Door*, Meg learns that she is a Namer, one who is called to identify who someone is meant to be and therefore help that person become his or her authentic self (80–81, 101). Her first major battle is to give up hating and instead Name the "dour" and "unattractive little" Mr. Jenkins (98, 130) as human and therefore lovable. This is a significant challenge because Mr. Jenkins, to her conscious knowledge, has nothing lovable about him. He is a cold and ineffectual school principal who does not intervene with school bullies, including those who regularly attack Charles Wallace (3, 12, 17–19), and who has regarded Meg as "the most belligerent, uncooperative," and "contumacious child" he has had the "misfortune to have in [his] office" (*Wrinkle* 26; *Wind* 121). Through Blajeny, Meg learns, however, that Mr. Jenkins is "a perfect host" for demonic takeover (*Wind* 130) because he has not, evidently, known love. Proginoskes affirms that when people like Mr. Jenkins are unloved, "the Echthroi can move in" (184).

Meg herself has known the power of un-Naming, and not only through Mr. Jenkins: she has regarded herself as a "delinquent" (*Wrinkle* 4) and a "monster" (6); she has suffered under the malicious village gossip that she and Charles Wallace are "subnormal" (9). She too was once imprisoned under the darkness of her own anger and resentment (189–90). However, the difference between the 'beast' Mr. Jenkins and the 'beast' Meg is that Meg was shown unconditional love, faith in her development, and forgiveness (9–10, 13, 189–90, 195, 201). And as Meg was restored to love and courage and therefore could lovingly Name the true Charles Wallace in order to save him from the demonic (207–9), she must now do the same for Mr. Jenkins. Were she to refuse, she would effectively participate in the Echthroi action of annihilating Mr. Jenkins. But with the help of the cherubim, Meg's subconscious memory reveals to her the man's compassionate gift of shoes for Calvin (*Wind* 119–20), and while she would prefer to continue hating him, she can no longer deny her own habit of "snarly" behaviour towards him (120) and instead chooses to affirm Mr. Jenkins as a fallible human like herself, and therefore lovable (125).

Meg must first love the beast within herself, choose goodness, and then, like Beauty in the fairy tale, offer that love to the 'beast' Mr. Jenkins: when she does, the beastliness of both is forgiven and transformed into beauty, and darkness is defeated (cf. *Rock* 251–54). The transformation in the Named Mr. Jenkins is extraordinary: he agrees to fight evil because "Margaret Named [him]" (*Wind* 131), and through a journey battling familiar feelings of inadequacy where he is "afraid to be[,]" a journey on which Meg again Names him with love (173), the once unimaginative, cowardly man becomes a resilient hero.

L'Engle's concept of Naming disabuses readers of the illusion that love requires feeling. When Meg struggles to Name Mr. Jenkins, the cherubim admonishes her to distinguish between transient human emotion and moral action: "Love isn't how you feel. It's what you do" (*Wind* 122). While warm feelings for Mr. Jenkins do follow her moral action, they are not nor should they be a prerequisite, nor do they necessarily follow as a consequence. Naming is informed by the command of Jesus to "Love your enemies, bless them that curse you, do good to them that hate you, and pray for them which despitefully use you, and persecute you."[12] In *An Acceptable Time*, the People of the Wind illustrate this principle in healing instead of killing their wounded enemies. As a result, Tav and others realize that human sacrifice is wrong and that love can prevail, the result being intertribal marriage between Klep and Anaral and restoration for all. But in the case of Polly's choice to save Zachary Gray although he was willing to have her sacrificed, when she chooses moral responsibility (273) she feels "nothing. No anger. No fear. No love" (338). Polly parts with Zachary in peace (342), knowing that she has helped give him the chance to change (336–40), and that is enough.

∿ Conclusion

THUS, L'ENGLE'S MYTHIC RE-ENCHANTMENT in the Time novels challenges and seeks to repair cultural rationalism with a holistic spirituality. Through her invitation into the world of the spirit—as Lewis has noted, "the only real 'other world' we know"—she strikes many chords, including the harmony of humility, interdependence, and Naming. Humble listening is life-giving. Acknowledgment of the need for interdependence results in community and healing. And lovingly Naming an individual is the truly caring, life-affirming response in every situation. Madeleine L'Engle's imaginative geography invites a Deepening into ancient, timeless piety, and therefore continued consideration in the emerging discourse of contemporary spirituality. She answers a world profoundly shaped by secularism with the gentle words of Blajeny in *A Wind in the Door*—"Believing takes practice" (139).

—————

Notes

1) For example, L'Engle cited the atomic physicist Friedrich Dessauer as speaking of humanity as being dependent "'entirely on revelation ... We / should not strive to impose the structures of / our / own mind/s/, / our / systems of thought upon reality'" but instead begin with "'humility'" (qtd. in "Subject to Change" 336). See Marek Oziewicz's excellent discussion of theoretical science and modern psychology as it is reflected in L'Engle's writings ("Integrating" 186–95), as well as his overview of literary criticism of her work (178–80).

2) See L'Engle, "George MacDonald: Nourishment for a Private World."

3) See John 8:32.

4) See Isaiah 55:8.

5) The question of whether universalism (or a Christian form of universalism) can be attributed to L'Engle is debated, and the debate is also connected with the issue of

George MacDonald's heterodox view of salvation. Marek Oziewicz has extended the conversation in his argument that L'Engle has aspired to a "new spirituality, new mythology and new paradigm" ("Envisioning" 78)—a "'mere Christian' theology" that is "viable" for emerging spirituality in the third millennium ("Integrating" 196–97). L'Engle disassociated herself from universalism in the interview "Allegorical Fantasy" and in *Walking on Water* (32). See note 7.

6) Colossians 1:15–17.

7) This passage is often cited as evidence of L'Engle's universalism (see note 5), but she denies that she equates Buddha and others with Jesus (*Walking* 32). In an interview with me, on 26 September 2000, L'Engle said that she intended this passage "to widen our understanding of good. We tend to be very narrow about it. 'Only those who believe the way I believe can be saved.' Well, I don't think that's true. [There are] many things that some people want to believe that I don't believe. That does not make me a heretic or a blasphemer."

8) See John 1:5.

9) See Psalm 96:1.

10) See Romans 12:21.

11) See I Corinthians 1:25.

12) Matthew 5:44.

Four

SPACE AND GENDER

15) Female Places in *Earthsea*

❦ PETER HYNES

＝＝＝＝＝

U rsula K. Le Guin's well-known Earthsea series takes place in a vividly imagined archipelago where each island has its own topography, its own human geography, and in many cases its own culture and
politics. The creation of fictional spaces, in other words, is crucial to the
books' elaboration of plot and character. Indeed, Le Guin has said on a
number of occasions that, for her, one basis for the creative imagination
is a sense of place. In an interview published in 2008, for example, she
explained that "'place' is obviously enormously important in all my work.
Look how many maps there are in my books. Often a map is one of the
first things I do" (*Conversations* 118). In 2012 she returned to this point in a
fresh afterword to *A Wizard of Earthsea:* "But *where* is as important in the
realms of pure imagination as it is here in mundanity. Before I started to
write the story, I got a big piece of posterboard and drew the map. I drew all
the islands of Earthsea, the Archipelago, the Kargad Lands, the Reaches"

(*Wizard* 219). Her sense of the centrality of imaginative geography to novel-writing in general comes clear in a comment from an essay on Tolkien: "Great novels offer us not only a series of events, but a place, a landscape of the imagination which we can inhabit and return to" (*Wave* 106). Finally, on the genesis of Earthsea Le Guin has written, "I am not an engineer, but an explorer. I discovered Earthsea" (*Language* 49–50).

This respect for the "landscape of the imagination" is amply evident in the Earthsea series, perhaps especially so in the first three novels (1968 to 1972). The series as a whole is named for its fantasy location, and maps form part of the apparatus. A map of the whole archipelago figures in each volume, and there are also close-ups at appropriate junctures to show exactly where the characters are. What is more, each volume of the initial trilogy begins with an extended description of a place. *A Wizard of Earthsea* sets the tone by introducing first of all the island of Gont, "famous for wizards," where "from the towns in its high valleys and the ports on its dark narrow bays many a Gontishman has gone forth to serve the Lords of the Archipelago in their cities as wizard or mage" (*Wizard* 1). Once the flavour of the entire island is established the novel settles in the village of Ten Alders, a remote hamlet where the hero, Ged, is beginning his career. The second book of the series, *The Tombs of Atuan*, is set up in a similar fashion: the Prologue presents the home farm of young Tenar, the protagonist, before changing locales to the Hall of the Throne where Tenar will be dedicated as the Eaten One, Arha, the Priestess of the Tombs of Atuan. *The Farthest Shore*, finally, focuses on the Court of the Fountain on Roke Island before enlarging its opening vision to include the Great House that encompasses the fountain, and the Isle of the Wise, which, in its turn, encompasses the house. The fountain is the central feature of the central building of the central island of the whole archipelago. It is impossible, in other words, even to begin reading any volume of *Earthsea* without being forcefully reminded that these books are about place.

Le Guin's techniques for rendering these spaces and places range from fantasy standbys like maps (although in most paper editions the book's spine annoyingly gets in the way of some of the most important central islands) to descriptive set pieces like the introductions and the scene setting that I mentioned above. These scenes are brought alive with a flair for visual detail that is surely unmatched in the field of contemporary prose in this genre. But to get to the main concern of this chapter, among their other traits these spaces large and small are frequently gendered. The island of Roke, for example, is given over to a central wizard school for boys and so is clearly marked as masculine space. The open sea is likewise a space where men can travel together: Ged and Vetch hunt the wizard's shadow far beyond the eastern boundaries of Earthsea in the first volume, and in the third Ged and his *protegé* Arren wander south and then west to the dragon lands. Not all of the trilogy's places, however, belong to men. In this essay I would like briefly to acknowledge the generally masculine ascriptions of place in the initial trilogy, but mostly to focus on a number of spaces that belong, explicitly or implicitly, to women, or at least throw an unusual light on the predominantly masculine feel of the books' geography. Examples I will discuss include domestic spaces like houses and castles as well as portentous major locations such as the underground dwellings of the young priestesses of the island of Atuan. Finally, I will discuss the peculiar status of the Land of the Dead that Ged visits in the final volume. I will try to show how these places are articulated with the predominantly male space of the novels, and then, in the interests of historical completeness, conclude with a brief discussion of Le Guin's later reassessment, in *Tehanu* (1990) and after, of the gender profiles represented in this early work.

In very general terms the articulation of men's and women's spaces has been the study of feminist geography, a trend in academic geographical work that dates from the revival of the women's movement in the later 1960s. Feminist geographers draw attention to the truth that space is rarely neutral: it belongs to different people in different ways, it is traversed or inhabited differently by men and by women, and its symbolic values lean in either masculine or feminine directions. In the words of Linda McDowell, "the idea that women have a particular place is the basis not only of the social organization of a whole range of institutions from the family to the workplace, from the shopping mall to political institutions, but it is also an essential feature of Western Enlightenment thought, the structure and division of knowledge and the subjects that might be studied within these divisions" (12). Such insights have been applied to work on spaces in fiction as well as real spaces; Stacey Alaimo, for example, has studied the meanings of natural spaces like forests and deserts for an array of American women writers, and Susan Brooker-Gross has taken a geographically informed look at the spaces of the Nancy Drew Young Adult mystery novels. My discussion of the Earthsea series may be seen as a contribution to this literature on women's spaces, in this case spaces as they appear in Young Adult fantasy.

To return to Earthsea: *A Wizard of Earthsea* is a coming-of-age tale whose protagonist, Ged, is a young man whose adventures rarely bring him into contact with women at all. His serious apprenticeship in magic takes place under the tutelage of Ogion the Silent, a solitary male mage, and Ged leaves Ogion's house only to move to the entirely masculine wizards' school on the island of Roke. Thereafter, his major quest involves chasing a mysterious shadow creature around Earthsea, sometimes alone, and sometimes in the company of his old school friend Vetch, another male.

Nonetheless, women are not entirely absent from the novel, and where they appear they are usually framed by their own kind of space. One of the first avowedly female places to be mentioned is the house of Ged's aunt in his childhood village of Ten Alders. Ged, whose mother died when he was only one year old, is brought up in his father's forge, a clearly masculine venue. There is brief mention of his six older brothers, as if to underscore the overwhelmingly male flavour of his childhood environments. Ged's aunt, on the other hand, is a minor sorcerer who teaches the young hero his first spells. Her house is described in these terms:

> She took him into her hut where she lived alone. She let no child enter there usually, and the children feared the place. It was low and dusky, windowless, fragrant with herbs that hung drying from the crosspole of the roof, mint and moly and thyme, yarrow and rushwash and paramal, kingsfoil, clovenfoot, tansy and bay. There his aunt sat cross-legged by the firepit. (3–4)

This minute rendering is interesting for a number of reasons. Darkness and fear are the predominant notes, to the point that even the conventional association between women and children is broken: the aunt has none herself, and the village children are afraid of her. It is worth observing also the sense of profusion that the description conveys. The list of hanging herbs goes beyond what is needed to render the essentials—that is, a dark home with stuff dangling from the ceiling—to specify each plant in detail. And they are a motley bunch indeed, encompassing real herbs like thyme and mint, inventions like "rushwash" (which is apparently used to make tea in Earthsea), and borrowings from J.R.R. Tolkien ("kingsfoil," also known as "athelas," is applied to Frodo's wounds from a Nazgul sword). The blending of very concrete elements contributes to the feeling that a woman's place must be close to the earth, fertile and also menacing.

The negative tone conveyed by this woman's space is only reinforced as the aunt takes a greater interest in her nephew's talents for magic. There are hints of shady self-interest in her sudden concern for her Ged, culminating in the novel's first articulation of the Earthsea principle that "woman's magic" is untrustworthy (6). Le Guin temporizes, to be sure, unwilling to cast the village witch as an evil character, yet the novel goes into some detail about the "dubious ends" of some of the sorcery applied by the aunt (6), and her inability to teach Ged properly and fully is emphasized.

By way of contrast, it is useful to look at the neighbouring descriptions of the house of Ogion the Silent, the true wizard who takes on Ged as an apprentice after his aunt can teach him no more. Austerity is the keynote here. Ogion's place is on the brow of a cliff overlooking the sea, buffeted by hard winds, and the decor is correspondingly abstract and spare:

> The mage's house, though large and soundly built of timber, with hearth and chimney rather than a firepit, was like the huts of Ten Alders village: all one room, with a goatshed built on to one side. There was a kind of alcove in the west wall of the room, where Ged slept. Over his pallet was a window that looked out to the sea, but most often the shutters must be closed against the great winds that blew all winter from the west and north. In the dark warmth of that house Ged spent the winter, hearing the rush of rain and wind outside or the silence of snowfall, learning to write and read the Six Hundred Runes of Hardic. (21)

In sum, the contrasting descriptions of place at the very beginning of the novel play into the larger contrast, very important to the world of Earthsea, between "woman's magic"—the object of proverbial mistrust—and the "true" magic practised and taught by men. Ged's move from his aunt's house to the house of Ogion is a story of progress, of leaving behind women's things and setting out on the path to manhood.

As its imaginary geography expands, *Wizard* complicates the schematism implied in the early contrast between the aunt's demarcated female space and the unmistakably masculine habitat of Ogion. The aunt's house may be depicted as a questionable place, but it is always clear that it is *her* place. There are, however, other places whose fundamental appurtenance is ambiguous or misleading. One major example of this kind of space is the house of Vetch the wizard on Iffish; another is the Court of the Terrenon where Ged recuperates after a harrowing fight with his Shadow.

Ged meets Vetch at school, but his travels around Earthsea lead him to reconnect with his old friend on his home island. Vetch lives there in a comfortable house with his younger brother and his sister Yarrow. It is useful to note first of all that the sequence leading up to Ged's visit has been carefully crafted to enhance the sense of the hero's solitary, threatened existence. It is winter, and he is travelling alone across the sea in search of his double, his Shadow. He has landed at islands where he has been made unwelcome and has settled for the mercenary hospitality of inns. His walk through the town of Ismay, described immediately before his meeting with Vetch in the street, stresses the exile's wistful glance into the warm, well-lit interiors of other people's domestic spaces.

It is from this dreary set of circumstances that Ged is retrieved by Vetch's invitation to come home. Indeed, Vetch repeatedly emphasizes the words "come" and "home" on the occasion: "'In trouble and in darkness you come to me, Ged, yet your coming is joy to me.' Then he went on in his Reach-accented Hardic, 'Come on, come home with us, we're going home, it's time to get in out of the dark'" (183).

Once the family party gets back to the house the locale and its contents are described in some detail. Says the narrator:

> The house was spacious and strong-beamed, with much homely wealth of pottery and fine weaving and vessels of bronze and brass on carven shelves and chests. A great Taonian harp stood in one corner of the main room, and Yarrow's tapestry-loom in another, its tall frame inlaid with ivory. There Vetch for all his plain quiet ways was both a powerful wizard and a lord in his own house. There were a couple of old servants, prospering along with the house, and the brother, a cheerful lad, and Yarrow, quick and silent as a little fish, who served the two friends their supper and ate with them, listening to their talk, and afterwards slipped off to her own room. All things here were peaceful, well-founded, and assured; and Ged looking about him at the firelit room said, "This is how a man should live" and sighed.
>
> "Well, it's one good way," said Vetch. "There are others." (184–85)

The interior with its "homely wealth" is most immediately noticeable, combining as it does the spoils of a sea-trading ancestor's voyages with an array of practical and aesthetic items that signify not so much the lordship of the wizard-owner as the domestic talents of his sister. The father may have been responsible for outfitting the home, but the details are either explicitly associated with the daughter ("Yarrow's tapestry-loom") or very likely included in the household repertoire as reflections of her character and activities—the novel does not include hints about the manliness of harp-playing, for example. On this showing, Vetch's house is very much Yarrow's space. But Yarrow serves, listens, and disappears: so a decor that appears tailored to fit her female habits and inclinations is revealed to be in fact a proof of the comfort and self-esteem of the ruling male. Here we find a man who is "lord in his own house," who lives as "a man should live."

Vetch's house, then, is ambiguous. It is filled with signs of female possession and management, but these represent a subordinate sphere. This pattern can be seen, albeit with important variations, in another domestic interior from *Wizard*: the castle of the Terrenon where Ged faces the temptation to consult an ancient stone of power.

The Court of the Terrenon, a refuge for Ged after a battle with his shadow on the forbidding island of Osskil, is presided over by Serret, an alluring woman whose aged husband, Benderesk, is a faded presence only nominally in charge of the household. Despite its isolation and its harsh natural setting, the castle abounds in unfamiliar luxury: down-filled mattresses, satin coverlets, silk tunics (128). This insistence on comfort and high living suggests conventional femininity, and indeed the Court is presented as answering above all to Serret's tastes and commands. She greets Ged, attends to his needs, does the honours of the house.

A jarring note is sounded, though, by the pervasive cold: the coddling and the luxury Ged experiences are mostly visual, and he finds himself chilled to the bone, depressed, and confused. There is enough literary precedent for the threats posed by beautiful and mysterious hostesses that readers must process both the luxury and the femininity of the castle not as signs of rescue or haven, but as a seductive danger. As the narrator points out of Ged's reaction to his time there: "The beauty of the Lady of the Keep confused his mind" (133). This confusion persists even when the lady makes her true plans for Ged more explicit: she would like him to call on the services of the Terrenon, an ancient paving stone imbued with overwhelming power that is hidden in the foundations of the Court.

As things turn out, Ged, the reader, and even Serret herself seriously overestimate her control of this space: her husband senses that she is plotting with Ged to exploit the powers of the Terrenon, and when she turns herself into a bird in order to escape Benderesk's awakened wrath, his own flying vengeance creatures hunt her down and kill her. For his part, Ged transforms into a hawk and flies to safety back in Gont, to the house of his old master Ogion. The conclusion about the Terrenon sequence, then, has to be that masculine power is eventually reasserted over female wiles. Benderesk, the antiquated lord of the Keep, has actually been pulling the strings all along.

The aunt's house, the peaceful home of Vetch and Yarrow, the Court of the Terrenon: these examples represent two major possibilities. First, there is the wholly female space, and second, there is the space that belongs in appearance to a woman but is in fact controlled by a man. To round out this tour of women and space in *Wizard* it may be instructive to add a third variant: the case where a woman shows up unexpectedly or inappropriately in a man's space. This kind of intrusion (for, as I will show, an intrusion it is) occurs overtly in the novel when an aristocratic wife dines at Roke School, and in a more oblique fashion when a young woman tempts Ged to make his first ill-judged spell of transformation.

The uniform masculinity of the wizard school is interrupted in Chapter 3 by a very rare spectacle. A visiting lord arrives with his wife, a young and beautiful woman:

> That night the Lord of O was a guest of the school, himself a sorcerer of renown. He had been a pupil of the Archmage, and returned sometimes to Roke for the Winter Festival or the Long Dance in summer. With him was his lady, slender and young, bright as new copper, her black hair crowned with opals. It was seldom that any woman sat in the halls of the Great House, and some of the old Masters looked at her sidelong, disapproving. But the young men looked at her with all their eyes. (59)

This female presence shakes the already unsteady peace among the students, as Ged's nascent rival Jasper takes advantage of the lady's attention to show off some of his best magic. She is charmed—the narrator records that she "cried out with pleasure" (60)—but Ged's jealousy is patent, and it becomes ever more clear that he and Jasper are on a collision course. Their conflict reaches a climax when Ged is induced to cast a powerful resurrection spell that, in addition to achieving its official purpose of bringing back the legendary lady Elfarran, releases his fateful Shadow into the world.

This big mistake has deep roots in Ged's character and in his past. Most important, the episode of the Shadow builds explicitly on a precursor incident from the period of Ged's apprenticeship with Ogion the Silent. While out gathering herbs for his master, Ged meets a girl, the daughter of the Lord of Re Albi, with whom he converses over a term of several days. The girl is curious about his magical background and teases him about the scope of his powers. In response to this teasing Ged foolishly attempts an advanced spell that brings forth a threatening shadow creature—a clear anticipation of the major disaster on Roke Knoll. Even though the conversations between the two young people have taken place in a public space, a meadow near Re Albi, it is fair to interpret the young woman as an interloper. Her will if not her body makes its way into Ogion's house as Ged retreats there to concoct the spell that she persuaded him to try, and the incipient error is set right only when Ogion comes back in thunder to reclaim his own hearth.

It is remarkable how consistent—or repetitive!—Le Guin's rendering of these temptations is. The encounter with the girl in the meadow is constructed almost exactly like the numerous cumulative rubs that Ged experiences with Jasper when he becomes a student on Roke. Le Guin's evident goal in both instances is to provide a narrative pretext for Ged's rash sorcery. Goading is the central issue here, as well as the ambiguous circumstances in which goading may be offered and understood. The represented dialogue, especially where Jasper is concerned, might appear innocent enough to a forgiving observer, but to Ged the words are always interpreted as challenges. In both cases—at Re Albi and on Roke—Le Guin plays with the qualities of the provocation but never loses sight of the essential conceit that makes Ged susceptible to it. His buttons are easily pushed, and the button-pushers he meets are very good at their job. The result in both instances is overreach and all its evil consequences.

The temptation to parade his skills in front of or on account of a woman, then, is depicted in *Wizard* as one of the most threatening obstacles in a young man's path to true wizardry, and such temptations arise mostly when women are not respectful of masculine space. The Lady of O is strictly speaking out of place in the man's place of the school, just as Elfarran is out of place in the world of the living and the daughter of the Lord has no place in the wizardly activities that occur in Ogion's house. Overall the effect of this interlinked sequence is to suggest that the maintenance of gendered space is crucial to the balance of Earthsea. Men have their place, women have their place, and it is not good for women at least to trespass on the territory of the other sex.

The ambiguity of these complex sequences is worth keeping in mind when considering the second volume of the trilogy, *The Tombs of Atuan*. Here the central location is the warren of underground tombs where Tenar, the chosen priestess of the local Kargish cult, presides. The Tombs are set away from the rest of Karg society by an extensive desert, and only women take active part in the religious observances there. Tenar, therefore, has only female companions, exercises a peculiar female cultic power, and dwells in an almost painfully symbolic space: a dark womb underground, inhabited by the Old Powers who predate even the masculine magic of Ged's people. "In all the world," says the narrator, "she knew only one place: the Place of the Tombs of Atuan" (11). And later on she refers to "the precincts of the Place—that was all the name it had or needed, for it was the most ancient and sacred of all places in the Four Lands of the Kargish Empire" (13-14). The repetition of the word "place" emphasizes the sheer centrality of location, a centrality that is enhanced by details like the deep connection between the Priestess and the Tombs. Unlike a wizard, for example, she cannot take her powers with her; her functions are not portable at all. She has to live right by the tombs and cannot leave for any reason. Worse, once renamed in the traditional dedication ritual Arha has no identity outside that of priestess dedicated to the place. Her birth name, Tenar, is taken away when she is consecrated to the Nameless Ones, and she recovers it only when Ged, diviner of names, restores it to her.

The plot of the novel turns on the coming of a male intruder to the tombs: Ged, on a quest to find the lost half of the Ring of Erreth-Akbe, whose retrieval will apparently regenerate the government of Earthsea. Tenar captures Ged and imprisons him in the tombs, only to be convinced eventually that he is her friend and that her best chance of a decent life is to side with him and escape. The requisite feats are successfully carried out, and Tenar is liberated: she and Ged make their way overseas to the Inner Lands and freedom.

What to make of this addition to the trilogy's repertoire of spaces? On the one hand, this second volume clearly marks an inversion of the spatial dominants of the first. It is a novel about a woman's space *par excellence*. Furthermore, it is told mostly from the point of view of Tenar, with only a brief late interlude where Ged explains to her how he came to her shrine and what his mission is. Other aspects of the novel's mythology play into this sense of women's prominence. The history of the Karg lands, as recounted to Tenar by her guardian Manan, is the story of how the once powerful Priestesses were the arbiters of disputes among kings and chieftains: effective secular rulers, in fact. Over time their influence has waned, and now the place of the Priestess of the Tombs is secondary to the sway of the masculine Godkings and their own religious conventions. As Manan puts it:

> "Long ago," he said, "you know, little one, before our four lands joined together, before there was a Godking over us all, there were a lot of lesser kings, princes, chiefs. They were always quarreling with each other. And they'd come here to settle their quarrels. That was how it was, they'd come from our land Atuan, and from Karego-At, and Atnini, and even from Hur-at-Hur, all the chiefs and princes with their servants and their armies. And they'd ask you what to do. And you'd go before the empty throne and give them the counsel of the Nameless Ones. Well, that was long ago. After a while the Priest-Kings came to rule all of Karego-At, and soon they were ruling Atuan; and now for four or five lifetimes of men the Godkings have ruled all the four lands together, and made them an empire. And so things are changed. The Godking can put down the unruly chiefs, and settle all the quarrels himself. And being a god, you see, he doesn't have to consult the Nameless Ones very often." (27)

The waning of female power in the Karg Lands is made evident not only in the relative unimportance of the Priestess, but also in the decrepitude of her stronghold: the Hall of the Throne where great ceremonies take place is falling apart and will eventually collapse altogether when Ged and Tenar leave the underworld for freedom. In Tenar's impatient description, the hall has "holes in the roof, and the dome cracking, and the walls full of mice, and owls, and bats" (28).

The psychological weight of the bond between Priestess and place runs parallel here to a very interesting stylistic feature of Le Guin's prose. Her passages about the interior of the labyrinth and the tunnels under the Tombs almost resemble descriptions from a *nouveau roman*: mathematical precision and the absence of visual cues are noteworthy features of this style. Overall it presents a real challenge to the reader's powers of visualization, as the topography is both immensely detailed and somewhat improbable. There is a complex network of tunnels and chambers directly underneath the tombs and the temples, and an equally complex system of doorways and spyholes corresponding to them on the surface. There are inner chambers upon inner chambers, sacred retreats within sacred retreats, and a variety of secret rooms whose connections with one another are never easy to realize. Perhaps this is why Le Guin usually has her protagonist negotiate the tunnels in darkness: it is actually easier to describe the number of paces Tenar takes than to deliver some sense of the different appearances offered to the eye. In any case, descriptions of the tunnels and caverns take the form of listing the number of steps the explorer must take to get from one room to another, and underscoring tactile features like iron door surfaces or telltale draughts that betray the proximity of a pit or an open space. Tenar's instructions to Ged about the way to safety supply a typical case: "Go back along the river wall to the second turn. The first turn right, miss one, then right again. At the Six Ways, right again. Then left, and right, and left, and right. Stay there in the Painted Room" (84). Since it seems unlikely that Le Guin wanted to regale her officially Young Adult audience with extensive homage to Alain Robbe-Grillet, one might conclude that the numerization of the space of the Tombs is intended to draw further attention to its spatiality *per se*, and to cement the connection between the female occupant and her place by adverting forcefully to Tenar's sheer mastery of the system. She has counted every step of her place, and owns it intimately.

Sympathetic critics like Holly Littlefield have built on the woman-centred foundation of this novel to argue that it is in some sense feminist. As Littlefield writes of Ged and Tenar:

> When they meet in the underground tombs it is she who is in control, not he. When a fifteen-year-old girl single-handedly manages to outwit, entrap, and control the most powerful wizard in the land, it should be obvious that she is not a simpering, helpless female needing some knight in shining armor to rescue her. In fact, her strength is in many ways equal to or even greater than Ged's.

This point is well put; and to Littlefield's appreciation of the importance of a major female character we may add some other details. Surrounding Tenar in the Tombs are an assortment of powerful and vividly described women: Kossil the skeptic, who serves the gods out of a sense that political power hangs on respecting the public worship; Penthe the pragmatic novice, who doubts the value of being a priestess at all; Thar, the aloof but well-meaning instructor of Tenar's apprenticeship. Like the other two volumes of the series, this one features a sea voyage; but while *Wizard* and *The Farthest Shore* unite male companions in this quintessentially masculine undertaking, in *The Tombs of Atuan* it is Ged and Tenar who take ship together, crossing over from the Karg lands to Havnor in his little boat. A small dent is made in the convention that boats and sea voyages are for men.

Still, this story obviously is a variant on the age-old theme of the rescued damsel, and it concludes with collapse of the cave system where Tenar presided as priestess, the undoing (presumably) of the more female-oriented Old Powers, and the appropriation of a Kargish woman to the territory and the purposes of the men of the Inner Lands. Even the talismanic ring that Ged rescues from the tombs has a masculinist overtone: it will serve to restore the long-departed King to his throne in Havnor. Atuan is in some senses the Court of the Terrenon writ large: a woman's space is central to the story, but it ends up belonging in essence to a man.

The Farthest Shore, the final volume of the initial trilogy, presents very different issues. The most prominent location in the novel is the boat *Lookfar*, home to Ged and his youthful ward Arren as they roam Earthsea looking for the cause of the lapse in vitality that forms the major obstacle or challenge for this plot. The spaces of the novel are in fact quite varied: there is the masculine Great House on Roke, the masculine boat, the more or less inclusive rafts of the ocean-dwelling Raft People, and the dragons' zone, which is hinted to be beyond gender entirely. Most of the characters are male, but the women include Akaren, a mad sorcerer from the silk island of Lorbanery, and the career-changing witch from Hort Town who has given over magic and now runs a haberdashery stall. The attribution of the new evil that afflicts Earthsea to a male antagonist helps soften the earlier volumes' suspicion of women's power, and overall we miss the sense that gender relations play an obtrusive part in the structure of the society and its magic. It is almost like a different world, as if Le Guin had been paying close attention to gendered space in the first two volumes and then made a deliberate switch in the third.

What might be the reason behind this apparent change in effect? A clue may lie in the sense of emptying out that comes as Cob's magic drains energy and life through what is vividly described as a hole in the world. All over Earthsea people, both male and female, are losing their powers, their distinctiveness, to a feeling of overwhelming ennui. Ostensibly this ennui is tied to the promise of eternal life circulated by Cob, but to the adult reader at least the novel's descriptions of depressed affect and drug addiction suggest less abstract afflictions. The foremost effect of the change is a kind of forgetfulness: wizards can no longer utter their spells, singers stop in mid-melody, craftspeople stare at their tools in incomprehension. Ged's quest is to discover why these changes are occurring; whatever their efficient cause, though, in *The Farthest Shore* the loss of real humanity is connected to the loss of the distinctions that define human difference. Furthermore, among these precious distinctions we must include gender itself.

This effacement of difference is seen most clearly when Ged and Arren find their way to the land of the dead, the "Dry Land" as it is known in Earthsea. It is telling that the climactic sequences of the novel involve a journey to a place, and that here death is represented in terms of an imaginary geography. The land of the dead in fact has walls, some sort of city, a dried riverbed, and, perhaps most important, a forbidding frontier range known as the Mountains of Pain. Ged and Arren get there by stepping over the wall, following Cob to the foothills, and, after Ged has dealt with the evil wizard's plan and sealed the hole in the world, climbing all the way up the mountains until they emerge in the dimension of the living, back on the Earthsea island of Selidor.

So the Dry Land does have its equivalent of an earthly topography. Nonetheless, when Le Guin tries to represent its metaphysical and emotional flavour she uses language that suggests the very opposite of an ordered and intelligible domain. Obscure and dreary, this place is described as "formless" (220). Moreover, "in this timeless dusk there was, in truth, neither forward nor backward, neither east nor west, no way to go" (223). "There was no horizon" (223). Repetitions of "no" multiply in the pages devoted to this land, flattening all binaries, sinking all the living difference of the upper world in an undistinguished monotone, signalling not positive torment but nothingness. Perhaps most significantly, people are as much affected as things: they are mostly described as a generic group, "figures," or simply "the dead" in their gender-neutral form. Passive constructions are called upon to convey not just the absence of activity but the absence of any connection between an activity and a human actor: "There was no buying and selling there, no gaining and spending. Nothing was used; nothing was made" (220). The personal distinctions among the dead have been removed: mothers no longer recognize their children, and "those who had died for love passed each other in the streets" (221).

It is fitting that the force responsible for this erasure is known not as the Destroyer but as "the Unmaker."[1] And it may be that one of the book's lessons concerns the positive status of gender difference as proof that at the very least a gendered human is alive and involved with the world. In connection with this point I would like to recall a sequence from Le Guin's science fiction novel *The Left Hand of Darkness*. In it the earthman Genly Ai is sent to a labour camp run by the Soviet-style alien nation of Orgoreyn. Normally the alien Gethenians are androgynous beings who go into an estrus-like state once a month, but in the labour camp the prisoners are given drugs to suppress the hormones that govern this process of change. The goal is to prevent the periodic emergence of sexual feeling among the prisoners, and more generally to keep their mood down. Genly is deeply impressed by the depression he sees around him and processes it as a departure from the vigorous androgyny of the Gethenians in their normal state (in one of his more obtuse moments, he even says that the prisoners remind him of women). In any case, it is noteworthy that in a text roughly contemporary with *The Farthest Shore* Le Guin entertained a similar vision of people reduced to listlessness by the removal of their own version of sexual difference. In both fictional worlds a space that belongs to no one, where the effacement of gender is part of a general suppression of all the living difference of the world, is Hell indeed.

One thing that many readers have concluded about gendered space in Earthsea, I think, is that these novels do open themselves to charges of sexism. In fact the response to all of Le Guin's work from her early period has included reservations about its limitations in that area, and it is now accepted that this feminist-inspired criticism profoundly affected Le Guin's career thereafter.[2] As Amy Clarke has pointed out, there is a pause in the writer's productivity following the wonder years of the late 1960s and the early 1970s, a pause characterized by a great deal of self-reflection particularly where questions of gender are concerned. So it is perhaps not surprising that in the wake of this process Le Guin should have decided to revisit Earthsea in order to recalibrate some of its more evident imbalances. The result was a fourth volume entitled *Tehanu*, which appeared in 1990, and then a collection of short stories (*Tales from Earthsea*) and a presumably final novel, *The Other Wind*. I do not have the space here to discuss this work fully, but will mention only *Tehanu*'s new approach to place in order to reinforce my point that the association of places and gender is indeed a meaningful part of even the original trilogy.

Tehanu is an astonishing palinode, a retraction of practically every significant characteristic attributed to Earthsea in the first three books. Le Guin set out to show the supposed underbelly of her imaginary society: kindly wizards transform into sadistic beasts, the loose organization of the Earthsea polity is recast as a cruel anarchy where bandits and warlords have their way, and romanticized conflict, like the Karg raid described at the beginning of *A Wizard of Earthsea*, is replaced by crimes as horrible as the rape and burning of a little girl by her own family. As the critical problems posed by the relations between this work and its predecessors are severe and complex I will not begin to rehearse them here: but I will round off my discussion of gendered place by noting that even in the specialized domain of spatial representations *Tehanu* turns things upside down. For example, Tenar, the young Priestess who left her Karg home in company with Ged at the end of *The Tombs of Atuan*, reappears as a major character in Tehanu. She has settled in Gont, married a farmer, and after her husband's death controls his house and lands. This looks like a reversal of the journey that led her away from the powerful ownership of place that characterized her girlhood and towards an exile's displacement in the Archipelago. At the same time, her control of the farm is threatened by the return of her son Spark, who has a legal claim to the land. Tenar and her protégée Therru travel freely around Gont, but one of the novel's most unusual features is the great suspicion with which Earthsea people regard wanderers of any sort. Open space belongs to bandits, thieves, and untrustworthy people generally, while Tenar's farm is also something of a fortress. It is in fact besieged at one point and has to be protected by force of arms. By the end of the novel, however, the process of change has gone far. Some inveterate male wizards have been defeated, a new King is doing wonders for public safety all over the Archipelago, Ged has united with Tenar in a truly egalitarian relationship, and the existing spatial hegemony is beginning to be shaken up. The most striking example of this kind of progress is to be seen in the fate of one domestic space: the house of Ogion the Silent. This austere and masculine home of Ged's first real lessons in wizardry is a major location for the new novel, and it goes through some significant changes in ownership. Ogion still lives there at the beginning of the story, but after his death he leaves his house to Tenar, her adopted daughter Therru, and the good-hearted witch Aunty Moss. The manly taciturnity of the wizard's study hall is enlivened by the clucking of chickens, the buzz of a full domestic life, and, most important, the recovered voices of women. They have asserted their place at last.

―――――

Notes

1) Coinages using "un" are a favourite with Le Guin, who described the "untrance" practised by religious Foretellers in *The Left Hand of Darkness* and expressed the ambition of her scientist-hero from *The Dispossessed* as the desire to "unbuild" the walls, both literal and figurative, that keep his people from achieving their full potential.

2) General accounts that encompass a range of Le Guin's work from the late 1960s through the 1970s, usually with a special emphasis on *The Left Hand of Darkness*, include Russ, Larbalestier, and Lefanu. Critics who focus more on Earthsea itself are Comolletti and Drout, Donaldson, Hatfield, Nodelman, Rawls, and Robinson. It is safe to state that when critics deal extensively with gender in the novels they are alert to the problems of gender politics that they exemplify. Critics whose emphasis lies mostly on the structure of the coming-of-age tale or on questions of audience—Elizabeth Cummins and Mike Cadden are examples—tend to pay less attention to gender issues.

16) Dancing and Hinting at Worlds in Theatre for Young Audiences

❧ HEATHER FITZSIMMONS FREY

The experience of being in the audience when a play or an opera is being performed is not simply passive. It's not like watching TV; it's not even like going to the cinema. Everyone in that big space is alive, and everyone is focused on one central activity. And everyone contributes. The actors and singers and musicians contribute their performance; the audience contribute their attention, their silence, their laughter, their respect. And they contribute their imagination, too. The theatre can't do what cinema does, and make everything seem to happen literally … But the limitations leave room for the audience to fill in the gaps. We pretend these things are real, so the story can happen. The very limitations of theatre allow the audience to share in the acting. In fact, they require the audience to pretend. It won't work if they don't. But the result of this imaginative joining-in is that the story becomes much more real, in a strange way. It belongs to everyone, instead of only to the performers under the lights. The audience in the dark are makers too. (Pullman, "Theatre")

One way theatre practitioners appeal to the imagination of the audience in order to transform the tangible geography of the performance space into an imaginative geography is through *dance*. Hundreds of thousands of Canadian children[1] see Theatre for Young Audiences (TYA) each year,[2] and the productions often include dance. Dancing, performing bodies are a tool directors may use to define and map out on-stage worlds for two major reasons: first, to construct an imaginary geography on stage that the audience could not otherwise access; and second, to help the audience navigate the fictional space. In Aïda Hudson's introduction to this book she explains that an imaginative geography relies on one's perspective and on what is *seen* in the mind's eye. She also notes that geography is "imaged earth writing"—a way of defining, describing space. I became curious as to how dance hints at a geographically other world that young spectators can see and even *experience* for the duration of the performance, and why dance has the potential to be effective not only at creating worlds but at dismantling barriers as well.

How could a dancing body express imaged earth writing? And how could young spectators access that dynamic way of mapping space? To learn more about ways artists believe dance functions in their productions, I conducted interviews with eighteen directors and choreographers who work with theatre for young audiences, from six different provinces (Fitz-simmons Frey, "Dance" 1).[3] Although the nature of the projects the artists were working on influenced the direction of the conversations we had, my interviews included questions such as these: "How would you describe the dance, the dance form, and the dance creation process in some of the shows you've worked on?"; "Why is dance included in those productions and how does dance function dramaturgically?"; "How do you feel children respond to dance aspects of productions?"; and "Explain your ideas about the value of dance in theatre for young audiences." The interviews revealed that in theatre for young audiences, directors, choreographers, and curators do not share a common definition of dance, physical theatre, or movement. Clearly this is another research direction. For the purposes of this chapter, if the artists I spoke to considered the work to be "dance" or to contain "elements of dance," I accepted their labelling.

Edward Said's concept of imaginative geography operates quite transparently on the theatrical stage. Through performative suggestion and an audience member's willingness to, as Pullman puts it, "pretend these things are real" for the duration of the play, bordered by the area that spectators and actors mutually agree is the performance space, "distinctive objects are made by the mind, and ... these objects, while appearing to exist objectively, have only a fictional reality" (Said, *Orientalism* 54). Said argues that the agreed upon fictional reality can establish "us" and "them" boundaries, designating difference and "Otherness." To illustrate ways that imaginative stage worlds built by dance have the potential to render previously imagined boundaries as porous, simultaneously constructing and complicating an audience's conception of "Otherness," I look to TYA dance moments that may be illuminated through ideas of kinaesthetic empathy and space as a process. One performance takes the audience on a journey; in a second, characters move in and out of a dream world. The third example comes from work by a theatre company that draws spectators into several very real but also distant and possibly unfamiliar geographies. The fourth piece uses movement in ways that challenge audiences to imagine ways of inhabiting place and space that goes beyond the specificities of time.

ᖇ How Might Children Experience Earth-Writing Through Dance?

THE THOUGHT EXPERIMENT of connecting elements of these theoretical concepts together suggests a balancing act to me, so I have imagined them as a tower of ideas, with each major idea represented by a single block. Blocks are placed one on top of the other so that each idea supports the next. The foundation of my tower is *empathy*—specifically, the way children empathize. According to aesthetics philosopher Ellen Handler Spitz, empathy is derived from ancient Greek. The prefix *em-* means "in," and the root *pathos* means "experience." "Em-" is transitive, so it functions like *into*, as in "the outside enters in." In other words, Spitz argues, if someone is empathetic, that person is entering *into* someone else's experience (Spitz 546). Idiomatically, from the acting and cognitive psychology perspective, Bruce McConachie simply says that empathy is stepping into someone else's shoes (McConachie 99) and immediately complicates (but does not eliminate) ideas of imaginary "Otherness." "Feeling overrides cognition," argues Spitz (549), and unlike an intellectual activity, "to empathize is to experience in a deeply physical way, one that causes metaphor to blossom into reality" (546).

Essential to understanding how children might experience performance as earth-writing, Spitz emphasizes that "empathy suggests immersion—saturation, if you will—in the experience of another person (or persons, creatures, or art works)" (549), and argues that that young children can have a kind of "primordial aesthetic empathy"(551).[4] Shifra Schonmann, an expert in theatre for young audiences, writes about this phenomenon in children watching theatre. She is concerned that unlike adults, children may not have developed the skill of "aesthetic distance" and may not be able to separate themselves adequately from the performance in order to appreciate it (65). Her belief is that good children's theatre respects children: it does not play them into a state of hysterical laughter, or terrify them to the point of screaming and sobbing, because those are simply indications that the children are *too* absorbed in the work and unable to appreciate it from an aesthetic point of view (65).[5] Schonmann also explains that aesthetic distance inhibits motor activity and that before children have acquired that skill, besides expressing physical responses like laughter or sobbing, they may be unable to control their own body movement (65). This means that as people learn aesthetic distance, they are better able to process their emotional responses to aesthetic experiences intellectually; until then, responses are often embodied. Regardless of whether a child's tendency to be absorbed *into* a performance is a positive feature of child audience members or whether it is negative, Spitz's and Schonmann's research projects indicate that children can be empathetic audience members—probably more so than adults. If the performance of abstract movement is respectful, and if overly taxed emotions do not factor into a child's empathetic experience, children have the potential to relate to performance in a way that is especially profound.

However, in this exploration, I am specifically interested in the way *dance* works in performance, since so many theatre for young audiences artists use dance as an earth-writing communication tool. Watching dance is different from watching some kinds of story-centred performance because it is primarily physical, and physical performance allows for the possibility of something more than *empathy*—in fact, *kinaesthetic* empathy. Researchers using neuroscience, cognitive science, and qualitative research methodologies have carried out numerous studies on dance spectatorship, demonstrating that watching dance allows spectators to simultaneously *feel* the work the dancer is doing.[6] Dance scholars like Susan Foster, dance therapy scholars like Cynthia Berrol, and dance spectatorship scholars like

Matthew Reason and Dee Reynolds have all approached these issues differently, and their insights are advancing this nascent and rapidly changing field of research. According to "mirror neurons" theory,[7] when spectators watch a physical movement, their brains synapse in the same way as if their muscles were actually jumping, spinning, contracting, or stretching. Reason and Reynolds have found that the more virtuosic the dance, the more embodied a connection to the performer some spectators experience (58). Of course, Reason and Reynolds remind us that "spectators' perceptions of dance are highly context-specific" (52); and for a wide variety of reasons related to past experience with dance, not all spectators will enjoy the same kinds of dance or enjoy dance in the same way. Nevertheless, the potential exists for a multisensory kinaesthesia that allows spectators to imaginatively enact the movement they see on stage.[8]

Connected to kinaesthetic empathy is "kinaesthetic contagion," existing between the spectator and the performer. Those mirror neurons may allow the spectator to experience and appreciate the physical sensation of *doing* a particular thing on stage, and furthermore, the connection between the brain and the body of the spectator to the brain and body of the performer means that when the spectator connects to the on-stage movement, he or she is more inclined to *feel* the emotional response that the performer is trying to portray through that physical gesture; this has the potential to break down barriers between self and Other, which is so important in Said's reading of Orientalist discourse. For example, Reason and Reynolds explain that contagion might describe those instances where the spectators passionately describe the feeling of joyful pleasure in uplifting and graceful movement (71). Spectators connect to moving bodies on stage because the movement and action can connect the spectator to emotions.

In addition, children may well be more likely to be open to an empathetic experience than most adult audience members, allowing them to be swept up by the movement and readily absorbed into the physical and emotional moment in a performance. The second block in the tower is made out of the kinaesthetic varieties of empathy. Building the tower by placing the kinaesthetic empathy block on top of the foundational empathy block, imagine that children inhabit the kinaesthetic empathy that dance can evoke more readily, and more profoundly, than adults.

Let us leave the spectators for the moment and move on to the performing body, which is the core of the next building block. Sally Ann Ness has written a provocative article, "The Inscription of Gesture: Inward

Migrations in Dance," in which she suggests that dance is not just an ephemeral art form, leaving traces in the air through which the dancer moves, but also that the gestures of dancing can be a kind of inscription or engraving on the dancer's body. As with the "em-" in empathy, Ness takes the "in-" of *inscription* to mean "into," and it is "place-seeking rather than place-being" (1). In her view, dance gestures get inscribed on the performers' body, moving into the tissues, the muscles, the bones, and are reawakened in every moment the performer draws on muscle memory to move. Ness's idea that the dancer can literally embody information reflects the bodies of professional dancers who have dedicated their lives to a rigid and technical dance form such as ballet or the East Indian classical dance Bharatanatyam. Such long-term training has a significant impact on the performing body; however, it may be possible to extend Ness's thoughts to the bodies of performers who are not classically trained but who engage in intense repetition during the rehearsal process and subsequent touring for a theatre for young audiences production, and who may also find that dance inscribes upon or at least etches upon their bodies. In this way, dance gestures become something more than a sign to be interpreted through semiotics or the spectator's imagination. As Ness says, the body is a "historically informed home for inward moving symbols, allowing [dancers] to be the ultimate interpretants of embodied knowledge" (Ness 22). Dance imprints ideas onto the body of the performer—and this engraving process on the performing body is the third block in the tower.

The final and most precarious part of my tower directly connects to the notion of imaginative geographies and demands that Ness's idea of inscribing on and into the moving body be related to the ideas of kinaesthetic empathy and space. Performance takes place in a specific real location and usually evokes an imaginary, fictitious space. Doreen Massey's book *For Space* passionately argues that space is made of relations constantly in process (9). The gathering together of multiple open-ended, interconnected trajectories produces what Massey terms a "sometimes happenstance, sometimes not-arrangement-in-relation-to-each-other" (111). She also suggests that space is *the* condition for the unexpected. Nowhere is this idea of space more true than in a theatre, which is always itself and all the geographic imaginings that are brought through it, to it, evoked in it, and hinted at there, not only by the performers but also by the complicity of the audience members whose imaginations, as Philip Pullman says, are essential to making the play happen (Pullman, "Theatre"). The special status of theatre *space* is eloquently described by Dennis Kennedy who writes: "We are bodies who occupy space and metaphorically are occupied by it ... Theatres, which are spaces separate from ordinary life by definition, affect us not only by their architecture and décor but also by the spatial relationship established between actor and spectator." (133). Artists engage audiences to reclaim and reinvent theatre space so that it is always multidimensional, multiple places at once, just like the imaginary worlds of children's play—defined, but easy to dissolve or change into something else.

In an effort to carefully balance my tower of idea blocks, I would like to explicitly connect these disparate theories. After an intense rehearsal process, the repetition of particular gestures and movement vocabulary might etch or inscribe the idea of a specific imaginary geography onto the body of the dancing performer. When the dancer performs in the theatre space, through processes of kinaesthetic empathy and contagion, spectators can experience the physical pleasures of moving by virtue of mirror neurons, as well as emotional connections associated with the intentions of the physical activity; and, perhaps, they can *experience* that geography that was engraved on the dancer after an intense period of repetition. Furthermore, through Massey's idea of space made of trajectories and relationships, the theatre space can, however temporarily, *become* that imaginary geography because of the imaginative collective creation carried out by the performers and the audience. As a result of the performer's physical movement embodying place, a spectator, especially an empathetic child, might be able to "be" there, in the world of the play. Or, to put it another way, if something abstract, like an imaginary world, can be engraved on a dancer's body, and if that imaginary world can be accessed by an empathetic child through experiencing *in* the body, empathizing with the movement, feeling the movement with the body, and, as such, perhaps *feeling* the space represented, that could allow young people to be surrounded, absorbed, sucked *in* to another space through movement on stage, facilitating child spectators' ability to complicitly create and experience another world.[9]

∾ *Dancing Imaginative Geographies*

I WOULD LIKE TO SUBJECT my tower of idea blocks to the vibrations caused by actual performance examples and the geographies they evoke. Allow me to start with dance that is also a journey, guiding the characters, and also the audience, into the imaginary world of the play. If we embark on a journey, we might imagine a map. But Doreen Massey argues that one of the problems with maps is that they are "a completed horizontality" and reduce observations to an ordered and clearly connected surface, in which dynamism of change is exorcised (Massey 12). Mapping tames space, but dance reminds us that space can be wild and that it can be inhabited wildly. As directors push the boundaries of where the body can seem to be in space, the space becomes different things for the bodies that are there. Some directors I spoke to acknowledged the literal limitations of the physical theatre space and explicitly created movement motifs to guide the audience into the imaginary world of the play and to define the space—imposing a kind of embodied map that expands the literal space of the theatre. In Cheshire Unicorn's production of "Wanda T. Grimsby, Detective Extraordinaire" (Toronto FringeKids 2011) the child detective, Wanda, has to travel to her boss's secret hideaway in order to learn about her mission. Director Melissa Major says, "we're working with a small space, but we want to create an epic journey to get to the boss's door. We had [imagined] a wall, a swinging rope, an elevator ... yeah, we put a few things in there. The journey helps children come play with us" (Major, personal interview). Yet Major did not really "put a few *things in* there"; rather, her performer defined the space and the associated geographic features with her body, by climbing, swinging, jumping, crawling, and rolling over invisible obstacles, inviting the young audience members to navigate the imaginary three-dimensional map with her, and, in the process, drawing them into the imaginary world. "[We were] using that movement sequence in particular in that way," Major says, "because it transports them there, it plants a seed that they can take in their imagination, and enter along with us into the world of the play" (Major, personal interview). Major sought to expand literal space and also to control and define space through the performers' bodies, and the movement created signposts pointing to particular imaginary spaces.[10] As mentioned, Massey suggests that space is *the* condition for the unexpected, so does establishing a temporary space map that simultaneously lives in the bodies and minds of the performers along with the imaginations of the spectators artificially combat the ambitions of space to constantly change? An imaginary "other world" space inside a theatre might actually *be* unexpected (it certainly defies literal thinking, although Kennedy suggests that theatre spaces are partly defined by the shared cultural attitudes of the builders). Furthermore, an embodied map is impossible to read without movement, and without a spectator's own historical trajectory—an embodied map defies a static horizontality and *is,* in fact, change. If child audience members can empathetically and kinaesthetically access that temporary map, the embodied map may allow them to feel more rooted in the story, to feel like they can know it with their bodies.[11]

Navigating space had a different meaning for Clare Preuss, the choreographer for "Monster Under the Bed" (Young People's Theatre 2010). She wanted to use moving bodies and embodiment rules to demonstrate when characters were in "the dream world" of the monster or in the more pedestrian real world (Fitzsimmons Frey, "Dance" 34). "Monster Under the Bed" by Kevin Dyer is about a boy who needs to face up to his fears. In the process of doing just that, he switches places with the child-monster living under his bed. Preuss's "science fiction" approach to choreography meant that she created rules about the ways bodies could interact with and within that monster dream world space: the texture of the air and even gravity had unusual qualities in the dream world. Performers moved completely differently depending on the world they were inhabiting, and the only physical clue to *locate* the performers was the performing bodies' movement style. Even though they were consistently heightened, movement qualities in the dream world contained a lot of variation. Performers worked with the idea of viscosity when they were "underwater," and with variations in tempo to explore different aspects of flight (slow, suspended movements for soaring, and light running through space to create the speed of flight). Besides being heightened, the movements were emotionally charged, variously evoking menace, wonder, and curiosity. Each actor had nuanced interactions with the dream world that incorporated the "rules" of the space and their understanding of their character. The actor playing the child explored a wide range of bodily extensions, and maintained a "sense of flow" or sustained quality to his movements that was reminiscent of martial arts; while the actor playing the father was heavier on his feet, and, as Preuss put it, there was "lots of weight and size to him with a punchy, direct feel" (Preuss, personal correspondence). Artistic director Allen MacInnis of Young People's Theatre initially wondered if the decision to define space through body movement might confuse young audiences, but in fact, audiences instantly connected to the characters and place through the style and quality of movement. He explains: "One of the advantages we have with young audiences is that they are ready to use their imaginations ... We have to allow them to see the world in a different way. So [for example] the body can show me where I am rather than a set change" (Fitzsimmons Frey, "Dance" 34). MacInnis goes on to muse that while young people are fascinated by the process of meaning-making and while the meaning they find in dance does not have to be literal, "if they start to suspect that their presence in this live event doesn't matter, they disengage. If making some kind of meaning out of it doesn't matter, they disengage. [I have felt] them give up" (35). But if they know that they need to be complicit in the way that Pullman describes in the opening quote, they can make meaning and even co-construct worlds as they connect with the performers. Evoking the idea of kinaesthetic empathy again, Preuss told me that "dance allows young people to use their imagination differently and to be more in their bodies" (37). For movement to act like a compass or locating device, it does not have to define and map space, as it does in a play that uses the idea of "journey" like "Wanda T. Grimsby." Alternatively, it can enhance the spectator's ability to imagine, experience, and even enter a world that characters inhabit in unusual ways because bodies operate differently there.

Staging journeys using movement to define the space can draw audiences along for the trip, and inviting audiences to enter an imaginary or dream world is like inviting a friend to play "let's pretend" as you collectively create a world together. But sometimes the landscape a director needs spectators to imagine may be a real, specific, but probably unfamiliar world, and dance is one tool some directors use to transport the audience to a faraway place. Using the dancing body to write the earth on stage in these contexts draws attention to complexities Said raises when he argues that the imaginative construct of the "Orient" is based on dichotomizing beliefs about "us" and "Other." Dance can be a marker for an unfamiliar place while simultaneously inviting audiences through, kinaesthetic empathy, to cross boundaries and become a part of it.

Director Lynda Hill regularly uses dance as an invitation to enter an unfamiliar world. For example, with choreographer Lata Pada, she included Bharatnatyam (classical Indian dance) in Theatre Direct's production of *Beneath the Banyan Tree*, which drew young audiences into Anjali's experience of difference as a new immigrant in Canada as well as into her reimaginings of stories from the *Panchatantra*. In this chapter, I will discuss two other examples from Hill's repertoire at Theatre Direct: *Sanctuary Song* and *Binti's Journey*.

"Theatre is storytelling but dance isn't only storytelling—it is much more," Hill explains (Fitzsimmons Frey, "Dance" 35). Theatre Direct's production of *Sanctuary Song* by Marjorie Chan (2008, 2012) is an opera based on a true story about an elephant named Sydney. In the story, Sydney's mother is probably killed by poachers, Sydney is separated from her herd, and she has an incredible journey that includes working for a circus, travelling on a ship that catches fire, living in a zoo, and finally reaching an elephant sanctuary in Tennessee, where she unexpectedly reconnects with her friend, Penny, from her original herd. Hill uses dance in a touching scene of "play" between two elephant friends, Sydney and Penny, when they were still living together in the wilds of Southeast Asia. The elephants dance to music composed by Abigail Richardson. Written in an operatic style inspired by elements of Asian music, it is a form that is probably aurally unfamiliar to most children. Hill explains that the dance in the production is "inspired primarily by the graceful movements of the elephant" and that the shapes of the arms, hands, and fingers are inspired by "the more ornamented vocabulary of Indian classical and Southeast Asian dance forms like Thai and Balinese" (Hill, personal correspondence). While many child audience members would not be familiar with these forms of dance, the scene is framed by and rooted in the idea of *play*—something with which children are entirely familiar, and which offers them a way to enter the space created by probably unfamiliar music referencing distant parts of the world. I might have expected the musical style to make some children restless and uncomfortable, but the day I saw the production (2012), children were rapt and attentive during the dance sequence: perhaps their sense of empathy was harnessed by the physicality of dance and enhanced by the way it brought the trajectories of other histories and cultures into the space. The ornamented dance vocabulary did not exoticize the characters or alienate the audience; instead the use of dance and the potential of kinaesthetic empathy may have simultaneously established an imaginary geography on stage, and then broken down the boundaries between here and there, us and them, allowing children to inhabit and experience an elephant's world.

When Theatre Direct's Lynda Hill directed *Binti's Journey*, she needed to reimagine the theatre space as the real but distant country of Malawi. The story is about Binti, a little girl who loses her mother and father to AIDS, who loses her status as a radio play star when she has to leave the city, and who even loses track of her sister. But eventually she joins her grandmother in the countryside, where she helps look after younger and even more vulnerable AIDS orphans. Each transition between scenes was a West or South African folk song and dance. The four performers sang, sometimes with arms up-raised or incorporating body percussion, and other times engaging in simple skipping that became the more complicated jumping footwork of schoolyard play. At first, I wondered if the unfamiliar singing and dancing exacerbated the distance between the Canadian spectators and the heartbreaking world of the play, in which a little girl's family experiences a reality of deprivation, poverty, and AIDS that was probably unfamiliar to many members of the audience—perhaps dance would heighten the feeling of "us" and "Other." I also wondered if the dancing and singing had the potential to exoticize the story and encourage young people to reinscribe stereotypes they might have about "poor people" who live "far away" even while continuing to be happy singing and dancing people. I saw the piece twice: once in 2012 and once in 2017. At the 2012 performance, comments during the talk-back were few, and I could not be sure what the silent children were thinking, but one child's comment—"It sure made me glad to live where I do"—suggests that child felt more pity than empathy for the characters who lived "over there." While Hill asserts that "dance is so absolutely universal. It transcends language even when it has cultural roots" (Fitzsimmons Frey, "Dance" 36), Matthew Reason's and Dee Reynolds' researches in dance and audience reception indicate that adult viewers tend to be less engaged or moved when they watch completely unfamiliar dance forms than dance with which they have some experience (57). With unfamiliar dance, adults involved in their study had more trouble accessing a sense of kinaesthetic empathy and kinaesthetic contagion, although they appreciated that the dancers were highly trained and even virtuosic. But in this case, Reason and Reynolds were not working with children, and each audience is different: during the post-show discussion I witnessed in 2017, children asked questions about AIDS, easily recalled specific lines of the script, and could describe movement happening on stage to locate points they were making. The children in the audiences for both productions I saw clearly enjoyed the rhythms and the movements of *Binti's Journey*, and some tried them out afterwards and while they were walking back to school. Through kinaesthetic empathy, did dance draw children into the world of the story? Were the dance and the music one more marker that served to make it seem as though the story were happening "somewhere else" and "far away," or did dance provide a bridge for children to access that distant world? Even if Schonmann and Spitz are correct in their assertion that children have fewer ego barriers and experiences getting in the way of their ability to become absorbed in a performance, it is possible that the movements expressed by the dancing performers evoked a distant and unfamiliar place for the spectators. Yet the intimacy of the narrative, and the probability that children would empathize with Binti, could counteract any tendency to make dance a site for exoticization. Together dance, story, and character could simultaneously locate *Binti's Journey* as distant, while potentially enabling children to temporarily inhabit Binti's world.

In my fourth example, on-stage movement and movement development processes make it possible for the director to challenge the audience to imagine ways of inhabiting place and space that go beyond the specificities of time. Artistic director Sandra Laronde of Red Sky Performance does not use the term kinaesthetic empathy, but her descriptions of her work make it clear that she anticipates audience responses to her on-stage work to engage with kinaesthetic empathy. She explains that a mark of success for her is when spectators "are so profoundly moved … [they] can't even go up to their brain to analyze, they have to stay in their bodies and in their hearts, in that feeling" (Laronde, personal interview). She believes that one reason people become moved in that way is because of a kind of knowledge transmission she calls "soul speak." She claims that "the soul speaks in pictures." Laronde is interested in how those pictures can reach an audience, and she calls this kind of dance "imagistic"—not mime, but abstract work that evokes a picture or an idea (Fitzsimmons Frey, "Interview" 2013). The *content* of those images can vary, but like mapping the space in the first examples, imagistic dance can establish imaginary geographies on stage. Laronde works with both movement and story, but she believes that movement dominates text in performance for two reasons: even in utero, we move before we experience language; and movement is *vitality* so spectators will always be drawn to that. "Vitality breathes in the body. So does culture. When [a performer] holds an image or the energy of an image in the body on stage, we see it. Even if [the audience] doesn't get exactly what it is they *feel* it" (Fitzsimmons Frey, "Interview" 92).

On stage, an imagistic connection between landscape and performers seems plausible, but Laronde says the advance process of making that connection is complicated: imagining and imaging land through the vitality of the performing body is possible, but inscribing *place* on a performing body is not obvious or immediate. Laronde's relationship with her home-land in Temagami, Northern Ontario, grounds a lot of her work, but she asks, "How do you reveal and translate that through the body? And even more interestingly, if I can transmit it through *my* body because I have lived it, experienced it, how does a dancer who has not had this experience, and perhaps, grew up in the city, how do they transmit that through their body, and can they?" (Fitzsimmons Frey, "Interview" 90). Patrice Pavis writes that actors "simultaneously reveal the culture and community where they have trained and lived, and the bodily technique they have acquired" (3). Other choreographers I spoke to who work in forms like jingju (also called Peking or Beijing Opera) and Bharatanatyam similarly mused about what might be possible where culturally specific, non-Western dance is used to create a world when a performer has only limited exposure to a particular dance form and, by extension, culture. William Lau explains that he could only hope for performers to "achieve the right look" (Fitzsimmons Frey, "Forbidden," 47), and Lata Pada says that with untrained performers she can offer only "a window onto the form" (Pada, personal interview). Laronde elaborates that since, in her experience, children recognize truth in performance, creative process cannot take shortcuts. Laronde creates an "artistic iceberg." She explains that the tip of the iceberg is all the produc-tion-oriented work like aesthetics, movement, even marketing. "And below the surface of the water lives the more expansive part of the iceberg where time is deep" (Fitzsimmons Frey, "Interview" 91). The creative process of that bigger part of the iceberg involves aspects related to space, culture, and landscape, including "a larger world connected to the unconscious, to the spirit world, and a deeper culture, where countless hours of research, experimentation ... site-specific investigation, and so forth reside" (Fitz-simmons Frey, "Interview" 91). She believes that extensive preparation is essential even if the audience doesn't see what is there, because they will connect with it on a deeper, subconscious level. The combination of "achieving the right look," opening "a window," and multi-layered prepara-tion might be able to bring culture and landscape into a performer's body. Whether or not Ness's idea of engraving (or at least etching) on the body is possible following an intense rehearsal process is pivotal to imagining how a location might be performed by bodies neither intimately nor per-sonally familiar with specific places and cultures.

For Laronde's production of Drew Hayden Taylor's *Raven Stole the Sun* (inspired by Sháa Tláa Maria Williams's [Tlingit] retelling of her father's recounting of this well-known story), performing a kind of hybridized culture is actually to the production's advantage because Laronde is exploring the idea of going beyond the specificities of time. When her company does productions for children, the work tends to be more literal than her work for adults, but she believes children love abstraction too: "audiences just need landmarks so that they know where they are" (Fitzsimmons Frey, Interview" 92). These landmarks evoke place and help the audience navigate that place, and even draw the audience *into* a "place," but the *land*marks are all inscribed on the body. *Raven Stole the Sun* describes how Raven lives in a world of darkness and is tired of bumping into everything. He meets an old man and his daughter, and through trickster cunning he discovers the man has hidden the sun, moon, and stars in boxes. His plan to release them means he has to get into their house—and to do that, he decides that the daughter should give birth to him, the Raven. The performers do traditional powwow dances associated with Indigenous peoples living on Turtle Island, and what political maps label Canada, but Raven also does "The Raven Hop"—a highly spirited dance, one that fuses the men's prairie dance "the crow hop" and the contemporary hip hop. Children love the "Raven Hop." They love the energy and athleticism, and when I saw the production in a school gym, many children had trouble keeping still as they enthusiastically bounced around on the floor. Several were unable to remain on their bottoms and went up onto their knees, some started to laugh with pleasure, and many turned to look at their friends with wide grins on their faces to see how others were reacting. Laronde's perspective is that "traditional [dance] contains both the traditional and the contemporary at the same time" (Fitzsimmons Frey, "Interview" 91). Competitive powwow dancers regularly borrow from hip hop when they compete, and since, as Laronde points out, Raven is a trickster, Raven's sense of place does not have to be in a particular time. In Raven's case, space can dominate time in a way that Massey suggests it rarely does. By making the dances both traditional and contemporary, children have a way in, and the story belongs to them in their own world today. Laronde positions *Raven Stole the Sun* in the Canadian/Turtle Island landscape, but it is not fixed or contained in a historical map. The spirit world and the world of folklore exist in timeless spaces, simultaneously coexisting and interweaving with people's daily lives, and also moving parallel to the pedestrian, quotidian world in which people live. Laronde's production discourages young audience members from imagining static identities for Indigenous peoples.

She also untames space, making Raven's space dynamic and connected to a larger political project that Massey (using Johannes Fabian) calls a recognition of "coevalness" where distance is suddenly "eradicated both spatially and temporally" and where "other" voices and real differences can no longer be eradicated from our (Eurocentric) time (70). Laronde's production works to dismantle structures of the "vision of political reality" that promote what Edward Said describes as the difference between the familiar and the strange (43). When Laronde's performers create Raven's mythical world on stage, they embody spaces of the past and the present, they use image, and they play. Even though Raven's story is traditional, Laronde's approach to choreography allows it to exist in a timeless and all-times space, drawing the young people into a world that they recognize partly as their own and that makes a meaningful impact because of the deep base of the iceberg—a creation process that draws on Indigenous knowledges and a profound connection to the land.

The idea of performers embodying imaginary geographies is appealing to me: if children are predisposed towards empathy, then they are especially attuned to embodied communication; and if bodies can be engraved with a cultural knowledge or memory, then they can move in a way that is rich with meaning. Massey's idea that space is constituted through its relations, and that without relations, space does not exist, is very useful for extending these ideas to bringing *space* into bodies. Space and, therefore, imaginary on-stage worlds, are possible only through relationships—relations between the performer and the subject, between the performer and the spectator, and perhaps between the members of the audience community. If space is relationships, and if everyone in the room contributes their imagination to the creation of the on-stage world, then the performing body can guide willing and eager spectators into an imaginary emotionally charged place. Some artists know from instinct and from the response of their audiences that dance is a way to speak deeply to children, and the examples I drew from demonstrate ways dance supports the audience as they navigate space vicariously through the performers, as they enter dream worlds, as they visit real but unfamiliar places, and as they engage with stories that inhabit worlds that exist beyond time. The process is certainly not guaranteed. Not all spectators will engage empathetically with dance, and not all performers will be able to etch a landscape and culture onto their bodies and share that with an audience. Conversations with young people about the ways they experience dance in performance, and better understanding how inscribing culture and place on the body may happen through a rehearsal process, may help dance function as a communication tool that brings young audiences, if not adults, into visceral understandings of imaginary geographies. As suggested in the introduction to this book, imaginative geography may refer perhaps to one's perspective and to what is *seen* in the mind's eye, but perhaps with dance, not only are on-stage dancing bodies establishing and embodying markers that *show* audiences another world, but also the earth-writing that the audiences engage in through imaginative and empathetic collaboration with performers on stage are *felt* with the *body* of the spectator.

Notes

I would like to thank Rebecca Barnstaple, Ben Fletcher-Watson, and especially Seika Boye for supporting this project.

1) I acknowledge that "child" is a culturally constructed concept, often referring to chronological age, or to students attending a particular grade level in school. Although people often assume they know what "child" means, there are many interpretations and constructions of the concept. One limitation of this study is that writers, directors, choreographers, and even scholars do not necessarily clarify how they are using the term. In this article, "child" means a person who is not yet a teenager, but some of the assertions might be more applicable to preschool children. Since the directors or choreographers I interviewed created their work for children, their idea of "child" usually refers to the ages of people they anticipate in their audiences, which might vary radically from project to project. In studies of dance and child spectatorship, developing a more consistent approach to the idea of "child" is certainly desirable, but further research is necessary.

2) Theatre for Young Audiences refers to professional theatre for children, usually (but not always) performed by adults. There are no officially maintained statistics on Theatre for Young Audience numbers in Canada. However, I base this number on a combination of Green Thumb Theatre's estimate that 62,000 children were a part of their 2010 season audience, and PACT's agreement that the number is probably similar for eighteen other member companies. Add to that the hundreds of professional TYA companies that are not PACT members, who tour to schools throughout Canada, and "hundreds of thousands" seems like a reasonable statement. (See Fitzsimmons

Frey, "Dance" 1). Theatre for Young Audiences also describes theatre that is not for "general," "family," or "adult" audiences. Yet the audience is regularly subdivided into age-specific subcategories such as "the very young" (0—24 months), "preschool" (2—5), "K-6," "ages 12—15," or "15 and up." As mentioned in note 1, the artists I spoke to did not necessarily differentiate between their various projects or the audiences they were catering to when they discussed their work.

3) The focus of my research to date has been on the artists, their intentions, and their perceptions of their audiences. However, it is essential to find ways to learn more about child spectator experiences as they relate to dance from a child's point of view, rather than from the viewpoint of artists.

4) Spitz cites eleven-year-old Margaret's childhood experience of watching Shaw's *Saint Joan,* and her being absolutely present in and inhabited by "some uncanny hybrid of actress and saint" (551) as Joan cries out to the spirits who have abandoned her. The child's knees ached from the imagined hardwood boards, her fingers trembled, and her heart pounded. Spitz argues that Margaret's empathetic experience occurred because "the integuments of self slip away and one lives naked in another carapace, another membrane" (552).

5) Theatre for young audiences practitioners have a responsibility to consider children's empathetic tendencies when creating their work. Although performers may find it exciting to experience wildly vocal reactions from children, especially laughter, Schonmann argues that those kinds of out of control reactions do not indicate that the production was a "success" in a meaningful or aesthetic sense.

6) Jola and colleagues describe the ways that conducting the research is challenging and complicated. Their project combined neuroscience and phenomenology in ways that had not been previously considered in terms of dance. They acknowledge controversy and debate concerning kinaesthetic empathy (20).

7) At this time, the idea of mirror neurons in humans is a theory: mirror neurons have not been proven to exist in humans, and literature that references them uses primate models, and an assumption that they must surely exist in people. Jola, Abedian-Amiri, and colleagues' dance spectatorship research, for example, appears to support the idea of a "mirror system," while Lingaua and colleagues' recent study argues against it. Without using the phrase "mirror neurons" Reason, Jola, and colleagues' 2016 study combined qualitative research with functional brain imaging to examine embodied human experiences to music and dance, and found that in some circumstances embodied responses were common across spectators.

8) Further research with children is essential to better understand a child's potential to acknowledge their own experiences of kinaesthetic empathy.

9) Reason and Reynold's research indicates that virtuosic performing bodies on stage can inspire kinaesthetic empathy in spectators who could not possibly do the things the performers do (60). Nevertheless, it would be interesting to extend my research to learn

about the potential for kinaesthetic empathy among children with disabled bodies, and also to examine differences with children and their willingness to empathetically identify with a performer who does not share his or her gender.

10) More deliberate research with children regarding embodied dance/theatre experience and memory is necessary, but I share this anecdote because it points to the probability of learning about other, similar experiences. Research in the field is growing. For example, in 2015 Matthew Reason noted that he launched a dance spectatorship research project with children that encouraged them to use the 3D material plasticine to communicate some of their responses. Visit https://matthewreason.com/portfolio/researching-with-plasticine.

11) As mentioned, Reason and Reynold's stress that in their research, response to dance is highly individual and is at least partly determined by exposure, experience, and familiarity, particularly when it comes to evaluating a dancer's skill. Knowing through and with the body could not be the same experience for all children, and an interesting extension of the research would be to learn about differences in the ways girls and boys respond, differences among children from different ethnocultural or socio-economic communities, and differences among children who live with physical disability, or who engage in various forms of physical training: dance forms, gymnastics, or athletics.

17) Following the Path of the Unconscious in the Owen Skye Books, and Others

❧ ALAN CUMYN

―――――――

Alan Cumyn is the Canadian award-winning author of thirteen
novels for adults, teenagers, and children. His literary canvasses
reflect different parts of the world. His adult novels include *Waiting for
Li Ming* (1993), set in a small town in China, *Man of Bone* (1998), about a
kidnapped diplomat on an island in the South Pacific, and *The Sojourn*
(2003), about a First World War soldier on the Western Front, on leave in
London, England. His work for the young includes a trilogy of children's
novels, *The Secret Life of Owen Skye* (2002), *After Sylvia* (2004), and *Dear Sylvia*
(2008) (the last being a collection of Owen's amusingly misspelled letters),
which explore a string of Owen's hilarious adventures born of a freedom
accorded to Canadian children growing up in the 1960s. These stories take
place in the imaginative geography of unnamed rural surroundings that
are based on the New Brunswick countryside, where Cumyn lived for a
number of years as a child. Cumyn's works for young adults include the

novella *All Night* (2005), about a poor, young, artsy couple in Toronto staying up late to examine their life together after a friend's funeral, as well as *Tilt* (2011) and *Hot Pterodactyl Boyfriend* (2016), both set around life at home and school and both about teenage angst, sex, and falling in love. The first of these is painfully yet humorously realistic; the second is a Kafkaesque collision between the high school mania for hotness and popularity and the seductive power of a new student, Pyke, the pterodactyl. Throughout his work, Alan Cumyn never ceases to surprise, never ceases to be innovative, as he plumbs the unconscious for his fictional geographies.

Cumyn is past Chair of the Writers' Union of Canada and is now Chair of the prestigious Writing for Children & Young Adults MFA program at the Vermont College of Fine Arts. He has an MA in English Literature and Creative Writing from the University of Windsor, where he studied under the late Alistair MacLeod. In 2016 he won the Vicky Metcalf Award for his body of work for young people. He has lived in China, Indonesia, and all across Canada; he now lives in Ottawa.

∽

One Sunday in November some years ago, I woke up in a beautiful old B&B in the town of Rothesay, outside Saint John, New Brunswick. I was scheduled to read from my adult novel *Losing It* the following night at the University of New Brunswick, and to save money the organizer—the poet Anne Compton—had flown me in Saturday night and was putting me up in this B&B outside of town.

I remember it as if in a movie. That morning I wandered outside. It was cold but there was no snow, just foggy rain that limited visibility. A few blocks away, quite unexpectedly, I came face-to-face with my first elementary school, which I had not seen for decades. It looked practically unchanged: the old hulking structure, foreboding brown bricks, asphalt playground. And not far away, also practically unchanged, was the Kennebecasis River where I had played as a child with my brothers. The railway tracks, the woods, and farther on, the old falling-down farmhouse where our parents had moved us when my father's job prospects were dimming … They were all there still. The two corner stores we used to frequent for Popsicles: not only were they still in place, but they too looked exactly the same, down to the same peeling paint. It looked like the Land that Time Forgot.

Unwittingly I had wandered back into the remembered and imagined landscape of my first novel for children, *The Secret Life of Owen Skye*, which would later be completed as a trilogy with *After Sylvia* and *Dear Sylvia*. I say "unwittingly" because I hadn't realized how little the area would have changed over the years. Also, I had used the landscape as the setting for these novels without being entirely conscious of what I was doing.

This last part will take some explanation. How is it that, as a novelist, I could simply not be aware of where my inspiration was coming from? Or is it possible that I was aware but only to a certain degree—that the knowledge was in the back of my head somewhere, in my unconscious, accessible during composition but somehow hidden from my conscious mind?

This chapter explores aspects of the setting of the Owen Skye novels, and some of my work for adults as well, relating to craft—how setting can be crucial in grounding a story to give it a fictional yet compelling sense of reality, and how writers might use their unconscious memories to create vivid and powerful geographies for their own fictional worlds.

oeo

Back to New Brunswick for the moment. My family moved there when I was about four. I was born in Ottawa, and some of my earliest memories are from the Parkwood Hills area, where my mom and dad as a young couple had bought a beautiful little brand-new bungalow. But something was not going well with my father's work. In his twenties, he had spent several exciting years in Canada's North working in mineral exploration, pushing back the frontier, as he called it—but when it came time to marry and settle down, he found the bush incompatible with family life, so he moved back to the city and got a job as an insurance salesman. But that work didn't suit him, so despite the charming house in the beautiful new neighbourhood with the park across the street, he and my mom decided to pull up stakes and move to New Brunswick.

Anybody who knows the country well realizes how counterintuitive this strategy would have been. Traditionally, people from New Brunswick move to Ontario to improve their economic prospects, not the other way around. But as a boy my father had been sent to an exclusive boarding school in New Brunswick, and he had a lot of good memories of the place, Rothesay Collegiate, where he was a star athlete and a good student.

Things only unravelled for him after Rothesay, when he spent a year at Royal Military College in Kingston, Ontario, before being kicked out for enjoying himself rather too much. He went to Carleton University in Ottawa for a time and played hockey there, but couldn't settle down. His father took him aside and said, "George, if you don't want to do it the easy way, you can do it the hard way." That led to a job in Sudbury working underground in a gold mine. My dad hated it. But he did eventually fall in love with exploration work above ground, and that was the career that captured his imagination as a young man. So when the office life back in the big city was not going so well, something must have twigged and he decided to move the family back to where things had been better. My mother had gone to school at Mount Allison University and so had some good associations, too, with the Maritimes. I believe it was a gut decision, not entirely rational. At the time, as a four-year-old, of course, I had no input and no understanding. Everything that happened seemed normal.

Away we went: two parents and three young boys with a big dog on one of those epic car trips halfway across the continent with kids barfing every twenty minutes along the road. We settled in Renforth, outside Saint John and just down the road from Rothesay, in a big white house shared with the riotous Fowler family upstairs. They always seemed to be screaming at one another. One night the Fowlers' car caught fire in the driveway under mysterious circumstances. I remember the stench and the drama.

We had the bottom floor. It had a very large front yard, with a garage down at the bottom and the highway just in front of it. Past the highway were the railway tracks, then the Kennebecasis River. My father settled into a job selling cash registers, but despite having gone to school in the area, he was perpetually "from away" and suffered a crucial lack of contacts. Years later I remember him describing to me how he would sit down with the manager of a business and say, "I can help you replace five accountants with this one machine," and the reply would be something like, "But those accountants are my friends and they have families. What will they do if they lose their jobs?" This sort of quaint, refreshingly old-school argument seems to have no place in today's global economy, but back then it absolutely stumped my dad.

It meant that within a year of living in the big white house shared with the Fowlers, to save money we moved down the road to the falling-down farmhouse. Right next to the farmhouse was the bulls' field surrounded by a rickety fence that did nothing to keep boys out. I remember sitting in an apple tree with my brothers after seeing *Those Magnificent Men in Their Flying Machines,* the funniest movie in the history of the universe. I was pretending to be flying in the cockpit of a plane when the branch broke and for a moment I had this dizzy feeling of the world spinning. I wasn't sure what was going on and then I was on the ground. Somehow my brothers were beside me, looking at me, and in the deeply empathetic way of brothers my older brother Richard said, "You fell on your head, doesn't it hurt?" When he put it that way I realized it must hurt and so I started to cry.

Most of my earliest childhood memories are from the Maritimes: lying with my brothers in the ditch beside the railway tracks covering our ears as the train roared past at a million miles an hour and looking up just in time to see the face of the conductor as he shook his fist at us and yelled something foul; sneaking down into the basement of the farmhouse through the old coal chute and shivering in the darkness while we listened for the breathing of the beast that would later become the Bog Man in *The Secret Life of Owen Skye*; playing in the woods and stumbling upon a haunted house, an abandoned building where we would climb up and down the unfinished walls and try to avoid stepping on rusty nails, and where we would pick up old knife blades and broken hammers and pieces of wood and try to explain to each other what was going on. What was this house doing there in the middle of nowhere?

Readers of the Owen Skye books will recognize the haunted house where the brothers, Andy, Owen, and Leonard spend Halloween night scaring themselves senseless looking for the Bog Man's wife. We read a lot of comic books in those days, and knowledge of comic books pervades the interests of the three boys. The Bog Man idea comes from a classic called *Swamp Thing,* which has now had several incarnations in both comic books and movies. Most important for me was the way the boys actively shape and retell the story, so that readers can watch the story morphing depending on what's going on. On their way to the haunted house, for example, oldest brother Andy stops the procession and insists on telling the story of the Bog Man's wife, a story that gets further refined before they reach the dreaded house where she might be waiting for them on the mysterious red couch.

The Owen Skye stories literally began as Christmas stories for my two daughters, in particular for Gwen, the eldest, who had just started to read. I sat down and wrote out a story called "Valentine's Day" for her. I wanted to give her some clues about boys so I wrote a story about young Owen crossing the classroom to deliver a Valentine to his secret true love Sylvia

Tull, in the process making a complete fool of himself. I wrote that story for Christmas 1996 and showed it to my wife Suzanne before wrapping it up. I was on leave at the time from a writing and research job with the Immigration and Refugee Board. I'd received a Canada Council grant to write a human rights novel for adults, and by Christmastime was both out of money and burnt out from the intense subject matter. So the little Owen Skye story for Gwen was hitting two birds with one stone: coming up with a low-cost Christmas present as well as giving myself a chance to write something lighthearted.

Suzanne liked it very much. She said, "Alan, this is good, but you have two daughters. You can't just write one story for one of them for Christmas. It wouldn't be fair. You have to write another story for Anna." She didn't say, "You have to give it to both of them," otherwise I might have stopped writing Owen Skye stories right there. Instead, for Anna, I wrote the Bog Man's wife story about Halloween night, and that's what started me off.

I have said that the imaginative setting for these stories comes from those couple of years in New Brunswick. But actually, to be precise, the Valentine's Day story, the part from my own disastrous personal romantic experience, dates a little later on, after the family had limped back to Ontario. I attended the school that is just a few blocks from where I still live. As an author, what was I up to? I was borrowing stories and geography from both places and mixing it all up in a kind of leftover stew. Most of the external geography came from the two places where I had lived in New Brunswick. A lot of the internal geography, especially concerning Owen's love for Sylvia, comes from later, in Ottawa.

Now I've said already that I wasn't entirely conscious of what I was doing. I do remember making a bit of a decision to mix settings from different places. I'd already been doing that in the human rights novel I'd been working on, *Man of Bone*. That book came out of years of research that I had done at the Immigration and Refugee Board writing about human rights conditions in different countries, the very worst of what we do to one another as human beings on this planet. The reports I was writing for the government had to be very factual, academic. The emotional part of the story, anything graphic or gut-wrenching, was edited out in large part. The material had to hold up in Federal Court and sometimes the fate of people's lives depended on what got into those reports, which were used as background for refugee hearings. When it came time for me to write a novel using that material, I didn't want to be tied to the literal truth. I didn't want to write about a real place, I didn't want to write about things that had actually happened to someone. I wanted to be free to write fiction, which in some ways can get us closer to human truth. So in a way I put myself in the position of Bill Burridge, the novel's protagonist.

Burridge is a reflection of me in my mid-thirties, of a course I might have taken. At one point I was considering a career in the foreign service. Bill Burridge has a deeply loved wife and son with him on his first posting in a little country I made up called Santa Irene. Three weeks in, he gets kidnapped. The story starts after he's been held for a few days and is freaking out. He is in a hood, in shackles, in a closet. He doesn't know who has taken him, if he's going to survive. It's some rebel group and he figures they've made a mistake, that they think he is American CIA. He actually knows nothing.

Fortunately I have never been kidnapped, have never been remotely close to such a situation. But through my work, my reading, people I'd met, and some travels of my own, I had got myself to a place where I felt I could make the journey imaginatively. I'd lived in China and Indonesia, had visited the Philippines, and had read an awful lot about other places— Guatemala and El Salvador, Pakistan and the Indian Punjab, Lebanon and Bangladesh—places where human rights can get trampled. The country that I created, Santa Irene, is very much a stew. I took elements from all kinds of places that I was thinking of, then put them together, and sometimes that involved a conscious decision, but often I was just pulling details from my own unconscious. I was using what felt right.

So what do I mean by that? I guess I mean that when I sit down to write imaginatively, when I'm writing fiction, I actually sink into a bit of a state. It's not a trance. I'm lucid. The phone rings, the spell breaks, I sound normal. But I'm also not entirely myself. I feel like I am slipping into the character, the voice, the point of view, the storytelling consciousness behind the project that I'm working on. In *Man of Bone* Bill Burridge was physically trapped. He wasn't going anywhere. Like many of us would, under torture he quickly folds. He wants to die, but they aren't killing him. He's slow to realize this, but when he does, his heroic action is to decide that if he's going to survive this ordeal, he wants to come through with a semblance of his personality, of himself, intact. That means getting through the day, the night, with aggressive use of flashback. He tries to mentally transport himself back to other times, other places, other memories. Life in Canada when he was happy: that's his refuge.

As I sat down to write him I would slip into his persona, his thoughts, his body, and try to write close to what his felt experience would be, using, as a starting place, the raw material of my own memories. So here's something that happens early on in the novel, after he's scared out of his mind but it's starting to become clear that they aren't going to kill him. He mentally goes back to childhood:

We're playing execution. Tony and his little brother Bob and James Meade, whose father is in the RCMP, and Brenda from behind the vacant lot, who isn't entirely a girl since she plays hockey better than some of us. And my brother Graham is here and Mario from across the street, who has blond curly hair and skin white and pudgy as a Greek statue.

I'm stunned to see them after so long. James Meade has the machine gun, a hockey stick broken in the shaft, not the blade, so it's too short but it could be used for a goalie stick ... James says, "Okay, Mario!" and then shoots him right on the lawn. Mario takes the first bullet in the shoulder and twists, jolted into the air, then is hit again so both hands fly back and he lands on his side, bounces on the grass, then moans and rolls away from his attacker. James stops shooting but Mario continues to get hit, the bullets making his body buck and sway in an almost sexual frenzy. Finally he's still, face down, legs splayed, grass stains all over his clothing.

"Tony!" James shouts and turns to shoot him down on the driveway. Tony screams in pain and surprise and heads toward the grass before the bullets take him down ... [But] he finds it hard to keep from laughing, even as his guts are oozing out of him, and then when he's supposed to be still he looks around from time to time at the others.

"Brenda!" James turns his wrath on her. She withstands a blast without flinching, advances despite the hail of bullets. She has long black hair and wide shoulders and wears a baggy black sweatshirt because of her breasts. Everyone can tell she likes James by the way she looks at him. He's tall and thin and has wavy brown hair and skin as fine as a girl's and can catch a football like a frog pulling in flies.

"I'm shooting you!" James yells, waving his machine gun, but Brenda simply advances, her eyes looking dangerous. When she's just a few feet away and by all rights should only be fragments of flesh James suddenly stops firing and cries, "Abandon stations! Abandon stations!" then turns to run. Brenda lurches for him, gets him by the waist but he twists out of

Space and Gender

her grasp and then takes off down the road. Brenda starts running too, and then we all howl and chase across lawns and over the curb after him.

This is a new game. We don't know what it is—it's just happening, which is the excitement of it. Brenda is a good runner but James is fast, his arms and legs working like oiled scissor blades. He opens up a twenty-yard lead, then ducks into McCloskey's backyard and across the Stephenson's deck. It's a daring thing to do. McCloskey is a grey-haired cactus of a man who hates us even playing on the road, and no one has ever gone in his yard before. But now we all charge across it screaming. And the Stephensons don't even have children, but we thump across their deck like it's ours, then keep on running, through the hedge and on.

Just running and running for the fun of it, for being able to, with air so fresh in our lungs it hurts and legs going faster faster, and there's no way to do this silently, we all have to yell and keep on yelling. (34-35)

So that's from *Man of Bone*, as dark a novel as I'm going to write. A Canadian diplomat being held and tortured has this childhood memory of playing execution, a game from my childhood. We were channelling the stories that we'd heard from the Second World War. That's the kind of thing that boys did back in the 1960s. And I was already journeying back when I was writing *Man of Bone*. In this dreamlike memory, execution turns into escape, running away, with sexual overtones as well. In the bubbling stew of his imagination, Burridge is figuring out ways to escape even while in real life escape is impossible. I don't remember being conscious of the various complex levels that memory works on for the story, for that moment in Burridge's ordeal. I was sitting at my desk, sinking into a state, and that's what came out.

Similarly, just a bit later, when I was writing "Valentine's Day," a story that on the surface couldn't be more different in tone, here's what came out:

At recess time they ran screaming from one end of the schoolyard to the other and back again. The girls chased the boys and kissed them if they caught them. The girls were bigger than the boys. There was a white line painted across the schoolyard to keep the girls from chasing the boys, but it didn't work. The girls took one look at that line and then ran right over it. And the teachers didn't care. They stayed in the staff room smoking at recess time. You could see the smoke puffing out of the window even though it was closed and the drapes were drawn. Those teachers didn't want anything to do with the kids at recess. Sometimes they forgot to ring the bell and the girls would chase and kiss boys for hours. (*The Secret Life of Owen Skye*, 44)

So there's a variation on the chase scene from *Man of Bone*—a girl takes off after a boy who has started running. I do remember in particular, in grade three, in the schoolyard that is just down the street from where I still live, we went through a phase when the white line that was painted down the middle of the playground became superfluous. We would meet up in the field above the playground, the girls and the boys together. The girls spontaneously started chasing the boys and the boys ran away because we didn't want to be caught. I forget who it was who suggested that if a boy got caught by a girl then the girl could kiss him. But somehow we all knew that that was going to be the penalty, and usually the boys didn't get caught. We were extremely motivated to stay fast and free, but sometimes the kissing happened and that just made it all the more exciting, and the fifteen minutes of recess would seem to go on and on and on. Afterwards, when we returned to the classroom, it was impossible to think about the times tables or why Holland was under water because of everything that had happened in the field.

As we ran around, chasing after one another, almost getting caught, almost kissing, we were playing in much the same way my brothers and I were playing in the apple tree after seeing *Those Magnificent Men in Their Flying Machines*. We were just doing what felt fun, what seemed dangerous and alluring at the same time. We did not talk about it, particularly; it all just happened, like water flooding over the banks of a river in spring and flowing wherever it feels like, wherever gravity pulls. That's the central image of an Alice Munro short story called "The Found Boat." A river overflows its banks in the spring, and two girls, Eva and Carol, come across a band of boys trying to make a raft. The girls find a wrecked old boat, which the boys resurrect, and the boys and girls go on an expedition together, and eventually it leads, somehow, to an abandoned, dark station house where inhibitions fall away. Soon they are all naked, splashing back in the water, laughing uncontrollably. This is the land of the unconscious, of Freud's id, which he famously placed within the realm of childhood. Here's Michael Egan writing about the id in *Peter Pan*, one of those children's classics that seems to be a veritable ocean of the unconscious:

If Hook is the Oedipal Father, however, then within the structure of the story Peter Pan himself must be his Son. In great part the tale's popularity derives from its dramatization, in symbolic terms, of the Oedipal Son's victory over the Father. When Peter defeats Hook, every son in the audience crows with glee.

If this reading seems a little forced, let us recall that once the children descend into the unconscious Peter and Wendy undergo what is in effect a marriage. They set up house; they have children; Peter goes out to work and Wendy darns socks. At the same time, however, their roles oscillate ambiguously. Wendy is now a mother, now a wife; Peter simultaneously her husband, son, and, as he insists on being called, "The Great White Father" (Barrie's original title for the play).

Freudians have gone to town for years on *Peter Pan* and its many passages swimming in unconscious childhood reveries, Oedipal or not. When I started writing the Owen Skye stories as gifts for my own children—children whose arrival had sparked in me a swelling of memories of my own childhood—my father was still alive. Extremely vigorous, in fact; I could not imagine that he would be dead of lung cancer within four years. I had killed him off, fictionally speaking, on the opening page of my 1993 novel for adults *Between Families and the Sky*, and he had forgiven me, although I'm not sure why. Here's how that page reads:

> The day of my father's funeral I stood in the doorway of the master bedroom and watched my mother weep. She was wearing a white slip, sitting on the bed with her long red hair tangled over her shoulders and her back to me. It is an astonishing thing to find your mother beautiful, in *that* way, at such a moment—I could not speak and I could not leave; all I seemed capable of was to watch with the silent, hawkish eyes of receding childhood.
>
> I don't remember how the scene ended, whether she turned to me and said something or I finally crept away. In a way the scene has never ended—in a part of my mind it is etched in acid, and I can smell my mother's powder, see her round white shoulder in the mirror and how rumpled the sheets are (not made in three days—she would have the whole bed carted off later, sentenced to oblivion). I can see too my father's clothes peeking out of the closet door as if waiting for him to come into the

room, and taste still the air charged with the burning electricity of fear and love and anguish. (9)

I read that passage now and the Oedipal ramifications are striking. I wasn't writing from life, of course—my father was still very much alive; my mother's hair was never red—but there it is on the page, somehow real in its own way. I didn't actually write a son-kills-father scene until my first Young Adult novel, *Tilt*. The showdown, between Stan Dart and his hopelessly failing father, takes place on the front porch late at night when the father is stealing away with his child from another marriage. Stan, who has been making up his own martial art, is ready to fight for the boy, Feldon. Instead, like a master fighter, he is able to simply persuade his father to hand Feldon over, by telling gentle lies that make it easier for the father to rationalize what he is doing. I had very little idea, on a conscious level, when I was writing that scene, that it was time to invoke Oedipus and have Stan metaphorically slay his father. If that had been my intention, I imagine the scene would not have worked at all. Instead, I was writing the way it felt like it had to come out; I believe I was giving my unconscious a chance to participate. It was only much later, after the book was published, that the literary critic part of my conscious brain started to realize what was behind the extra power of that scene.

Michael Egan contends that J.M. Barrie wasn't entirely conscious of what he was doing when he wrote *Peter Pan*, either. Here's a note from Barrie's journal in 1922, in his sixties: "It is as if long after writing P. Pan its true meaning came to me—Desperate attempt to grow up but can't" (Egan 38, quoting Andrew Birkin, *J.M. Barrie and the Lost Boys* [New York: Clarkson N. Potter, 1979], 297).

I don't know about you, but I'm starting to realize that I actually live a lot of my life in my unconscious. Six to eight hours a night I am asleep, journeying all over the place. Recently, for example, in a dream I spent a very strange time visiting my brother some place he has never lived. I seemed to be staying in the attic, and to get down the stairs for some breakfast I had to pass him and his former lover of many years ago who were squeezed into a medieval hay bed that closely resembled one showcased recently on television by the eccentric and delightful historian Lucy Worsley in her series *If Walls Could Talk*.

I don't know what that dream was about, really. But I do know that writers who pay attention to the unconscious of their characters have a chance to make the world they are creating seem more real, more distinctly human. Indeed, while we don't see it, the unconscious is the greater part of most of us. In *Tilt* it felt right to follow Stan into his own dream world sometimes, especially late at night, when his conscious mind is exhausted from fighting off the sexual demons that seem to be surrounding him. In a dream Stan as a young boy—the language has suddenly become simpler—is running in a swampy area, looking for his parents:

> There they were, his parents. He almost missed them! He almost ran past them hiding off the path in the green and brown wall where they'd thrown down the blanket.
>
> He pushed through to them and said, "Where's the swimming?"
>
> It was so hot they'd taken off some of their clothes. It was so hot down there on the blanket in the green and brown wall that Daddy and Mommy were squirming. Their underpants were at their ankles!
>
> Stan said again, "Where is the swimming?" because his sneakers and his legs were mucky.
>
> And Mommy turned her face to him. She looked like she'd been dreaming. Daddy had his face hiding into her neck.
>
> "What are you doing?" Stan asked.
>
> That's when his father lifted up his face, too. He was lying right on top of her in the hot hot and his face looked like he'd just been swimming himself.
>
> "We're planting your little brother, Sport," he said. And Mommy hit him—it wasn't much of a slap. Her arms were mostly trapped in his.
>
> So Stan ran to where the swimming was. It wasn't far at all and he did splash water on the sneakers and his legs till all the muck came off. Later when he looked up his parents were on the blanket on the beach in their swimsuits. His mother was reading a book and his father was sitting in his dark glasses staring at something far away. (64-65)

You can run from your unconscious, but you cannot hide. It's a bumbling, unruly, often seemingly incoherent stewpot of our desires, fears, ambitions, hatreds, loves. It's the nine-tenths of the iceberg below the surface, the silent hulking mass that, every so often, we run our own personal *Titanic* into.

Maybe it's a middle-age or older-age thing, the sudden need to pay attention to what's below the surface. The realization that those unresolved conflicts you had with your father or your mother or your siblings, those instinctive patterns of crouching for safety or hiding from demons laid down when you were too young to think about things consciously—those are the things that are sinking your passenger liner now, just when you should be sailing free, enjoying the champagne. I do believe that children and young people are closer to the unconscious, that it's at the root of play, which many adults forget how to do. The physician Gabor Maté[1] has written, to my mind, convincingly about the patterns of behaviour laid down in childhood and how, when one is unprotected at a young age, either physically or psychologically, the wounds run deep, and can be at the root of many illnesses later in life.

I know a lot of people's memories of childhood are fraught with worry, fear, a deep sense of unease. This period of instability in my family's life that I've spoken of, the misadventure of roaring off to the Maritimes when my family was very young, could easily have left me with similar scars. The financial worries were hard on my parents; I must have picked up on some of that. But by and large those were my parents' worries, they were not mine. I don't remember ever having a feeling that food would not be available, or we wouldn't have a place to live. It was an era when children were let out the back door and left to roam in ways that seem unimaginable to young parents now. There was a highway at the end of the property, railway tracks after that, a river after that. We fell out of trees, we crashed our bicycles from riding too fast down the hill, towards that highway, we made a fort on top of the garage door that never closed and one day, when I was climbing the cable to get up on that door, the door did close and my finger got caught between the cable and the pulley, the tip sliced off. A man was killed on the railway track opposite our house when he was riding the rails in a hand car with his son, whom he saved by throwing him aside at the last second.

Many of those episodes made it into the Owen Skye stories, in one way or another, stirred around in the stew of my unconscious and then served in a different form. *The Secret Life of Owen Skye* was published a couple of years after my father's death, but he had read earlier versions of the stories, and of all my writing he liked those best. The father doesn't appear very often in the early stories; it's just the boys getting into scrapes, and Owen figuring out what to do with his heart. When I woke up that morning in Rothesay, unwittingly pulled into the foggy land of my own unconscious, it was just a couple of years after my father's death. My host, Anne

Compton, took me to Rothesay Collegiate where my father had been a star. There was his name still on the wall for his boxing victories, his track and field accomplishments, there was the academic trophy with his name. I had last seen these records of his triumphs when I was six, just before we headed back to Ontario. Not far from where we stood, down the hill, the Kennebecasis River wound its way in reality and in my mind. I remembered that once my father was done for good with the world of selling and office jobs, when he had found his calling as a classroom teacher, he wrote to his old headmaster at Rothesay to tell him the good news of where his feet had finally settled.

As an adult, the more I think about that whole Maritimes misadventure, the more I shake my head in wonder. On a hunch, when things were hard for his young family, my father had brought us all back to a place that had felt right for him when he was young. I think he'd felt the pull of his own iceberg, his own unconscious. It was a radical thing to do; difficult; it shook us up. Ultimately, I believe those years were among the wealthiest of our lives as a family—if memories are wealth, if good feelings are bankable, if the unconscious is something that can be said to live and breathe and feed us in its own delightful ways. We were wealthy then. I feel wealthy now, sitting on top of such a rich pile of memories.

———

Note

1) See, for example, Gabor Maté, *When the Body Says No: The Hidden Cost of Stress*; and *In the Realm of Hungry Ghosts: Close Encounters with Addiction*.

Works Cited

Abel, Elizabeth, Marianne Hirsch, and Elizabeth Langland, editors. *The Voyage In: Fictions of Female Development*. UP of New England, 1983.

Akenside, Mark. *The Pleasures of the Imagination*. n.p., 1744.

Alaimo, Stacey. *Undomesticated Ground: Recasting Nature as Feminist Space*. Cornell UP, 2000.

Alanus ab Insulis. *Omnis Mundi Creatura*. Edited by Turba Delirantium and Gunter Krebs, 2003. turbadelirantium.skyrocket.de/bibliotheca.

Alcott, Louisa May. *Little Women*. Roberts Brothers, 1868.

Andersen, Hans Christian. "The Snow Queen." *Hans Andersen's Fairy Tales*, translated by Naomi Lewis, Puffin Classics, 1981, pp. 103-42.

Anderson, Gary A. *Sin: A History*. Yale UP, 2009.

Anon. *The Child's Irish Song Book*. Southway Junior Irish Literary Club, n.d.

Armstrong, Jeannettte. *Whispering in Shadows: A Novel*. Theytus, 2004.

Ashley, Kathleen M., editor. *Victor Turner and the Construction of Cultural Criticism*. Indiana UP, 1990.

Atwood, Margaret. *Survival: A Thematic Guide to Canadian Literature*. Anansi, 1976.

Axtell, James. "The White Indians of Colonial America." *William and Mary Quarterly*, 3rd series, vol. 32, no. 1, January 1975, pp. 55-88 (published by the Omohundro Institute of Early American History and Culture), www.jstor.org/stable/1922594.

Avery, Gillian. "The Beginnings of Children's Reading to c 1700." *Children's Literature: An Illustrated History*, edited by Peter Hunt, Oxford UP, 1995, pp. 1-25.

Bacon, Francis. "Of Gardens." *The Faber Book of Gardens*, Routledge, 2011, p. 61.

Bachelard, Gaston. *The Poetics of Space*. Translated by Maria Jolas, Beacon Press, 1969, repr. 1994.

Baker, Deirdre. *Becca at Sea*. Groundwood, 2007.

Ballantyne, R.M. *Ungava: A Tale of the Esquimaux*. Thomas Nelson and Sons, 1858.

Bancroft, Hubert Howe. *The Book of the Fair*. Bancroft Co., 1893.

Barrie, J.M. *Peter Pan and Peter Pan in Kensington Gardens*. Wordsworth Editions, 2007.

Barrow, John. *A Chronological History of Voyages into the Arctic Regions*. John Murray, 1818.

Bastian, Dawn E., and Judy K. Mitchell. *Handbook of Native American Mythology*. ABC-Clio, 2004. www.eso-garden.com/specials/handbook_of_native_american_mythology.pdf.

Baum, L. Frank. *The Wizard of Oz*. Rand McNally, 1956.

Bavidge, Jenny. "Stories in Space: The Geographies of Children's Literature." *Children's Geographies*, vol. 4, no. 3, December 2006, pp. 319-30.

——. "Vital Victims: Senses of Children in the Urban." *Children in Culture, Revisited: Further Approaches to Childhood*, edited by Karin Lesnik-Oberstein, Palgrave Macmillan, 2011, pp. 208-22.

Bayne, Marie. *Tales of Ireland for Irish Children*. Dublin: Fallon Brothers, n.d.

Bennett, Jane. *The Enchantment of Modern Life: Attachments, Crossings, and Ethics*. Princeton UP, 2001.

Berrol, Cynthia. "Neuroscience Meets Dance/Movement Therapy: Mirror Neurons, the Therapeutic Process and Empathy." *The Arts in Psychotherapy* 33, 2006, pp. 302-15.

Berry, Dave. "The Treatment of Mythology in Children's Fantasy." *The Looking Glass: New Perspectives on Children's Literature*, vol. 9, no. 3, 2005, n.pag. www.lib.latrobe.edu.au/ojs/index.php/tlg/article/view/29/34.

Berton. Pierre. *The Secret World of Og*. McClelland and Stewart, 1961.

Bettelheim, Bruno. *The Uses of Enchantment: The Meaning and Importance of Fairy Tales*. Vintage Books, 2010.

The Bible. King James Version, Hendrickson, 2011.

Bixler, Phyllis. *Frances Hodgson Burnett*. Twayne, 1984.

Black, Marilynne V., and Ronald Jobe. "Are Children Gaining a Sense of Place from Canadian Historical Picture Books?" *Picture Window*, vol. 9, no. 3, n.pag. www.lib.latrobe.edu.au/ojs/index.php/tlg/article/view/35/40.

Bohls, Elizabeth A. "Standards of Taste, Discourses of 'Race,' and the Aesthetic Education of a Monster: Critique of Empire in *Frankenstein*." *Eighteenth-Century Life*, vol. 18, no. 3, November 1994, pp. 25-36.

Bradford, Clare, Kerry Mallan, John Stephens, and Robyn McCallum. "Reweaving Nature and Culture: Reading Ecocritically." *New World Orders in Contemporary Children's Literature: Utopian Transformations*, Palgrave Macmillan, 2008, pp. 79-104.

Briggs, Katherine Mary. *The Personnel of Fairyland*. Alden Press, 1969.

Briggs, Melody, and Richard S. Briggs. "Stepping into the Gap: Contemporary Children's Fantasy Literature as a Doorway to Spirituality." *Towards or Back to Human Values? Spiritual and Moral Dimensions of Contemporary Fantasy*, edited by Justyna Deszcz-Tryhubczak and Marek Oziewicz, Cambridge Scholars Publishing, 2006, pp. 30-47.

Brontë, Charlotte. *Jane Eyre*. Oxford UP, 1975 [1993].

Brooker-Gross, Susan R. "Landscape and Social Values in Popular Children's Literature: Nancy Drew Mysteries." *Journal of Geography*, vol. 80, no. 2, 1981, pp. 59-64. doi.org/10.1080/00221348108980236.

Brown, Peter. "Author/Illustrator Interview: Peter Brown." By Peteredmundlucy. *Into the Wardrobe*, 19 July 2009, peteredmundlucy7.blogspot.com/2009/07/authorillustrator-interview-peter-brown.html.

——. *The Curious Garden*. Little, Brown, 2009.

Brown, Terence. *The Literature of Ireland: Criticism and Culture*. Cambridge UP, 2010.

Bruchac, Joseph. *Our Stories Remember: American Indian History, Culture, and Values through Storytelling*. Fulcrum, 2003.

Buckland, Corinne. "Fantasy and the Recovery of the Numinous." *Towards or Back to Human Values? Spiritual and Moral Dimensions of Contemporary Fantasy*, edited by Justyna Deszcz-Tryhubczak and Marek Oziewicz, Cambridge Scholars, 2006, pp. 17–29.

Buell, Lawrence. "Ecocriticism: Some Emerging Trends." *Qui Parle*, vol. 19, no. 2 (Spring-Summer 2011): 87–115.

——. *The Environmental Imagination: Thoreau, Nature Writing, and the Formation of American Culture*. Belknap Press, 1996.

——. "Environmental Writing for Children: A Selected Reconnaissance of Heritages, Emphases, Horizons." *The Oxford Handbook of Ecocriticism*, edited by Greg Garrard, Oxford UP, 2014. dash.harvard.edu/bitstream/handle/1/14555069/Buell_Environ mentalWriting.pdf?sequence=1.

——. *The Future of Environmental Criticism: Environmental Crisis and Literary Imagination*. Blackwell, 2005.

Bunyan, John. *The Pilgrim's Progress*. Edited by W.R. Owens, Oxford UP, 2003.

——. *Book for Boys and Girls or, Country Rhymes for Children. Being a Facsimile of the unique first edition, published in 1686, deposited in the British Museum*, edited by John Brown, Elliot Stock, 1890.

Burke, Edmund. *A Philosophical Enquiry Into Our Ideas of the Sublime and the Beautiful* [1757]. Edited by Adam Phillips, Oxford UP, 1990.

Burnett, Frances Hodgson. *The Secret Garden*. Original Illustrations by Charles Robinson. Chartwell Books, 2009.

——. *Two Little Pilgrims' Progress*. Scribners, 1895.

Burnford, Sheila. *The Incredible Journey*. Hodder and Soughton, 1961.

Byrd, Jodi A. "Red Dead Conventions: American Indian Transgeneric Fictions." *The Oxford Handbook of Indigenous American Literature*, edited by James H. Cox and Daniel Heath Justice, Oxford UP, 2014, pp. 344–58.

Cadden, Mike. *Ursula K. Le Guin Beyond Genre*. Routledge, 2005.

Cajete, Gregory. "Challenging the Chance of the Cheshire Cat's Smile." www.banffcentre.ca/articles/challenging-chance-cheshire-cats-smile.

Cameron, William E. *The World's Fair, Being a Pictorial History of the Columbian Exposition*. National Publishing, 1893.

Cariou, Warren. "An Athabasca Story." *Read, Listen, Tell: Indigenous Stories from Turtle Island*, edited by Sophie McCall et al., Wilfrid Laurier UP, 2017, pp. 98–103.

——. "Tarhands: A Messy Manifesto." *Imaginations Journal* 3, no. 2 (2012), pp. 17–34. dx.doi .org/10.17742/IMAGE.sightoil.3-2.3.

Carola, Leslie, editor. *The Irish: A Treasury of Art and Literature*. Hugh Lauter Levin, 1993.

Carpenter, Humphrey. *Secret Gardens: A Study of the Golden Age of Children's Literature*. Houghton Mifflin, 1985.

Carr, Emily. *Klee Wyck.* gutenberg.net.au/ebooks01/0100131.txt%3egutenberg.net.au/ebooks01/
0100131.txt.

Carroll, Jane Suzanne. "Death and the Landscape in *The Dark Is Rising* and Its Adaptations."
What Do We Tell the Children? Critical Essays on Children's Literature, edited by Patricia
Kennon and Ciara Ní Bhroin, Cambridge Scholars, 2012, pp. 74–89.

——. "The Green Topos: Gardens, Farms, Wilderness." *Landscape in Children's Literature.*
Routledge, 2011, pp. 49–89.

Carroll, Lewis. *Alice's Adventures in Wonderland* and *Through the Looking Glass.* Edited by Hugh
Haughton, Penguin, 1998.

——. *The Annotated Alice: Alice's Adventures in Wonderland* and *Through the Looking Glass.* Edited
by Martin Gardner, Penguin Books, 1970.

Chesterton, G.K. "On Mr. Rudyard Kipling and Making the World Small." 1905. *Heretics/
Orthodoxy.* Thomas Nelson, 2000, pp. 17–24.

Clark, Catherine Anthony. *The Golden Pine Cone.* Harbour, 1995.

Clarke, Amy M. *Ursula K. Le Guin's Journey to Post-Feminism.* McFarland, 2010.

Coleridge, Samuel Taylor. "Frost at Midnight." *English Romantic Writers*, edited by David
Perkins, Harcourt, 1967, pp. 422–23.

Comoletti, Laura B., and Michael D.C. Drout. "How They Do Things with Words: Language, Power, Gender, and the Priestly Wizards of Ursula K. Le Guin's *Earthsea* Books."
Children's Literature 29, 2001, pp. 113–41.

Cooper, Susan. *The Dark Is Rising.* [1973]. Margaret K. McElderry, 2013.

Cornum, Lindsey Catherine. "The Creation Story Is a Spaceship: Indigenous Futurism and
Decolonial Deep Space." www.vozavoz.ca/feature/lindsay-catherine-cornum.

Coughlan, Lynn, editor. *A Child's Famine: Poetry and Prose for Children by Children.* Bradshaw,
1995.

Cowell, Cressida. *How to Train Your Dragon.* Hodder Children's, 2017.

Craig, John. *The Long Return.* Bobbs-Merrill, 1959.

Cummins, Elizabeth. *Understanding Ursula K. Le Guin.* 2nd ed., U of South Carolina P, 1993.

Cumyn, Alan. *Between Families and the Sky.* Goose Lane Editions. 1995.

——. *Man of Bone.* Goose Lane Editions. 1998.

——. *The Secret Life of Owen Skye.* Groundwood Books. 2002.

——. *Tilt.* Groundwood Books. 2011.

Cutter-Mackenzie, Amy, Phillip G. Payne, and Alan Reid. *Experiencing Environment and Place
through Children's Literature.* Routledge, 2011.

Daley, Jason. "DNA Analysis Could Identify the Sailors (Including Women) of the Doomed
Franklin Expedition." www.smithsonianmag.com/smart-news/dna-extracted-doomed
-franklin-expedition-sailors-180963031.

Darton, F.J. Harvey. *Children's Books in England.* 3rd ed., edited by Brian Alderson, Cambridge
UP, 1982.

——. *Children's Books in England: Five Centuries of Social Life* [1932]. Cambridge UP, 2011.

Denisoff, Dennis. "Pleasurable Subjectivities and Temporal Conflation in Stevenson's
Aesthetics." *Journal of Stevenson Studies* 4, 2007, pp. 227–46.

Dillon, Grace L., editor. *Walking the Clouds: An Anthology of Indigenous Science Fiction*. U of Arizona P, 2012.

Dobrin, Sidney I. "Through Green Eyes: Complex Visual Culture and Post-Literacy." *Environmental Education Research*, vol. 16, nos. 3-4, June-August 2010, pp. 265-78.

Dobrin, Sidney I., and Kenneth B. Kidd, editors. *Wild Things: Children's Culture and Ecocriticism*. Wayne State UP, 2004.

Donaldson, Eileen. "Accessing the 'Other Wind': Feminine Time in Ursula K. Le Guin's *Earthsea* Series." *English Academy Review* 30, 2013, pp. 39-51.

Donnelly, Jim. *The Irish Famine*. kenanfellows.org/kfp-cp-sites/cp01/cp01/sites/kfp-cp-sites .localhost.com.cp01/files/LP3_BBC%20Irish%20Famine%20Article%20for%20Lab.pdf.

Donnison, T.E. "The Siren of the Pole." *The Idler* 10 (August 1896-January 1897), p. 757.

During, Simon. *Modern Enchantments: The Cultural Power of Secular Magic*. Harvard UP, 2002.

Duvar, John Hunter. "The Emigration of the Fairies." *Hernewood: The Personal Diary of Col. John Hunter Duvar June 6 to September 17, 1857*, edited by L. George Dewar and M.D. O'Leary, Williams and Crew, 1979.

Dyer, Richard. *Pastiche*. Routledge, 2007.

"East of the Sun and West of the Moon." Collected by Asbjornsen and Moe. Translated by George Webbe Dasent. *The Blue Fairy Book*, edited by Andrew Lang, Dover, 1965, pp. 19-29.

Ede, Charles. *The Home Amid the Snow or, Warm Hearts in Cold Regions: A Tale of Arctic Life*. T. Nelson, 1882.

Edwards, Gail, and Judith Saltman. *Picturing Canada: A History of Canadian Illustrated Books and Publishing*. U of Toronto P, 2010.

Egan, Michael. "The Neverland of ID: Barrie, Peter Pan, and Freud." *Children's Literature* 10 (1982). muse.jhu.edu/article/246035.

Egoff, Sheila, G.T. Stubbs, and L.F. Ashley, editors. *Only Connect: Readings on Children's Literature*. Oxford UP, 1969.

Elam, Diane. *Romancing the Postmodern*. Routledge, 1992.

Emberley, Julia. "In/Hospitable 'Aboriginalities' in Contemporary Indigenous Women's Writing." *The Oxford Handbook of Canadian Literature*, edited by Cynthia Sugars, Oxford UP, 2016, pp. 209-23.

Epple, Colette. "'Wild Irish with a Vengeance': Definitions of Irishness in Katharine Tynan's Children's Literature." *Divided Worlds: Studies in Children's Literature*, edited by Mary Shine Thompson and Valerie Coghlan, Four Courts Press, 2007, pp. 32-40.

Erdrich, Louise. *The Birchbark House*. Hyperion, 1999.

——. *The Game of Silence*. HarperCollins, 2006.

——. *The Porcupine Year*. HarperCollins, 2008.

Falconer, E. "Killarney." *The Child's Irish Song Book*. Anon. n.p.: Southway Junior Irish Literary Club, n.d.

Ferguson, Mary Anne. "The Female Novel of Development and the Myth of Psyche." *The Voyage In: Fictions of Female Development*, edited by Elizabeth Abel, Marianne Hirsch, and Elizabeth Langland, Dartmouth, 1983, pp. 228-43.

Fish, Stanley. "One University under God?" *Chronicle of Higher Education*, 7 January 2005. www.chronicle.com/article/One-University-Under-God-/45077.

Fisher, Philip. *Wonder, the Rainbow, and the Aesthetics of Rare Experiences*. Harvard UP, 1998.

Fitzsimmons Frey, Heather. "Dance, Dramaturgy, and Theatre for Young Audiences." *Society for Dance History Scholars Conference Proceedings*, 2011.

——. "Forbidden Phoenix and Anime." *Canadian Theatre Review* 139, Summer 2009, pp. 42–49.

——. "An Interview with Sandra Laronde." *Canadian Theatre Review*, 2013.

Flett, Donna Elwood. "Deepening the Reading Experience of Drew Hayden Taylor's Vampire Novel for Adolescents." *Knowing Their Place? Identity and Space in Children's Literature*, edited by Terri Doughty and Dawn Thompson, Cambridge Scholars, 2011, pp. 25–41.

Forster, E.M. "The Story of a Panic" [1902]. *Selected Stories*, edited by David Leavitt and Mark Mitchell, Penguin, 2001, pp. 1–23.

Foster, Susan. *Choreographing Empathy: Kinesthesia in Performance*. Routledge, 2011.

Fowke, Edith. *Folklore of Canada: Tall Tales, Songs, Stories, Rhymes, Legends, and Jokes from Every Corner of Canada*. McClelland and Stewart, 1976.

——. "Folktales and Folk Song." *Literary History of Canada: Canadian Literature in English*, edited by Carl F. Klinck, U of Toronto P, 1965, reprinted 1973.

Franklin, Colleen M. *The Strange and Dangerous Voyage of Captain Thomas James: A Critical Edition*. McGill-Queen's UP, 2014.

Frehan, Padraic. *Education and Celtic Myth: National Self-Image and Schoolbooks in 20th Century Ireland*. Rodopi, 2012.

Frye, Northrop. *Anatomy of Criticism: Four Essays*. Princeton UP, 1957.

——. "Conclusion." *Literary History of Canada*, edited by Carl F. Klinck, U of Toronto P, 1965, reprinted 1973.

——. *The Secular Scripture: A Study of the Structure of Romance*. Harvard UP, 1976.

Gaard, Greta. "Children's Environmental Literature: Ecocriticism and Ecopedagogy." *Neohelicon* 36, 2009, pp. 321–34.

Gaiman, Neil. "Introduction." *Fragile Things: Short Fiction and Wonders*. Perennial Edition, HarperCollins, 2007, pp. xiii–xxxiii.

Gardner, John. *The Art of Fiction: Notes on Craft for Young Writers*. Vintage, 1991.

Garth, John. *Tolkien and the Great War: The Threshold of Middle-earth*. Houghton Mifflin, 2003.

Gatenby, Greg, editor. *The Wild Is Always There: Canada Through the Eyes of Foreign Writers*. A.A. Knopf, 1993.

Gerzina, Gretchen. *Frances Hodgson Burnett*. Chatto and Windus, 2004.

Gilbert, Kevin, editor. *Inside Black Australia: An Anthology of Aboriginal Poetry*. Penguin Books Australia, 1988.

Glotfelty, Cheryll, and Harold Fromm. *The Ecocriticism Reader: Landmarks in Literary Ecology*. U of Georgia P, 1996.

Goddard, John. *The Last Stand of the Lubicon Cree*. Douglas & McIntyre, 1991.

Goldstone, Bette. "Postmodern Experiments." *Children's Literature: Approaches and Territories*, edited by Janet Maybin and Nicola J. Watson, Palgrave Macmillan, 2009, pp. 320–29.

Gordon, Jon. *Unsustainable Oil: Facts, Counterfacts, and Fictions.* U of Alberta P, 2015.

Gould, Glenn. *The Idea of North. Ideas.* CBC Radio, 28 December 1967, www.cbc.ca/player/ Radio/More+Shows/Glenn+Gould++The+CBC+Legacy/Audio/1960s/ID/2110447480.

Graham, Judith. "Reading Contemporary Picturebooks." *Modern Children's Literature: An Introduction,* edited by Kimberley Reynolds, Palgrave Macmillan, 2005, pp. 209-26.

Graham, Kathryn V. "The Devil's Own Art: Topiary in Children's Fiction." *Children's Literature* 33, 2005, pp. 94-114.

Grahame, Kenneth. *The Annotated Wind in the Willows by Kenneth Grahame.* Edited and annotated by Annie Gauger, W.W. Norton, 2009.

——. *The Wind in the Willows* [1908]. Edited by Peter Hunt, Oxford UP, 2010.

——. "By a Northern Furrow" [1888]. *Paths to the River Bank.* Edited by Peter Haining, Blandford Press, 1993, pp. 75-79.

——. "The Lost Centaur" [1893]. *Paths to the River Bank,* pp. 92-94.

——. "The Romance of the Road" [1891]. *Paths to the River Bank,* pp. 45-49.

——. "The Rural Pan" [1891]. *Paths to the River Bank,* pp. 29-34.

Green, Peter. *Beyond the Wild Wood.* London: Grange, 1982.

Greenway, Betty. "Introduction: The Greening of Children's Literature." *Children's Literature Association Quarterly,* vol. 19, no. 4, Winter 1994, pp. 146-47.

Gregory, Derek. "Imaginative Geographies." *Progress in Human Geography* 19, pp. 447-85.

Greiman, Liela Rumbaugh. "William Ernest Henley and *The Magazine of Art.*" *Victorian Periodicals Review,* vol. 16, no. 2, Summer 1983, pp. 53-64.

Grianna, Maire ni. "Memories of the Famine." *The Irish: A Treasury of Art and Literature,* edited by Leslie C. Carola, Hugh Lauter Levin, 1993, pp. 153-54.

Griffin, William D. "The Distant Land." *The Irish: a Treasury of Art and Literature,* edited by Leslie C. Carola, Hugh Lauter Levin, 1993, pp. 185-89.

Groen, Rick. "Harry Potter rings hollow without Hogwarts." *Globe and Mail,* 18 November 2010. www.theglobeandmail.com/arts/film/harry-potter-rings-hollow-without-hogwarts/ article1461702/.

Guroian, Vigen. *Tending the Heart of Virtue: How Classic Stories Awaken a Child's Moral Imagination.* Oxford UP, 1998.

Hack, Maria. *Winter Evenings; or, Tales of Travellers.* 4 vols., Darton, Harvey, and Darton, 1818.

Haestrup, Jørgen. *European Resistance Movements, 1939–1945: A Complete History.* Westport and Meckler, 1981.

Hatfield, Len. "From Master to Brother: Shifting the Balance of Authority in Ursula K. Le Guin's *Farthest Shore* and *Tehanu.*" *Children's Literature* 21, 1993, pp. 43-65.

Hein, Rolland. *Christian Mythmakers.* Cornerstone, 1998.

Hey, David, editor. *The Oxford Companion to Family and Local History.* Oxford UP, 2009. www .oxfordreference.com/view/10.1093/acref/9780199532988.001.0001/acref-9780199532988.

Higgins, Edward, and Tom Johnson. "The Enemy Church: Pullman's Agenda in 'Compass' and Beyond." *Christian Century,* vol. 125, no. 1, 2008, pp. 28-31. www.christiancentury .org/article/2008-01/ememy-church.

Higgins, Michael D., and Declan Kiberd. "Culture and Exile: The Global Irish." *New Hibernia Review*, vol. 1, no. 3, 1997, pp. 9–22.

Hitchens, Christopher. "The Boy Who Lived." *New York Times*, 12 August 2007. www.nytimes .com/2007/08/12/books/review/Hitchens-t.html.

Hong, Jackie, and Jessie Winter. "HMS Terror, second ship from doomed Franklin Expedition, found in Terror Bay." www.thestar.com/news/canada/2016/09/12/hms-terror -from-doomed-franklin-expedition-found.html.

Hutton, Ronald. *The Triumph of the Moon: A History of Modern Pagan Witchcraft.* Oxford UP, 2001.

Irish Monthly, vol. 35, no. 414 (1907), p. 701. https://www.jstor.org/stable/i20501258.

James, Lynette. "Children of Change, Not Doom: Indigenous Futurist Heroines in YA." *Extrapolation*, vol. 57, nos. 1–2, 2016, pp. 151–76.

Janeway, James. *A Token for Children: Being an Exact Account of the Conversion, Holy and Exemplary Lives, and Joyful Deaths of Several Young Children.* T. Norris and A. Bettesworth, 1691.

Jeffares, A. Norman, and Antony Kamm. *An Irish Childhood.* Collins, 1987.

Jefferies, Richard. *After London or Wild England* [1885]. Oxford UP, 1982.

Johansen, K.V. *Beyond Window-Dressing? Canadian Children's Fantasy at the Millennium.* Sybertooth, 2007.

Jola, Corinne, Ali Abedian-Amiri, Annapoorna Kuppuswamy, Frank E. Pollick, and Marie-Hélène Grosbas. "Motor simulation without motor expertise: enhanced corticospinal excitability in visually experienced dance spectators." *PLoS One*, vol. 7, no. 3, March 2012. doi.org/10.1371/journal.pone.0033343.

Jola, Corrine, Shantel Ehrenberg, and Dee Reynolds. "The Experience of Watching Dance: Phenomenological-Neuroscience Duets." *Phenomenology and Cognitive Sciences* 11, 2012, pp. 17–37.

Justice, Daniel Heath. "'Go Away, Water!' Kinship Criticism and the Decolonization Imperative." *Reasoning Together: The Native Critics Collective*, edited by Craig S. Womack et al., U of Oklahoma P, 2008, pp. 147–68.

Kealy, J. Kieran. "The Flame-Lighter Woman: Catherine Anthony Clark's Fantasies." *Canadian Literature*, 2 May 2013. canlit.ca/article/the-flame-lighter-woman/.

Kennedy, Dennis. "The Body of the Spectator." *The Spectator and the Spectacle: Audiences in Modernity and Postmodernity.* Cambridge UP, 2011.

Kernaghan, Eileen. *The Snow Queen.* Thistledown Press, 2000.

Kiberd, Declan. "Foreword." Nancy Watson, *The Politics and Poetics of Irish Children's Literature.* Irish Academic Press, 2009.

——. *Inventing Ireland: The Literature of a Modern Nation.* Jonathan Cape, 1995.

——. "School Stories." *Studies in Children's Literature 1500–2000*, edited by Celia Keenan and Mary Shine Thompson, Four Courts Press, 2004, pp. 54–69.

Kingsley, Charles. *The Water Babies.* Bridlington: Priory Books, n.d.

Kipling, Rudyard. *Kim.* Edited by Alan Sandison, Oxford UP, 1998.

——. *Puck of Pook's Hill.* Doubleday, Page, 1906.

Kirby, W.F., translator. *Kalevala: The Land of Heroes.* 2 vols. J.M. Dent & Co., 1907.

Kirkby, Mandy. *A Victorian Flower Dictionary: The Language of Flowers Companion.* Random House, 2011.

Kutzer, M. Daphne. *Empire's Children: Empire and Imperialism in Classic British Children's Books.* Garland, 2000.

Kwaymullina, Ambelin. "Edges, Centres, and Futures: Reflections on Being an Indigenous Speculative Fiction Writer." *Kill Your Darlings* 18, 2014, pp. 22-33.

Landy, Joshua, and Michael Saler. "Introduction: The Varieties of Modern Enchantment." *The Re-Enchantment of the World: Secular Magic in a Rational Age*, edited by Joshua Landy and Michael Saler, Stanford UP, 2009, pp. 1-14.

Lang, Andrew, editor. *The Blue Fairy Book.* Dover, 1965.

Larbalestier, Justine. *The Battle of the Sexes in Science Fiction.* Wesleyan UP, 2002.

Laronde, Sandra, Personal interview. 28 March 2012.

Larsen, Andrew. *The Imaginary Garden.* Illustrated by Irene Luxbacher, Kids Can Press, 2009.

Larsen, Erik. *The Devil in the White City: Murder, Magic, and Madness at the Fair That Changed America.* Vintage Books, 2003.

Lefanu, Sarah. *In the Chinks of the World Machine: Feminism and Science Fiction.* Women's Press, 1988.

Le Faye, Deirdre, editor. *Jane Austen's Letters.* 4th ed., Oxford UP, 2011.

"The Legend of White Buffalo Calf Woman. An interview with Bill Means." www.youtube .com/watch?v=ezNKgRbnVPY.

Le Guin, Ursula K. *Conversations with Ursula K. Le Guin.* Editor. Carl Freedman. UP of Mississippi, 2008.

——. *The Dispossessed.* Harper, 1974.

——. *The Farthest Shore* [1972]. Simon and Schuster, 2012.

——. *The Language of the Night: Essays on Fantasy and Science Fiction.* Edited by Susan Wood, Putnam, 1979.

——. *The Left Hand of Darkness.* Ace Books, 1969.

——. *The Other Wind* [2001]. Houghton Mifflin, 2012.

——. *Tales from Earthsea* [2001]. Houghton Mifflin, 2012.

——. *Tehanu* [1990]. Simon and Schuster, 2012.

——. *The Tombs of Atuan* [1970]. Simon and Schuster, 2012.

——. *The Wave in the Mind.* Shambhala, 2004.

——. *A Wizard of Earthsea* [1968]. Houghton Mifflin, 2012.

L'Engle, Madeleine. *An Acceptable Time.* [1989]. Bantam Doubleday Bell, 1990.

——. "Allegorical Fantasy: Mortal Dealings with Cosmic Questions." Interview with Madeline L'Engle. *Christianity Today*, 8 June 1979. www.christianitytoday.com/ct/2007/ septemberweb-only/136-52.0.html.

——. "Before Babel." *Horn Book Magazine* 42, December 1966, pp. 661-70.

——. "Childlike Wonder and the Truths of Science Fiction." *Children's Literature* 10, 1982, pp. 102-10.

——. *A Circle of Quiet.* HarperSanFrancisco, 1972.

——. "Do I Dare Disturb the Universe?" *Horn Book Magazine* 59, 1983, pp. 673-82.

——. "George MacDonald: Nourishment for a Private World." *Reality and the Vision*, edited by Philip Yancey. *Word*, 1990, pp. 111-21.

——. "The Heroic in Literature and in Living." *The Lion and the Unicorn* 13, 1989, pp. 120-28.

——. "Kerlan Award Lecture." *Kerlan Collection*, Fall 1990, pp. 1-7.

——. "Knowing Things Ahead of Time." *Things in Heaven and Earth: Exploring the Supernatural*, edited by Harold Fickett. *Paraclete*, 1998, pp. 3-9.

——. *Many Waters* [1986]. Dell, 1991.

——. *Penguins and Golden Calves: Icons and Idols*. Harold Shaw, 1996.

——. *The Rock That Is Higher*. Harold Shaw, 1993.

——. "Subject to Change without Notice." *Theory into Practice*, vol. 21, no. 4, 1982, pp. 332-36.

——. *A Swiftly Tilting Planet* [1978]. Dell, 1980.

——. Telephone interview with Monika Hilder, 26 September 2000.

——. *Walking on Water: Reflections on Faith and Art*. Harold Shaw, 1980.

——. "What Is Real?" *Language Arts*, vol. 55, no. 4, 1978, pp. 447-51.

——. *A Wind in the Door* [1973]. Square Fish, 2007.

——. *A Wrinkle in Time* [1962]. Dell, 1985.

L'Engle, Madeleine with Avery Brooke. *Trailing Clouds of Glory: Spiritual Values in Children's Literature*. Westminster, 1985.

Lévi-Strauss, Claude. *The Raw and the Cooked*. Translated by John and Doreen Weightman, Harper Colphon, 1975.

Lewis, C.S. "On Stories" [1947]. *On Stories and Other Essays on Literature*. Harcourt, 1982, pp. 3-20.

——. *The Silver Chair* [1953]. HarperTrophy, 2000.

——. "The Weight of Glory." [1941/1949]. *The Weight of Glory and Other Addresses*, 1977, pp. 1-15.

Lightman, Bernard. "Does the History of Science and Religion Change Depending on the Narrator? Some Atheist and Agnostic Perspectives." *Science and Christian Belief*, vol. 24, no. 2, October 2012, pp. 149-68.

Lingaua, Angelika, Benno Gesiericha, and Alfonso Caramazzaa. "Asymmetric fMRI Adaptation Reveals No Evidence for Mirror Neurons in Humans." *Proceedings of the National Academy of Sciences*, vol. 106, no. 24, pp. 9925-30, doi:10.1073/pnas.0902262106.

Little Bear, Leroy. "Leroy Little Bear: Native Science and Western Science." lib.asu.edu/librarychannel/2011/05/16/ep114_littlebear.

Littlefield, Holly. "Unlearning Patriarchy: Ursula Le Guin's Feminist Consciousness in *The Tombs of Atuan* and *Tehanu*." *Extrapolation*, vol. 36, no. 3, 1995, pp. 244+. connection.ebscohost.com/c/literary-criticism/9510294427/unlearning-patriarchy-ursula-le-guins-feminist-consciousness-tombs-atuan-tehanu.

"Living Myths. Native American Myths." www.livingmyths.com/Native.htm.

Lochhead, Marion. *The Renaissance of Wonder in Children's Literature*. Canongate, 1977.

Loeber, Rolf, and Magda Loeber. *A Guide to Irish Fiction, 1650–1900*. Four Courts Press, 2006.

Loomis, Chauncey. "The Arctic Sublime." *Nature and the Victorian Imagination*, edited by U.U.C. Knoepflmacher and G.B. Tennyson, U of California P, 1977, pp. 95-112.

Lurgan Ancestry. "The Great Famine." www.lurganancestry.com/famine.htm.

Luxbacher, Irene. "The Imaginary Garden: Meet-the-Author Book Reading." www.teaching
-books.net/tb.cgi?aid=11455.

MacDonald, George. *At the Back of the North Wind* [1871]. Broadview, 2011.

———. "The Fantastic Imagination" [1893]. *The Complete Fairy Tales*, Penguin, 1999, pp. 5–10.

———. *Phantastes: A Faerie Romance for Men and Women*. London, 1858. ftp.mirrorservice.org/
sites/ftp.ibiblio.org/pub/docs/books/gutenberg/3/2/325/325-h/325-h.htm.

———. *The Princess and Curdie* [1883]. ftp://ftp.mirrorservice.org/sites/ftp.ibiblio.org/pub/
docs/books/gutenberg/7/0/709/709-h/709-h.htm.

———. *The Princess and the Goblin* [1872]. www.mirrorservice.org/sites/ftp.ibiblio.org/pub/
docs/books/gutenberg/7/0/708/708-h/708-h.htm.

MacDonald, Robert H. *The Language of Empire: Myths and Metaphors of Popular Imperialism,
1880–1918*. Manchester UP, 1994.

MacDonald, Ruth K. *Christian's Children: The Influence of John Bunyan's The Pilgrim's Progress on
American Children's Literature*. Peter Lang, 1989.

Macfarlane, Robert. *The Wild Places*. Penguin, 2008.

MacLennan, Hugh. *Scotsman's Return and Other Essays*. Heinemann, 1961.

MacMillan, Cyrus. *Canadian Wonder Tales*. John Lane, The Bodley Head, 1920. www.gutenberg
.ca/ebooks/macmillan-wonder/macmillan-wonder-00-h-dir/macmillan-wonder
-00-h.html.

MacSorley, Catherine Mary. *Nora: An Irish Story*. SPCK, n.d.

———. *An Irish Cousin*. SPCK, n.d.

Maeterlinck, Maurice. *The Blue Bird: A Fairy Play in Six Acts*. Translated by Alexander Teixeira
de Mattos, Dodd Mead, 1911. www.gutenberg.org/cache/epub/8606/pg8606.html.

Mah, D.B. "Australian Landscape: Its Relationship to Culture and Identity." Master of
Visual Arts thesis, University of Western Sydney, 1997. researchdirect.westernsydney
.edu.au/islandora/object/uws:257.

Major, Melissa. Personal interview, 20 September 2012.

Maracle, Lee. "The Other Side of Me." *Canadian Cultural Studies: A Reader*, edited by Sourayan
Mookerjea, Imre Szeman, and Gail Faurschou, Duke UP, 2009, pp. 383–404.

"Margaret Mahy." Obituary. *The Guardian*, 26 July 2012. n.pag. www.theguardian.com/
books/2012/jul/26/margaret-mahy.

Massey, Doreen. *For Space*. SAGE Publications, 2005.

McCarthy, Pat. "Raven Stole the Sun Study Guide." Red Sky Performance, 2008. www
.redskyperformance.com/wpcontent/uploads/2017/05/RedSky_Raven_StudyGuide
.pdf.

McClure, John. *Late Imperial Romance*. Verso, 1994.

McConachie, Bruce. *Engaging Audiences: A Cognitive Approach to Spectating in the Theatre*.
Palgrave Macmillan, 2008.

McDowell, Linda. *Gender, Identity, and Place: Understanding Feminist Geographies*. U of Minnesota P, 1999.

McDowell, Marjorie. "Children's Books." *Literary History of Canada*, edited by Carl F. Klinck,
U of Toronto P, 1965, reprinted 1973.

M.E.T. *Exiled from Erin: A Story of Irish Peasant Life*. Dublin: James Duffy and Sons, n.d.

Meunier, Christophe. "The Cartographic Eye in Children's Picturebooks: Between Maps and Narratives." *Children's Literature in Education*, vol. 48, no. 1, 2017, pp. 21-38. link .springer.com/article/10.1007/s10583-016-9302-6.

Millar, George. *Maquis* [1945]. Cedric Chivers, 1973.

Miller, Mary Jane. *Outside Looking in: Viewing First Nations Peoples in Canadian Dramatic Television Series*. McGill-Queen's UP, 2008.

Milloy, John S. *"A National Crime": The Canadian Government and the Residential School System, 1879 to 1986*. U of Manitoba P, 1999.

Milne, A.A. *Winnie-the-Pooh*. 1926. Dell, 1982.

Moebius, William. "Picturebook Codes." *Children's Literature: Approaches and Territories*, edited by Janet Maybin and Nicola J. Watson, Palgrave Macmillan, 2009, pp. 311-20.

Molson, Francis J. *"Two Little Pilgrims' Progress*: The 1892 Chicago Columbian Exposition as Celestial City." *Markham Review*, 1978, pp. 55-59.

Monaghan, James. "Song of an Exile." *Young Ireland*, vol. 12, no. 5, 1886, pp. 73.

Monk, Patricia. *Mud and Magic Shows: Robertson Davies's Fifth Business*. ECW Press, 1992.

Montgomery, L.M. *Anne of Green Gables*. McGraw-Hill Ryerson, 1968.

——. *Emily of New Moon*. Frederick Stokes, 1923.

Moore, Thomas. "Let Erin Remember the Days of Old." *Gems for the Young from Favourite Poets*, edited by Rosa Mulholland, M.H. Gill and Son, 1891.

Morgenstern, Joe. "'Harry Potter' and the Endless Ending." *Wall Street Journal*, 18 November 2010. www.wsj.com/articles/SB10001424052748704104104575622452675925176.

Morin, Karen M. "Edward W. Said." *Key Thinkers on Space and Place*, edited by Phil Hubbard, Rob Kitchin, and Gill Valentine, SAGE Publications, 2004, pp. 237-44.

Mowat, Farley. *Lost in the Barrens*. McClelland and Stewart, 1956.

Muller-Funk, Wolfgang. "On a Narratology of Cultural and Collective Memory." *Journal of Narrative Theory*, vol. 33, no. 2, Summer 2003, pp. 207-27. muse.jhu.edu/article/376012.

Mulholland, Rosa. *The Return of Mary O'Murrough*. Phoenix, n.d.

Mulholland, Rosa, editor. *Gems for the Young from Favourite Poets*. M.H Gill and Son, 1891.

Murray, Heather, "The Geography of the Imagination: the Fantastic Frontier of Catherine Anthony Clark." *Children's Literature Association Quarterly*, vol. 8, no. 4, Winter 1983, pp. 23-25. muse.jhu.edu/journals/chq/summary/v008/8.4.murray.html.

Nagy, Joseph Falaky. "Liminality and Knowledge in the Irish Tradition." *Studia Celtica* 16-17, 1981-82, pp. 135-43.

Ness, Sally Ann. "The Inscription of Gesture: Inward Migrations in Dance." *Migrations of Gesture*, edited by Carrie Noland and Sally Ann Ness, U of Minnesota P, 2008.

Nodelman, Perry. "Reinventing the Past: Gender in Ursula K. Le Guin's *Tehanu* and the *Earthsea* 'Trilogy.'" *Children's Literature* 23, 1995, pp. 179-201.

Noonan, Mark. "From Century to St. Nick, or How Mary Mapes Dodge Came to Fame Editing the Infamous Frances Hodgson Burnett." *Popular Nineteenth-Century American Women Writers and the Literary Marketplace*, edited by Earl Yarington and Mary De Jong, Cambridge Scholars, 2007, pp. 367-87.

Oppel, Kenneth. *Airborn*. HarperCollins, 2004.

Orr, James. "The Irishman." *The Child's Irish Song Book,* Southway Junior Irish Literary Club, n.d.

Ortner, Sherry. "Is Female to Male as Nature Is to Culture?" *Women, Culture and Society,* edited by Michelle Zimbalist Rosaldo and Louise Lamphere, Stanford UP, 1974, pp. 67-87.

Oziewicz, Marek. "Envisioning Spirituality in a New Paradigm: Madeleine L'Engle's Time Quartet and the Hope for Humanity's Survival." *Towards or Back to Human Values? Spiritual and Moral Dimensions of Contemporary Fantasy,* edited by Justyna Deszcz-Tryhubczak and Marek Oziewicz, Cambridge Scholars Press, 2006, pp. 62-81.

——. "Integrating Science and Spirituality: Madeleine L'Engle's Time Quartet (1962-1986)." *One Earth, One People: The Mythopoeic Fantasy Series of Ursula K. Le Guin, Lloyd Alexander, Madeleine L'Engle, and Orson Scott Card,* McFarland, 2008, pp. 171-97.

Padley, Jonathan, and Kenneth Padley. "'A Heaven of Hell, a Hell of Heaven': *His Dark Materials,* Inverted Theology, and the End of Philip Pullman's Authority." *Children's Literature in Education* 37, 2006, pp. 325-34. doi.org/10.1007/s10583-006-9022-4.

Paquette, Aaron. *Lightfinder*. Kegedonce Press, 2014.

Pavis, Patrice. "Introduction: Towards a Theory of Interculturalism and Theatre." *The Intercultural Performance Reader,* edited by Patrice Pavis. Routledge, 1996, pp. 1-19.

Posesorski, Sheri. "Urban Nature Boy." Review of *The Curious Garden* by Peter Brown. *New York Times Sunday Book Review.* www.nytimes.com/2009/05/10/books/review/Posesorski-t.html.

Pratt, Annis. *Archetypal Patterns in Women's Fiction.* Indiana UP, 1981.

Prickett, Stephen. "Fictions and Metafictions." *The Gold Thread: Essays on George MacDonald,* edited by William Raeper, Edinburgh UP, 1990, pp. 109-25.

Pullman, Philip. *The Golden Compass.* Yearling-Random House, 2001.

——. "Theatre - The True Key to Stage." *The Guardian,* 30 March 2004. www.theguardian.com/books/2004/mar/30/booksforchildrenandteenagers.schools.

Rand, McNally & Co.'s Handbook to the World's Columbian Exposition. Rand, McNally, 1893.

Raposa, Michael L. *Boredom and the Religious Imagination.* UP of Virginia, 1999.

Rawls, Melanie A. "Witches, Wives, and Dragons: The Evolution of Women in Ursula K. Le Guin's *Earthsea* - An Overview." *Mythlore,* vol. 26, nos. 3-4, 2008, pp. 129-49.

Read, Alan. "Fear, Fur, Fauna, Falling: Into Bed with The Lion, the Witch and the Wardrobe." *Papers,* vol. 12, no. 1, 2002, pp. 20-29.

Reason, Matthew. *The Young Audience: Exploring and Enhancing Children's Experience of Theatre.* Trentham Books, 2010.

Reason, Matthew, and Dee Reynolds. "Kinesthesia, Empathy, and Related Pleasures: An Inquiry into Audience Experiences of Watching Dance." *Dance Research Journal,* vol. 42, no. 2, 2010, pp. 49-75.

Reason, M., C. Jola, R. Kay, D. Reynolds, J.-P. Kauppi, M.-H. Grobras ... F.E. Pollick. 2016. "Spectators' aesthetic experience of sound and movement in dance performance:

A transdisciplinary investigation." *Psychology of Aesthetics, Creativity, and the Arts*, vol. 10, no. 1, pp. 42–55. dx.doi.org/10.1037/a0040032.

Reeve, Philip. *Larklight*. London: Bloomsbury, 2007.

Reimer, Mavis, editor. *Home Words: Discourses of Children's Literature in Canada*. Wilfrid Laurier UP, 2008.

Reimer, Mavis, and Bradford, Clare. "Home, Homelessness, and Liminal Spaces: The Uses of Postcolonial Theory for Reading (National) Children's Literatures." *Children's Literature Global and Local: Social and Aesthetic Perspectives*, edited by E. O'Sullivan and R. Romoren Kristiansand, Novus, 2005, pp. 200–217.

Richards, Thomas. *The Imperial Archive: Knowledge and the Fantasy of Empire*. Verso, 1993.

Ricou, Laurie. "So Big about Green." *Canadian Literature* 130, Autumn 1991, pp. 3–6.

Robinson, Christopher. "The Violence of the Name: Patronymy in *Earthsea*." *Extrapolation*, no. 49, no. 3, 2008, pp. 385–409.

Rosebury, Brian. *Tolkien: A Critical Assessment*. St. Martin's Press, 1992.

Ross, Malcolm. *Poets of the Confederation: Carman/Lampman/Roberts/Scott*. McClelland and Stewart, 1960.

Rowling, J.K. *Harry Potter and the Deathly Hallows*. Raincoast Books, 2007.

——. *Harry Potter and the Philosopher's Stone*. Bloomsbury, 1997.

Russ, Joanna. "The Image of Women in Science Fiction." *The Country You Have Never Seen*. Liverpool UP, 2007, pp. 205–18.

Sadlier, Mrs. J. *Bessie Conway; or, The Irish Girl in America*. P.J. Kennedy and Sons, n.d.

Said, Edward W. *Orientalism*. Vintage, 1978.

——. *Culture and Imperialism*. A.A. Knopf, 1993.

Sale, Roger. *Fairy Tales and After: From Snow White to E.B. White*. Harvard UP, 1978.

Saler, Michael. *As If: Modern Enchantment and the Literary Prehistory of Virtual Reality*. Oxford UP, 2012.

Sandner, David. *The Fantastic Sublime: Romanticism and Transcendence in Nineteenth-century Children's Fantasy Literature*. Greenwood Press, 1996.

Sangster, Charles. *The St. Lawrence and the Saguenay and Other Poems: Hesperus and Other Poems and Lyrics*. U of Toronto P, 1972.

Schonmann, Shifra. *Theatre as a Medium for Children and Young People*. Springer, 2006.

Scutter, Heather. *Displaced Fictions: Contemporary Australian Books for Teenagers and Young Adults*. Melbourne UP, 1999.

Selby, Joan. "The Creation of Fantasy: The Fiction of Catherine Anthony Clark." *Canadian Literature* 8, December 2011. Originally appeared in *Canadian Literature* 11, Winter 1962, pp. 39–45.

Seuss, Dr. *The Lorax*. Random House, 1971.

Showalter, Elaine. *Sister's Choice: Tradition and Change in American Women's Writing*. Clarendon, 1991.

Sigler, Carolyn. "Wonderland to Wasteland: Toward Historicizing Environmental Activism in Children's Literature." *Children's Literature Association Quarterly*, vol. 19, no. 4, Winter 1994, pp. 148–53.

Silkenat, David. "Workers in the White City: Working Class Culture at the Columbian World Exposition of 1893." *Journal of the Illinois State Historical Society*, vol. 104, no. 4, 2011, pp. 266-300.

Simpson, Leanne. *As We Have Always Done*. U of Minnesota P, 2017.

Singh, Rashna B. *Goodly Is Our Heritage: Children's Literature, Empire, and the Certitude of Character*. Scarecrow, 2004.

Slipperjack, Ruby. *Little Voice*. Coteau Books, 2001.

Smith, David E. *John Bunyan in America*. Indiana UP, 1966.

Smith, Duane, Karen A. Vendl, and Mark A. Vendl. *Colorado Goes to the Fair: World's Columbian Exposition*. U of New Mexico P, 2011.

Smith, Lane. "Grandpa Green: Meet-the-Author Book Reading." www.teachingbooks.net/tb.cgi?tid=25358.

——. *Grandpa Green*. Roaring Book Press, 2011.

Sorfleet, John R. "The Nature of Canadian Children's Literature: A Commentary." *Windows and Words: A Look at Canadian Children's Literature in English*, edited by Aïda Hudson and Susan-Ann Cooper, U of Ottawa P, 2003, pp. 219-25.

South Australian Weekly Chronicle [Adelaide], 4 May 1867. trove.nla.gov.au/newspaper/page/8388555.

Spitz, Ellen Handler. "Empathy, Sympathy, Aesthetics, and Childhood: Fledgling Thoughts." *American Imago*, vol. 64, no. 4, 2008, pp. 545-59.

Stafford, Margot. "Journeys Through Bookland's Imaginative Geography: Pleasure, Pedagogy and the Child Reader." *Space and Place in Children's Literature, 1789 to the Present*, edited by Maria Sachiko Cecire, Hannah Field, and Malini Roy, Ashgate, 2015, pp. 147-64.

Stevenson, Robert Louis. *A Child's Garden of Verses*. Charles Scribner and Sons; London: John Lane, 1895.

Sullivan, Timothy Daniel. "Songs from the Backwoods." *The Child's Irish Song Book*, Southway Junior Irish Literary Club, n.d.

Tally, Robert T., Jr., editor. *Spatiality*. Routledge, 2013.

Taylor, Drew Hayden. *The Night Wanderer: A Native Gothic Novel*. Annick Press, 2007.

——. "The Night Wanderer." www.youtube.com/watch?v=xpSo1H81c04.

Taylor, Ken. "Landscape and Memory." Paper given at 16th ICOMOS General Assembly and International Symposium, 2008, pp. 1-14. openarchive.icomos.org/139/.

Thomas, Joyce. "Woods and Castles, Towers and Huts: Aspects of Setting in the Fairy Tale." *Only Connect: Readings on Children's Literature*, 3rd ed., edited by Sheila Egoff et al., Oxford UP, 1996, pp. 122-29.

Thomas, Trudelle H. "Spiritual Practices Children Understand: An Analysis of Madeleine L'Engle's Fantasy, *A Wind in the Door*." *International Journal of Children's Spirituality*, vol. 13, no. 2, 2008, pp. 157-69. eric.ed.gov/?id=EJ811384.

Thomson, James. *Winter*. Edinburgh, 1726.

Thwaite, Ann. *Frances Hodgson Burnett*. Boston: Twayne, 1984.

——. *Waiting for the Party: The Life of Frances Hodgson Burnett, 1849-1924*. Charles Scribner, 1974.

"Tjurunga." *Encyclopedia Britannica*. www.britannica.com/topic/tjurunga.

Tolkien, J.R.R. *The Hobbit*. [1937]. George Allen and Unwin, 1966.

———. *The Lord of the Rings*. [1954–55]. Illustrated by Alan Lee, Houghton Mifflin, 2004.

———. "On Fairy-Stories." *Tree and Leaf Including the Poem Mythopoeia*, HarperCollins, 2001, pp. 3–81.

Toohey, Peter. *Boredom: A Lively History*. Yale UP, 2011.

Tourism British Columbia. Official Travel Website of BC Canada, n.d. www.hellobc.com.

Traill, Catherine Parr. *The Backwoods of Canada, Being Letters from the Wife of an Emigrant Officer, Illustrative of the Domestic Economy of British America*. Charles Knight, 1836. gutenberg .org/catalog/world/readfile?fk_files=3486025.

Trevor, William. *A Writer's Ireland: Landscape in Literature*. Thames and Hudson, 1984.

———. *Canadian Crusoes: A Tale of the Rice Lake Plains*. Edited by Agnes Strickland, Arthur Hall, 1852. www.gutenberg.org/files/8382/8382-h/8382-h.htm#2H_PREF.

Trites, Roberta Seelinger. *Disturbing the Universe*. U of Iowa P, 2000.

Truth and Reconciliation Commission of Canada. *Canada's Residential Schools: The History*, Pt. 1: *Origins to 1939*. Truth and Reconcilation Commission. McGill-Queen's UP, 2015. www .myrobust.com/websites/trcinstitution/File/Reports/Volume_1_History_Part_1_ English_Web.pdf.

———. *Canada's Residential Schools: The History*, Pt. 2: *1939 to 2000*. Truth and Reconciliation Commission. McGill-Queen's UP, 2015.

———. *Honouring the Truth, Reconciling for the Future: Summary of the Final Report of the Truth and Reconciliation Commission of Canada*. Truth and Reconciliation Commission. McGill-Queen's UP, 2015. www.trc.ca/websites/trcinstitution/File/2015/Findings/Exec _Summary_2015_05_31_web_o.pdf.

———. *The Survivors Speak*. Truth and Reconciliation Commission. McGill-Queen's UP, 2015. www.trc.ca/websites/trcinstitution/File/2015/Findings/Survivors_Speak_2015_05 _30_web_o.pdf.

Turner, Victor. *The Ritual Process: Structure and Anti-structure*. Routledge & Kegan Paul, 1969.

Turner, Victor, and Edith Turner. *Image and Pilgrimage in Christian Culture: Anthropological Perspectives*. Columbia UP, 1978.

Tynan, Kathleen. *Peeps at Many Lands: Ireland*. Adam and Charles Black, 1909.

Upton, Bertha. *The Golliwogg's Polar Adventure*. Illustrated by Florence Upton, Longmans, Green and Co., 1900.

von Slatt, Jake. "At the Intersection of Technology and Romance." *Steampunk II: Steampunk Reloaded*, edited by Ann and Jeff Vandermeer, Tachyon, 2010, pp. 404–8.

Wadsworth, Sarah, and Wayne A. Wiegand. *Right Here I See My Own Books: The Woman's Building Library at the World's Columbian Exposition*. U of Massachusetts P, 2012.

Wartofsky, Alona. "The Last Word: Philip Pullman's Trilogy for Young Adults Ends with God's Death, and Remarkably Few Critics." *Washington Post*, 19 February 2001, C01. Academic OneFile. www.highbeam.com/doc/1P2-418825.html.

Weaver, Roslyn. *Apocalypse in Australian Fiction and Film: A Critical Study*. McFarland, 2011.

Wells, H.G. *The Island of Dr. Moreau* [1896]. Edited by Brian Aldiss, Dent, 1993.

Welty, Eudora. "Place in Fiction." *Collected Essays*, 1994. xroads.virginia.edu/~DRBR/welty.txt.

Westerfeld, Scott. *Leviathan*. Simon and Schuster, 2009.

White, Andrea. "Adventure Fiction: A Special Case." *Joseph Conrad and the Adventure Tradition*, Cambridge UP, 1993, pp. 39-61.

Wilde, Lady ["Speranza"]. *Ancient Legends, Mystic Charms, and Superstitions of Ireland, with Sketches of the Irish Past*. Ward and Downey, 1888, Facsimile ed., 1971.

Wolf, Casey. "Revisiting the Snow Queen: An E-mail Interview with Eileen Kernaghan." www.eileenkernaghan.ca/kernaghaninterviewreturnofsnowqueen.html.

Wolff, Doris, and Paul DePasquale. "Home and Native Land: A Study of Canadian Aboriginal Picture books by Aboriginal Authors." *Home Words: Discourses of Children's Literature in Canada*, edited by Mavis Reimer, Wilfrid Laurier UP, 2008, pp. 87-105.

World Columbian Exposition: Rand, McNally & Co.'s Sketch Book, Illustrating and Describing the Principal Buildings with their Locations, Dimensions, Const. etc. Rand, McNally, 1893.

Zenz, Aaron. "Interview #16: Lane Smith." *Bookie Woogie: Some Creative Kids and Their Illustrator Dad Just Talkin' about Books*. bookiewoogie.blogspot.com/2012/07/interview-16-lane-smith.html.

About the Contributors

‖

DEIRDRE F. BAKER is an Assistant Professor of English at the University of Toronto specializing in children's' literature. She is the author of *Becca at Sea* and its sequel *Becca Fair and Foul*. She is the children's book reviewer for the *Toronto Star* and other publications since 1998. She has published numerous scholarly articles and is co-author of *A Guide to Canadian Children's Books*.

CHRISTINE BOLUS-REICHERT is an Associate Professor of English at the University of Toronto and author of *The Age of Eclecticism: Literature and Culture in Britain, 1815–1885* (2009), as well as articles and papers on William Morris, George MacDonald, Daphne du Maurier, and, Peter O'Donnell. Her research focuses on Victorian and popular romance.

ALAN CUMYN is the award-winning author of thirteen wide-ranging literary novels for adults and for the young. His works for children and for young adults include the Owen Skye trilogy, *Tilt, All Night,* and *Hot Pteradactyl Boyfriend*. He is Chair of the Writing for Children and Young Adults MFA program at the Vermont College of Fine Arts.

PETRA FACHINGER is a Professor of English at Queen's University with a PhD in Comparative Literature from UBC. She is the author of *Rewriting Germany from the Margins*, as well as articles in diaspora and transnational studies, Indigenous literatures, and the environmental humanities.

JOANNE FINDON teaches medieval English literature, children's literature and creative writing at Trent University. She is the author of the time-travel fantasy novel *When Night Eats the Moon*, three picture books, numerous short stories, and most recently *Seeking Our Eden: The Dreams and Migrations of Sarah Jameson Craig,* a biography of her feisty great-grandmother, who was a radical dress reformer.

COLLEEN M. FRANKLIN a retired professor of English Literature. She has published numerous articles on British literature and northern exploration. Her book *The Strange and Dangerous Voyage of Captaine Thomas James: A Critical Edition* was published by McGill-Queen's University Press in 2014.

HEATHER FITZSIMMONS FREY is a Banting Post-Doctoral Fellow at York University, Toronto. A theatre director, dramaturge, and educator, her current research interests are theatre for young audiences, intersections between dance and theatre, performing marginalized identities, and Victorian children's amateur theatricals.

MONIKA B. HILDER is a Professor of English at Trinity Western University, where she teaches children's and fantasy literature. She is co-founder and co-director of the Inklings Institute of Canada and the author of a three-volume study of C.S. Lewis and gender, including *Surprised by the Feminine: A Rereading of C.S. Lewis and Gender* (2013). She has published articles on C.S. Lewis, George MacDonald, L.M. Montgomery, Madeleine L'Engle, and literature as ethical imagination.

MARGOT HILLEL OAM is a professor and Chair of the Academic Board at Australian Catholic University and is the co-author of *Child, Nation, Race and Empire: Child Rescue Discourse, England, Canada, and Australia, 1850–1915* and co-editor of *The Sands of Time; Children's Literature, Culture, Politics and Identity.* She has published numerous chapters and articles on subjects as varied as child rescue, concepts of childhood and children in Victorian literature in Britain and Australia, and the history of children's literature. Her chapter is based on research conducted while she was a Visiting Fellow at Trinity College, Dublin.

AÏDA HUDSON is a lecturer at the University of Ottawa and is co-editor of *Windows and Words: A Look at Canadian Children's Literature in English.*

PETER HYNES is an Associate Professor with the Department of English at the University of Saskatchewan; he publishes mostly in eighteenth-century studies.

LINDA KNOWLES is a Canadian independent scholar who has lived in the United Kingdom for the past forty years. A member of the British Association for Canadian Studies, she edited *CanText*, the BACS Literature Group Newsletter, and researches various aspects of Canadian literature, such as the novels of Carol Shields and Robertson Davies and the Scottish origins of the song, "Farewell to Nova Scotia." Linda has a PhD from the University of St. Andrews.

MELISSA LI SHEUNG YING holds a PhD in English from Queen's University and an MA in Comparative Literature from the University of Alberta. Her research specialties include Canadian literature, children's literature, and literature and the environment. She teaches at MacEwan University and the University of Alberta.

JANET LUNN CM, O Ont., was an award-winning historical novelist and picture book author for children. She also wrote numerous historical works and biographies, including a biography of Lucy Maud Montgomery. Her books include the well-known trilogy *Shadow in Hawthorn Bay*, *The Root Cellar*, and *The Hollow Tree*.

SHANNON MURRAY is a Professor at the University of Prince Edward Island, where she teaches Early Modern and Children's Literature. She has published on John Bunyan, John Milton, and children's adaptations of canonical literature. She is a 3M National Teaching Fellow.

CORY SAMPSON is a PhD student at the University of Ottawa who is studying emotional labour and women in the workforce in Victorian literature. His other academic interests include the effects of colonialism on British subjects in the nineteenth century and the effects of privilege on literary production.

ALAN WEST is a lecturer at the University of Ottawa and specializes in teaching children's literature, the mystery novel, and utopian and dystopian fiction.

SARAH FIONA WINTERS is an Associate Professor at Nipissing University. She is the co-editor of *Marvellous Codes: The Fiction of Margaret Mahy*. Her published articles include studies of C. S. Lewis, J. K. Rowling, and Philip Pullman.

Index

Abrahams, Wilbur, 38–39

acedia, and boredom, 219–20

A Child's Garden of Verse (Stevenson), 51–52; "The Northwest Passage," 52–57

adventure narratives: and imperialism, 42; of Late Victorian Britain, 25; as literary genre, 30; non-British adults as "Other," 31; polar exploration, 28–30, 45–46; "sublime quest" of the North, 46–48, 51, 64n3; as surrogate character-builder, 30–32, 42; and transgeneric fiction, 125–26; of travel and exploration, 48; *Winter Evenings* (Hack), 48–51

After Sylvia (Cumyn), 301

Airborn (Oppel), 13, 19, 229, 236–37; flying as sailing, 230; remythologizing of flight, 237

Akenside, Mark, *Pleasures of the Imagination*, 49

Alaimo, Stacey, 265

Alanus ab Insulis (Alan of Lille), 189; "Omnis mundi creatura," 186–87

Alcott, Louisa May, *Little Women*, 95, 99n9

Allingham, William, "Tenants at Will," 76

The Amber Spyglass (Pullman), 153

An Acceptable Time (L'Engle), 244, 247, 252, 256, 258–59

Andersen, Hans Christian, *The Snow Queen*, 14, 18, 197, 210

Anderson, Gary, 218

animals, as magical beings, 109, 110, 113, 117, 120n4

Anne of Green Gables (Montgomery), 104

anthropocentric relationships, 157, 168n3

anthropomorphism, *The Wind in the Willows*, 170–71, 173–75, 176

antinomial theorists of modernity, 231–32; and wonder, 239–41

apocalypse, and apocalyptic themes: *After London* (Jefferies), 173, 183n2; native apocalyptic storytelling, 136–37, 137nn3–5; as speculative fiction, 123–24

Armstrong, Jeannette, *Whispering in Shadows*, 131

Art of Fiction (Gardner), 1–2

A Swiftly Tilting Planet (L'Engle), 243; compatibility of religion and science, 246, 259n1;

interdependence, 255; listening, as spiritual principle, 253, 254; moral choice, 247; naming, as spiritual principle, 257; spirituality, 252

At the Back of the North Wind (MacDonald), 57–60, 65n9, 65n10

Austen, Jane, *Mansfield Park*, 5, 7

Australia: and the "Dreamtime," 126; Indigenous dystopian novels, 123; as *terra nullius*, 123

A Wind in the Door (L'Engle), 243; compatibility of religion and science, 245, 259n1; interdependence, 255; listening, as spiritual principle, 254; moral choice, 247, 250; mythic re-enchantment, 244; naming, as spiritual principle, 257–58; spirituality, 252

A Wizard of Earthsea (Le Guin), 263–64; as coming-of-age tale, 266; female places of, 266–67; gendered space, maintenance of, 271–72; masculine spaces, 270, 271–72; solitary existence of hero, 268; subordinate sphere of female possession, 267–69; woman's magic, 266–67

A Wrinkle in Time (L'Engle), 243, 246, 247, 252

Bachelard, Gaston, 202

Baker, Deirdre, 185–94; antiphony, use of, 191–92; background, 185–86; on imaginative geography, 193–94; nouns/verbs, use of, 193; place, as the story, 18; as regional writer, 187–88

Baker, Deirdre – WORKS: *Becca at Sea*, 186, 190–92; *Becca Fair and Foul*, 186

Ballantyne, R.M., 48; *Ungava*, 61

Barrie, J.M., 107, 312

Barrow, Sir John, *A Chronological History of Voyages into the Arctic Regions*, 47

Bavidge, Jenny, 157

Bayne, Marie, *Tales of Ireland for Irish Children*, 69, 83n2

Belacqua, Lyra, 14

Bennett, Jane: *The Enchantment of Modern Life*, 232–33; ethical relevance of enchantment, 237–38; and re-enchantment, 231, 234, 236, 240

Berrol, Cynthia, 284

Berry, Dave, 112

Berton, Pierre, *The Secret World of Og*, 102

Bessy Conway (Sadlier), 81, 83n7

Bettelheim, Bruno, realistic stories, and exposure to fairy tales, 106

Bildungsroman: Deathly Hallows, 216, 222–23; *The Snow Queen* (Kernaghan), 198, 212n4

Binti's Journey, 290, 291–93

biophilia, and place-attachment, 10, 21

Birch, Reginald, 86

Black, Marilynne V., 105

Black Beauty (Sewell), 10

Blyton, Enid, 61

Bolus-Reichert, Christine, 19, 229–41

bordem: of *acedia*, 219–20; existential boredom, 220

boy's adventure narratives. *See* adventure narratives

Bradford, Clare, 162

Briggs, K.M., 73

Briggs, Melody, 246

Briggs, Richard S., 246

British Columbia: as celebrity geography, 189; geology of, 188; marine life, 188; physical features, 188–89; questions of representation, 187–88

British Empire: Anglo-European superiority over the "other," 31, 32-33, 34-36; character, qualities of, 30-32, 42; geographic expansionism of, 26-27; Heroic Age of Exploration, 28-29; and imperialism, 14; polar exploration, 28-30, 44n3; preoccupation with Canadian North, 45-46; and scientific exploration, 209, 211; theoretical construct of "empire," 26-27, 43n1

Brooker-Gross, Susan R.: Nancy Drew series, 169-70, 265

Brown, Peter, *The Curious Garden*, 154, 161-64

Brown, Terence, 70

Bruchac, Joseph, 134

Buber, Martin, 248

Buchan, Susan, on Canadian wilderness, 101-2

Buckland, Corinne, 245

Buell, Lawrence: ecocriticism as multidisciplinary field, 155-56; "Environmental Writing for Children," 9-10, 12; *The Future of Environmentalism*, 5, 8-9; imaged earth-writing, 8-9

Bunyan, John, 189; *The Pilgrim's Progress*, 15, 85, 186; *Pilgrim's Progress*, 216. *See also Two Little Pilgrim's Progress* (Burnett)

Burke, Edmund, *A Philosophical Enquiry into the Origin of our Ideas on the Sublime and the Beautiful*, 47

Burnett, Frances Hodgson: about, 89-90; *Little Lord Fauntleroy*, 89, 92; *A Little Princess*, 89, 92; as progressive feminist, 87; *The Secret Garden*, 7-8, 89, 153; *Two Little Pilgrim's Progress*, 15,

85-99; Woman's Building library contribution, 89, 98n1

Burnham, Daniel, 90

Byrd, Jodi A., 125-26

Cameron, Scott, 142, 143

Canada: bias against fantasy, 104-5; Canadianism of wilderness, 105-6; development of Canadian children's literature, 101-3; fantasy literature, lack of, 103-4; Oil Sands ecocide, 122, 137nn3-4; survival tales, 109

Canadian Crusoes (Traill), 102, 105, 110, 111, 120n2; Spirit World creatures, 114

Canadian North, 13, 14; Franklin expedition, 46, 51, 198, 212n1; of Kernaghan, 18-19; and the Northwest Passage, 14-15, 45-46; *The Idea of the North* (Gould), 29, 44n3, 198

Canadian Wonder Tales (Macmillan), 110, 112

Cariou, Warren: "An Athabasca Story," 135, 138n11; "Tarhands," 135

Carpenter, Humphrey, 102; *Secret Gardens*, 108

Carr, Emily, *Klee Wyck*, 110

Carroll, Jane Suzanne, 87

Carroll, Lewis: *Alice's Adventures in Wonderland*, 126; Alice's imagined geography, 3

Carson, Rachel, *Silent Spring*, 9

Cecire, Maria S., *Space and Place in Children's Literature*, 21

Chan, Marjorie, *Sanctuary Song*, 291

Chesterton, G.K., 250

Chicago World's Fair: association with Bunyan's Celestial City, 95-97; nationalistic purpose of, 90-91, 92; souvenir guidebooks,

89–90, 99n3; Woman's Building
library, 89, 98n1
childhood: importance of place in, 9;
realistic stories, and exposure to
fairy tales, 106
children: colonization of, 14, 38–42,
44n5; as culturally constructed
concept, 282, 297n1; empathy, and
imaginary "Otherness," 283–87,
298n3, 298n4; environmental
awareness, development of, 165;
influence of natural environment
on, 158, 168n4; place-attachment,
and identity formation, 10, 11
Children's Literature Association Quarterly,
"Ecology and the Child," 156
The Child's Irish Songbook, 67, 79, 83n1
Christian heroism: construct of in
epic fantasy, 19, 216, 218–20
Christian theology: story of
redemption, 37
Clark, Catherine Anthony, 109–10;
The Golden Pine Cone, 15, 102
Clarke, Amy, 277
Coleridge, Samuel Taylor, "Frost at
Midnight," 172
colonialism, and capitalism, 121,
132–33
coming-of-age novels: *A Wizard of
Earthsea*, 266; and transgeneric
fiction, 125–26
Compton, Anne, 302, 315
Conan Doyle, Arthur, 107
Connor, Ralph, 63
Conrad, Joseph, *Heart of Darkness*, 230
consumption, and corporate logic of
growth, 132–36
Cooper, Susan, *The Dark is Rising*, 221
Cornum, Lindsey Catherine, 125
Cowell, Cressida, *How to Train Your
Dragon*, 109
Craig, John, *The Long Return*, 103

Crawford, Isabella Valancy, 110
creative process: antiphony, use of,
191–92; childhood influences
on writing, 303–5, 310;
interconnectedness of family,
148–49; Janet Lunn, 16–17, 139–49;
nouns/verbs, use of, 193; role
of the unconscious, 20–21, 302,
310–14; use of flashback, 307–8
Croker, Crofton, 74
Cuchulain of Muithemne (Gregory), 70
cultural appropriation, 32–34
cultural survival literature, 12–13;
and Indigenous futurism, 124–25,
127–28, 137nn5–6; as protest
literature, 12
culture, and spatialization of social
theory, 6
Culture and Imperialism (Said):
imperialism, and nostalgia for
empire, 27, 43; *Mansfield Park*
(Austen), 5, 7
Cumyn, Alan, 301–15; childhood
influences on writing, 303–5, 310,
314–15; role of the unconscious,
20–21, 302, 310–14; Young Adult
novels, 302, 312
Cumyn, Alan – works: *Dear Sylvia*, 301;
Between Families and the Sky, 311–
12; *Man of Bone*, 306–9; *The Secret
Life of Owen Skye*, 13, 301, 305, 314;
After Sylvia, 301; *Tilt*, 302, 312, 313;
"Valentine's Day," 309–10
The Curious Garden (Brown), 154; desire
to shape nature, 162–64, 168n7;
urban landscapes, greening of,
161–64, 168n6

dance. *See* theatre
Darton, F.J. Harvey, 46
Davies, Robertson, on spiritual
adventures, 106

Dear Sylvia (Cumyn), 301
death: as imaginary geography, 276;
    and landscape of the green topos,
    87, 160; *The Snow Queen* as, 203–4
defamiliarization, process of, 233
Denisoff, Denis, 52
DePasquale, Paul, 12–13, 127
Dessauer, Friedrich, 259n1
Dillon, Grace L., 125; *Walking the
    Clouds*, 136
The Disappearance of Ember Crow
    (Kwaymullina), 123
Dobrin, Sidney I., 154, 166; *Wild Things*,
    10, 21, 157–58
Donnison, T.E., 57
Doolittle, Bev, *The Earth Is My Mother*, 10
Doremus, Robert, *The Long Return*, 103
Doughty, Terri, *Knowing Their Place?*, 21
Drabble, Margaret, 81
dromomania, 220
During, Simon, 231
Dyer, Kevin, 289
Dyer, Richard, 230
dystopian novels: battle between
    good and evil, 126–27; *Lightfinder*
    (Paquette), 16, 122–23, 128–32,
    136–37, 137n2; and transgeneric
    fiction, 125–26

The Earth Is My Mother (Doolittle), 10
Earthsea series (Le Guin). *See* Le
    Guin, Ursula
ecocriticism: and anthropocentric
    relations, 157, 168n3; development
    of, 154–58; earth-centred approach
    to literary studies, 154; eco-
    critical study of literature, 17; and
    environmental awareness, 154; as
    multidisciplinary field, 155–56,
    168n1; shift towards children's
    literature, 156–58, 168n2

The Ecocriticism Reader (Glotfelty,
    Fromm), 154–55
Ede, Charles, *The Home Amid the Snow,
    or, Warm Hearts in Cold Regions*, 61
Edison, Thomas, 90
Egan, Michael, 310–11, 312
Egoff, Sheila, 112
Elam, Diane, *Romancing the
    Postmodern*, 236
Emberley, Julia, 124, 128
Emigration of the Fairies
    (Hunter Duvar), 104
empathetic collaboration between
    performer/audience, 20, 283–87,
    298nn3–4
The Enchantment of Modern Life
    (Bennett), 232–33
environmentalism: and activism,
    9–10; Alberta Oil Sands, 122–24,
    135, 138n11; anthropocentric
    relationships, 157, 168n3;
    biophilia, 10, 21; and dystopian
    novels, 122–24, 137nn2–4; and
    ecocriticism, 154–58; Glotfelty's
    definition of, 9; holistic images
    of earth-writing, 12–13; and
    imaginative geography, 3, 5–6;
    and Indigenous knowledge,
    136–37; and place-attachment,
    8–11, 17; scholarly contributions
    towards, 154–58, 168n1
Environment Award of Children's
    Literature, 9
epic fantasy, and construct
    of Christian heroism, 19
Epple, Colette, 72
Erdrich, Louise: *The Birchbark House*,
    131; *The Game of Silence*, 131;
    The Porcupine Year, 131
Exiled from Erin (M.E.T.), 79–80, 82
existential boredom, 220

*Experiencing Environment and Place
through Children's Literature*
(Cutter-Mackenzie, Payne,
and Reid), 157

Fachinger, Petra, 16, 121–38
fairies, scarcity of in New/Old World,
107–9
Falconer, E., "Killarney," 71
fantasy: bias against, 104–5; and
Canadian wilderness, 15; fairies,
scarcity of in New/Old World,
107–9; lack of in Canadian
literature, 103–4; methods of
plausibility, 108–9, 120n4; Old
World "magical displacement,"
106; place, and myth, 11–13; and
realism, 2–3, 232, 247; role in
inventing geography, 186–87;
as Secondary World, 11–12;
as a "spell," 245; Spirit World
creatures, 113–15, 116–17; talisman
as magical object, 109; and
wonder, 239–41
The Farthest Shore (Le Guin), 264; death
as imaginary geography, 276;
gendered spaces of, 275–77, 279n1;
repetitions of "no," 276–77
female characters: heroism of, 207–8,
210–12, 213n8; representation of in
adventure narratives, 60–63, 235
Ferris, George, 90
fiction: character development, 2;
importance of "place" in, 1–2,
21–22
Field, Eugene, 89
Field, Hannah, *Space and Place in
Children's Literature*, 21
Findlay, Timothy, 145
Findon, Joanne, 18–19, 197–213
First Nations: intergenerational
trauma, 125, 127–28; residential

school system, 14, 127–28, 137–38n5;
social issues affecting, 127–28
Fish, Stanley, 243
Fisher, Philip, 234; *Wonder, the Rainbow,
and the Aesthetics of Rare Experiences*,
239–40, 241
Fitzsimmons Frey, Heather, 20, 281–99
flight, and age of sail, 19, 230, 233
folklore: place and myth, 11–13
Folklore of Canada (Fowke), 104
forests, 221–22, 226
The Foretelling of Georgie Spider
(Kwaymullina), 123
Forster, E.M., 149; "The Story of a
Panic," 176
Foster, Susan, 284
Fowke, Edith, *Folklore of Canada*, 104
Fragile Things (Gaiman), 21, 22n2
Franklin, Colleen M., 14–15, 45–65;
*The Strange and Dangerous Voyage
of Captain Thomas James*, 46
Fromm, Harold, *The Ecocriticism
Reader*, 154–55
Frye, Northrop: Canadianism of
wilderness, 105–6; fiction as
"conscious mythology," 103;
The Secular Scripture, 232
The Future of Environmentalism (Buell),
5; imaged earth-writing, 8–9

Gaard, Greta, 156, 168n2
Gaiman, Neil, *Fragile Things*, 21, 22n2
gardens, and green places:
balance between instruction
and entertainment, 157–58;
cyclical nature of, 153–54,
160–61, 163; and ecocriticism,
154–58; environmental awareness,
development of, 165–67; renewal,
and spiritual regeneration, 165;
and self-cultivation, 167; topiary,
158–61, 162–64, 168n7; as *topoi*,

17–18; urban environment as
imaginative geography, 164;
urban landscapes, greening of,
17, 21, 161–64, 168n6

Gardner, John: *Art of Fiction*, 1–2;
similarity between "fables"
and realism, 2

Garner, Alan, *The Owl Service*, 120

Garth, John, 227

gay sexuality: and Pan, 176; "The Lost
Centaur" (Grahame), 182–83; "The
Story of a Panic" (Forster), 176

Gems for the Young from Favourite Poets
(Mulholland), 70

gender: feminist geography, 265–66;
male/female places, 20, 265–66;
subversion of roles, 198, 199–200,
201, 212n5, 213n6

geographic patriotism: in Irish
children's literature, 68; as kind
of folk memory, 69–70; and
national identity, 70–73

geography: of boredom, 216, 228;
concrete geography, 186–87;
conveyance of immanent
presence, 189–91; function of
in historical fiction, 140–41;
questions of representation,
187–88; role of fantasy in, 186–87

Gerzina, Gretchen, 89

Gilbert, Kevin, 130–31

Gilson, Rob, 227

Glotfelty, Cheryll, 9, 157; *The
Ecocriticism Reader*, 154–55

Godolphin, Mary, 99n7

The Golden Compass (Pullman), 13;
imperialism, and the Other, 14;
imperialist themes, 27, 38–43;
parallels to residential school
system, 38–42; stylization of
Victorian Britain, 25–27

The Golden Pine Cone (Clark), 15, 102;
magical device of, 111–12; mythic
time, 112–13; relationship of
characters, 115–17; Tekontha,
comparison to Galadriel, 117–18,
120n7; use of First Nations
myths, 110–11

Goldsmith, Oliver, 48

Goldstone, Bette, 166

The Golliwogg's Polar Adventure
(Upton), 61–63

Goodly Is Our Heritage (Singh), 30

Gordon, Jon, irrationality as mode
of engagement, 135–36

Gould, Glenn, *The Idea of the North*,
29, 44n3

Graham, Judith, 159, 168n4

Grahame, Kenneth: on the English
countryside, 172–73, 175; "The
Lost Centaur," 174, 175, 182–83;
"The Rural Pan," 176–77; *The
Wind in the Willows*, 10, 13, 17–18,
169–83

Grandpa Green (Smith), 154, 168n5;
allusive artistry of topiary,
158–61; life-and-death metaphor,
160; role reversal, 161, 168n5

Grant, George, 145

Greeley, Adolphus, 52

Green, Peter, 174, 176

Green Book Award for Children's
Literature, 9

Gregory, Derek, "Imaginative
Geographies," 6

Gregory, Lady Augusta, *Cuchulain
of Muithemne*, 70

Griffin, William, 77

Groen, Rick, 215, 220

Group of Seven, 63

Guroian, Vigen, 250

Hack, Maria, *Winter Evenings*, 48–51, 60–61

Hakluyt, Richard, 48

Haldane, J.B.S., 244

Hall, Mr and Mrs S.C., 74

Halley, Edmund, 48

Harry Potter and the Deathly Hallows (film), 215–16, 224

Harry Potter and the Deathly Hallows (Rowling): as *Bildungsroman*, 216, 222–23; "camping section," 215; comparison to *Lord of the Rings*, 216, 217–19; construct of Christian heroism, 218–20; contrasts to *Lord of the Rings*, 220–21; domestic claustrophobia, 222–24; forests, 221–22, 226; genres of quest narrative, 216; geography of boredom, 19, 219–20; parallel worlds of, 3; and Partisan Resistance, 216, 224–26, 228

Hartley, L.P., 229

Hauff, Wilhelm, "Heart of Stone," 115

Hein, Rolland, literary realism vs. literary myth, 244–45

Henty, George Alfred, 30, 48

Herodotus, 57

Higgins, Edward, 37

Higgins, Michael, "Culture and Exile," 80

Hilder, Monika, 2, 19–20, 243–60

Hill, Lynda, 290

Hillel, Margot, 15, 67–83

His Dark Materials (Pullman), 3, 25; criticisms of, 36–38

historical fiction, function of geography in, 140–41

Hitchens, Christopher, 215

The Hobbit (Tolkien), 11, 216, 218

Holland, Patricia, 82

The Hollow Tree (Lunn), 17, 139, 140, 141–42, 146, 147

Holmes, H.H., 99n10

The Home Amid the Snow, or, Warm Hearts in Cold Regions (Ede), 61

home-away-home narratives, 81

Hudson, Aïda, 1–22; conversation with Janet Lunn, 16–17, 139–49; perspective of mind's eye, 282

Hughes, Arthur, 57, 58

Hunter, Molly, 147

Hunter Duvar, John, *Emigration of the Fairies*, 104

Hutton, Ronald, 176

Hynes, Peter, 20, 263–79

identity: and geographic patriotism, 70–72

illustrators: Reginald Birch, 86; T.E. Donnison, 57; Arthur Hughes, 57, 58; Irene Luxbacher, 154, 165, 167; Charles Robinson, 52, 54, 55; Lane Smith, 154, 158–61, 168n5

The Imaginary Garden (Larsen), 154, 165–67

imaginative geography, 1–2, 31, 60, 80, 106, 127, 162, 164, 193, 198, 214; Australian desert and, 123; Bowen and, 22n1; Cumyn and, 301; and dance, 282, 287–90; environmentalism, 8–11; as imaged earth-writing, 6, 7–8; Le Guin and, 264; L'Engle's Time Quintet and, 243–60; *Lightfinder* and, 125, 131; and mimetic truth, 22; myths, and the mythical, 11–13; place-attachment, 8–11; postcolonialism of Said, 3–6; roots of, 1; steampunk and, 229, 231; theatrical embodiment, 282–83; travels and, 6–7; Victorian literature of the North and, 48–50; Welty and, 21; *Wind in the Willows* and, 169 imperialism: British character, qualities of, 30-32; in children's literature, 6; class system, 7-8; and cultural appropriation, 32-34; in *Lord of the Rings*, 12; *Mansfield Park* (Austen), 5, 7; nostalgia for empire, 27; Pullman and, 26; religious imperialism, 38-42; *The Secret Garden* (Burnett), 7; and the "sublime quest," 47-48, 64n3; West/non-West dualism of "ours"/"theirs," 4-5; of *Winter Evenings*, 49-51

The Incredible Journey (Burnford), 103

Indigenous Australian traditions: Dreamtime, 130-31, 138n9; dystopian novels, 123

Indigenous futurism, 124-25, 137nn5-6; and *biskaabiiyang*, 136; and transgeneric fiction, 127-28

Indigenous knowledge: kinship relations, 133-34; new generation of Elders, 128-29, 131-32, 134-35, 138n10; principles of, 127; totem, and clan, 134

Indigenous myth, 104; *The Golden Pine Cone* (Clark), 15, 110-20; kinship relations, 124; *Lightfinder* (Paquette), 16; and spirit, 121; Wendigo, 129; White Buffalo Calf Woman, 118, 120n7, 122, 124, 129, 136

"inscape" of place, 19-20, 244

intergenerational trauma: and colonial violence, 125; decolonizing strategies, 127-28; new generation of Elders, 129-30, 138n8

The Interrogation of Ashala Wolf (Kwaymullina), 123

Ireland: *The Child's Irish Songbook*, 67, 79, 83n1; The Famine, and emigration, 75-79, 83nn4-6;

geographic patriotism, 68, 70-72, 77-80; Irish Literary Revival, 70; land as feminized ideal, 71; nationalistic literature, 67-68; nostalgic patriotism of emigrants, 15, 68-70, 72-73, 76-79, 82, 83n2, 83n6; oral tradition, 73-74; saints and fairies, 73-74; "The Potato Digger's Song" (Irwin), 74-75, 77; *Young Ireland* journal, 78-79

Irish diaspora, and cultural memory, 68-70, 83n2

"The Irishman" (Orr), 67, 83n1

irrationality as mode of engagement, 135-36

Irwin, Thomas, "The Potato Digger's Song," 74-75, 77

James, Lynette, 124-25, 137n6

Janeway, James, *A Token for Children*, 95

Jefferies, Richard, *After London*, 172, 183n2

Jobe, Ronald, 105

Johansen, K.V., 119

Johnson, Pauline, 110

Johnson, Samuel, 48

Johnson, Tom, 37

Justice, Daniel Heath, kinship relations, 133-34

Kealy, J. Kieran, 111

Kennedy, Dennis, 287, 288

Kernaghan, Eileen, 212n2; the idea of North, 202; *The Snow Queen*, 14, 18-19

Kiberd, Declan, 68, 72; "Culture and Exile," 80

Kidd, Kenneth B., 154, 166; *Wild Things*, 10, 21, 157-58

"Killarney" (Falconer), 71

kinaesthetic empathy, 284–86, 287, 290, 293–94, 298n5, 298n8

King, Thomas, *The Back of the Turtle*, 138n8

Kingsley, Charles, *The Water Babies*, 114

Kipling, Rudyard, 230; *Kim*, 32–33; *Puck of Pook's Hill*, 107–8, 109

Knowing Their Place? (Doughty, Thompson), 21

Knowles, Linda, 15, 101–20

Kutzer, M. Daphne, 28, 35

Kwaymullina, Ambelin, Tribe series, 123–24

kything, 253–54

land, as feminized ideal, 71

Landy, Joshua, *The Re-Enchantment of the World*, 231, 232, 233

Larklight (Reeve), 19, 229; aether-ships, 230–31; British Empire, and outer space, 233–35; interspecies crossings, 238–39; and pirate romance, 234; remythologizing of flight, 237; textual allusions, 235

Laronde, Sandra, 293–94

Larsen, Andrew, *The Imaginary Garden*, 154, 165–67

Larson, Erik, 99n10

Last Leaf First Snowflakes to Fall (Yerxa), 13

Late Imperial Romance (McClure), 19, 230, 231

Lau, William, 294

Lawrence, D.H., 174

Lee, Dennis, *Alligator Pie*, 3

The Left Hand of Darkness (Le Guin), 277, 279n1

Le Guin, Ursula: Earthsea series, 13, 20; gendered spaces of Earthsea, 264–65; sense of place, role in creative imagination, 263–64; sexism of gendered space, 277, 279n2

Le Guin, Ursula – WORKS: *The Farthest Shore*, 264, 275–77; *The Left Hand of Darkness*, 277; *Tehanu*, 265, 277–79; *The Tombs of Atuan*, 264, 272–75; *A Wizard of Earthsea*, 263–64, 266–72

L'Engle, Madeleine: compatibility of religion and science, 245–46, 259n1; human choice, and moral vision, 250–51, 260n7; as incarnationalist, 248, 259–60n5; "inscape" of place, 19–20, 244; interdependence, as spiritual principle, 254–56, 259; listening, as spiritual principle, 253–54, 259; matter, moral significance of, 248–50; mythic re-enchantment, 244, 245, 259; naming, as spiritual principle, 256–59; powers of good vs. evil, 250–51, 252; rationalistic labelling, 256–57; spirituality, 247, 252–53; symbiosis of cosmic harmony, 255–56, 259; and universalism, 244, 250–51, 259–60n5, 260n7

L'Engle, Madeleine – WORKS: *An Acceptable Time*, 244, 247, 252, 256, 258–59; "Before Babel," 244; *Many Waters*, 243, 244, 246, 247, 253, 254, 255; *A Swiftly Tilting Planet*, 243, 246, 247, 252, 253, 254, 255, 257; Time Quintet, 3, 19–20, 243–44; *A Wind in the Door*, 243, 244, 245, 247, 250, 252, 254, 255, 257–58, 259n1; *A Wrinkle in Time*, 243, 246, 247, 252

Le Pan, Doublas, 145

"Let Erin remember the days of old" (Moore), 69–70, 83n3

Leviathan (Westerfeld), 19, 229, 230; Darwinists vs. Clankers, 231, 238; remythologizing of flight, 237

Lewis, C.S., 36, 119, 247–48; on
  imaginative fantasy, 241; *The Lion,
  the Witch and the Wardrobe*, 109,
  110; *The Silver Chair*, 219–20
Lightfinder (Paquette), 13, 15;
  colonialism, and capitalism,
  121, 132–33; as dystopian novel,
  122–23, 137nn2–4; Indigenous
  Australian traditions, 130–31,
  138n9; and Indigenous futurism,
  124–25, 127–28, 137nn5–6; new
  generation of Elders, 129–30, 131–
  32, 134–35, 138n10; relationship to
  the land, 128; spiritual rebirth,
  134–35; transmission of cultural
  knowledge, 128–29, 131–32
Lightman, Bernard, 246
The Lion, the Witch and the Wardrobe
  (C.S. Lewis), 109
The Lion and The Unicorn, "Green
  Worlds," 156
Li Sheung Ying, Melissa, 9, 17, 153–68
listening, as spiritual principle, 253–54
Literary History of Canada (Frye), 105–6
The Literary History of Canada
  (McDowell), 102
literary realism vs. literary myth,
  244–45
Little Bear, Leroy, 127
Littlefield, Holly, 274
Little Lord Fauntleroy (Burnett), 89, 92
Little Voice (Slipperjack), 12, 131
Lochhead, Marion, *Renaissance of
  Wonder*, 248
Locke, John, 48
The Long Return (Craig, Doremus), 103
The Lord of the Rings (Tolkien), 11, 12;
  comparison to *Deathly Hallows*,
  216, 217–19; and construct of
  Christian heroism, 19, 218–20;
  contrasts to *Deathly Hallows*,

220–21; as World War I allegory,
  226–28
"The Lost Centaur" (Grahame), 174,
  175, 182–83
Lost in the Barrens (Mowat), 103
Lunn, Janet, 139–49; awards and
  honours, 140; creative process,
  16–17; environmental concerns,
  148–49; function of geography
  in historical fiction, 140–41, 148;
  interconnectedness of family,
  148, 149; on shadows and ghosts,
  142–43
Lunn, Janet – WORKS: *Dear Canada, A
  Rebel's Daughter*, 141; *The Hollow
  Tree*, 17, 139, 140, 141–42, 146, 147;
  The Root Cellar, 17, 139–40, 142,
  143–44; *Shadow in Hawthorn Bay*,
  16–17, 139, 141, 144–46, 147; *The
  Story of Canada*, 140
Luxbacher, Irene, 154, 165, 167

MacDonald, George, 246, 248; *At the
  Back of the North Wind*, 57–60,
  65nn9–10; *The Princess and the
  Goblin*, 60, 109; "The Fantastic
  Imagination," 232, 240
MacDonald, Ruth, 99n7
Macfarlane, Robert, *The Wild Places*, 172
MacInnis, Allen, 289–90
MacLennan, Hugh, 107
Macmillan, Cyrus, *Canadian Wonder
  Tales*, 110, 112
MacSorley, Catherine: *An Irish Cousin*,
  77–78; *Nora*, 82
Maddox, Matthew, 255
Maeterlinck, Maurice, *The Blue Bird*, 113
The Magazine of Art, 52
magical devices: *The Golden Pine Cone*
  (Clark), 111–12; use of in fantasy
  narratives, 109, 120n4

Mah, D.B., landscape and national identity, 70–72

Mahy, Margaret, Old World "magical displacement," 106

Major, Kevin, 140

Major, Melissa, 288

Man of Bone (Cumyn), 306–9

Mansfield Park (Austen), 5, 7

Many Waters (L'Engle), 243; compatibility of religion and science, 246, 259n1; interdependence, 255; listening, as spiritual principle, 253, 254; moral choice, 247; mythic re-enchantment, 244

Maquis (Millar), 225–26

Maracle, Lee: "The Other Side of Me," 12–13; *Will's Garden*, 12

Marlowe, Christopher, 46

Massey, Doreen, *For Space*, 286, 288, 296

Maté, Gabor, 314

matter, moral significance of, 248–50

McClure, John, *Late Imperial Romance*, 19, 230, 231

McConachie, Bruce, 283

McDowell, Linda, 265

McDowell, Marjorie, *The Literary History of Canada*, 102

Millar, George, *Maquis*, 225–26

Miller, Mary Jane, 111

Milloy, John S., 39, 41

Milne, A.A., *Winnie the Pooh*, 63

Milton, John, 46

mirror neurons theory, 285, 298n6

Mitchell, W.O., 140

Molson, Francis, 97

Monaghan, James, "Song of an Exile," 78–79

Montgomery, L.M.: *Anne of Green Gables*, 104; *Emily of New Moon*, 95, 99n9

Moore, Christopher, *The Story of Canada*, 140

Moore, Thomas, "Let Erin remember the days of old," 69–70, 83n3

moral choice, powers of good vs. evil, 250–51, 260n7

Morin, Karen, 5

Morris, Alexander, 39

Morris, Desmond, *The Human Zoo*, 146

Mowat, Farley, *Lost in the Barrens*, 103, 105, 120n2

Mulholland, Rosa: age of intended readership, 72; *Gems for the Young from Favourite Poets*, 70; *The Return of Mary O'Murrough*, 72–73, 76, 78, 80

Muller-Funk, Wolfgang, 68

Munro, Alice, "The Found Boat," 310

mythic re-enchantment, and rationalism, 244, 245, 259

myths, and the mythical: beingness of the North, 13; in Creation stories, 126; First Nations myths, 109–10, 112–13, 118, 120n7; Indigenous myth, 15, 16, 104; Irish myths, 70; the *Kalevala*, 18; literary realism vs. literary myth, 244–45; loss of, 145–46; mythic time, 112–13; power of, 212n2; and re-enchantment, 18–20; relation to place, 3, 11–13, 21–22; remythologizing of flight, 237; Wendigo, 129; White Buffalo Calf Woman, 118, 120n7, 122, 124, 129, 136

Nancy Drew series, 265; compared to *Wind in the Willows*, 169–70

National Film Board of Canada (NFB), 110

nationalistic literature, and geographic patriotism, 67–68

Nesbit, E., 108; *Five Children and It*, 109; *The Story of the Amulet*, 109

Ness, Sally Ann, 294; "The Inscription of Gesture," 285–86

The Night Wanderer (Taylor), 124

Nodelman, Perry, 81

Noonan, Mark, 87

Nora (MacSorley), 82

the North: beingness of, 13; British representations of, 46–48; the Canadian North, 13, 14, 18–19; female representations of, 57–60, 61–63, 64n7; as idea, 197; polar exploration, 28–30, 44n3; public fascination with, 46; shift in British representations of, 63; as the "sublime quest," 47–48, 51, 57, 64n3. *See also* Northwest Passage

The Northern Lights (Pullman). *See The Golden Compass* (Pullman)

Northwest Passage: and British imperial ambition, 14–15, 45–46; Franklin expedition, 46, 51, 198, 212n1; "The Northwest Passage" (Stevenson), 52–57, 61

nostalgic patriotism, of Irish emigrants, 15, 68–70, 72–73, 76–79, 82, 83n2, 83n6

O'Grady, Standish, *In the Gates of the North*, 70

"On Fairy-Stories" (Tolkien), 11, 108, 119, 232, 240, 241; on imaginative fantasy, 245

Oppel, Kenneth: *Airborn*, 13, 19, 229, 230, 236–37; female characters, 235

Orientalism (Said): global reach of, 6; postcolonial discourse of, 3, 4–5

Orr, James, "The Irishman," 67, 83n1

the Other, and Otherness: *The Golden Compass* (Pullman), 14; and imperialism, 14; of Indigenous literatures, 12–13; non-British adults as "Other," 31; "our land-barbarian land" duality, 9, 17–18, 21; West/non-West dualism of "ours"/"theirs," 4–5, 21

Oziewicz, Marek, 252

Pada, Lata, 290

Padley, Jonathan, 37

Padley, Kenneth, 37

Pan, as literary symbol, 176

Paquette, Aaron, *Lightfinder*, 13, 16, 122–23, 137nn2–4

Pavis, Patrice, 294

Peri Hypsous (Longinus), 47

perspective: of mind's eye, 1–2, 282; West/non-West dualism of "ours"/"theirs," 4–5

Peter Pan (Barrie), 107, 312; Freudian interpretations of, 310–11

picture books: garden as positive landscape, 154; *The Golliwogg's Polar Adventure* (Upton), 61–63; Indigenous picture books, 12–13; as introduction to a garden, 158, 168n4; portrayals of Aboriginal children, 127; postmodern, 166

The Pilgrim's Progress (Bunyan), 15, 85, 186; Burnett's secularization of, 97–98; comparison to *Deathly Hallows*, 216–17; as spiritual guidebook, 95, 99n9; versions for children, 94–95, 99n6, 99n7; World's Fair, association with Celestial City, 95–97. *See also Two Little Pilgrim's Progress* (Burnett)

place: and fiction, 1–2, 21–22; importance of in childhood, 9; "inscape" of, 19–20; mental colonization, 18; and myth, 3, 11–13; as reflection of imagined geography, 2–3; role in creative imagination, 263–64; spatial representations, 278; variation in scale, 3

place-attachment: and environmentalism, 8–11, 17, 21; sentimentalizing of, 18

Pleasures of the Imagination (Akenside), 49

Politics and Poetics of Irish Children's Literature (Watson), 68

Posesorski, Sherie, 164

postcolonialism: cultural survival literature, 12–13; and imaginative geography, 3–6; imperialism in children's literature, 6

postmodern culture, and fantasy literature, 245

postmodern romance, juxtaposition of historical periods, 236

"The Potato Digger's Song" (Irwin), 74–75, 77

Pratt, Annis, 212n4

Preuss, Clare, 289, 290

Prickett, Stephen, on realism and fantasy, 2, 247

Pullman, Philip: alternate universe of, 26–27, 43n1; comparison of Lyra to Kipling's Kim, 32–33; fictional reality of theatre, 282–83, 286–87, 290; geographic inquiry, 26; Lyra's universe, 25, 29; polar exploration, 29–30; theological knowledge, criticism of, 36–38

Pullman, Philip – WORKS: *The Amber Spyglass*, 153; *The Golden Compass*, 13, 14, 25, 32–33; *His Dark Materials*, 3, 25, 36–38

Purchas, Samuel, 48; *Purchas his Pilgrimage*, 47

quest narratives: construct of Christian heroism, 19, 216, 218–20; and epic fantasy, 19; Harry

Potter, 215–16; inner growth of characters, 208–9, 213n8; as religious allegory, 216; *The Snow Queen* (Kernaghan), 201–2, 205–7; unexpected results, 209–10

Raposa, David, 228

Raposa, Michael, 216, 219, 223

rationalism, and mythic re-enchantment, 244, 245

readers: age of intended readership, 72; manipulation of, 68; perspective of mind's eye, 1–2

realism: in Canadian children's literature, 101–3; and fantasy, 2–3, 232, 247

Reason, Matthew, 285, 292, 299n10

Red Sky Performance, 293–94

re-enchantment: and antinomial theorists of modernity, 231–32, 239–40; ethical relevance of enchantment, 237–38; of flight, 19, 229–31; hierarchy of significance, 233; kinds of, 231; mythic re-enchantment, 244–48; remythologizing of flight, 237; and renewal, 240–41; restoration of magic and mystery, 230, 231; and secondary worlds, 18–20, 21–22; as strategy of modernity, 231; and wonder, 239–41

The Re-Enchantment of the World (Landy, Saler), 231, 232, 233

Reeve, Philip, *Larklight*, 19, 229, 230–31, 233–35, 238–39

Reimer, Mavis, 81

religion, compatibility with science, 245–46, 259n1

residential school system (Canada): and colonization of children, 14; intergenerational trauma, 16, 127–28, 137–38n5; parallels to *The*

Golden Compass (Pullman), 38–42; as religious imperialism, 38–42, 44nn4–5; survivors' accounts, 38–39; Truth and Reconciliation Commission, 38, 44n4

The Return of Mary O'Murrough (Mulholland), 72–73, 76, 78, 80

Reynolds, Dee, 285, 292

Richards, Thomas, 26; geographic expeditions, and knowledge control, 28

Richardson, Abigail, 291

Robbe-Grillet, Alain, 274

Robinson, Charles, 52, 54, 55

Rolfe, Frederick William (Baron Corvo), 182–83

The Root Cellar (Lunn), 17, 142; time travel, 139–40, 143–44

Rosebury, Brian, *Tolkien*, 11–12

Rowling, J.K.: comparison to *Lord of the Rings*, 216, 217–19; contrasts to *Lord of the Rings*, 220–21; *Harry Potter and the Deathly Hallows*, 19, 215–28; Harry Potter books, 3

Roy, Malini, *Space and Place in Children's Literature*, 21

Rueckert, William, 154

Sadlier, Mrs J., *Bessy Conway*, 81, 83n7

Said, Edward: fictional reality of theatre, 282–83; imaginative construct of the "Orient," 290; imaginative geography, 3–6, 8; imperialism, and nostalgia for empire, 27; on *Kim* (Kipling), 32; "our land-barbarian land" duality, 4–5, 9, 17–18, 21, 169, 296; and postcolonial imperialism, 6

Said, Edward – WORKS: *Culture and Imperialism*, 5, 27, 43; *Orientalism*, 3–6

Sale, Roger, 197

Saler, Michael: *As If*, 232; *The Re-Enchantment of the World*, 231, 232, 233, 240

Sampson, Cory, 14, 25–44

Sanctuary Song, 290, 291

Sandner, David, 172

Sangster, Charles, *The St. Lawrence and the Saguenay*, 103–4

Scandinavia, 64n3

scapegoat, ritual of, 218

Schonmann, Shifra, 284, 293

science, compatibility with religion, 245–46, 259n1

science fiction: and transgeneric fiction, 125–26. *See also* steampunk fiction

Scotland, 65n9

Scott, Duncan Campbell, 110; "Powassan's Drum," 115

Scutter, Heather, 72

secondary worlds, and re-enchantment, 18–20

The Secret Garden (Burnett), 89, 153; double-worlding in, 7–8; and imperialism, 7

The Secret Life of Owen Skye (Cumyn), 13, 301, 305, 314

The Secret World of Og (Berton), 102

self-conquest, trope of, 15

Selina, Helen, "Lament of the Irish Emigrant," 76–77

Sendak, Maurice, *Where the Wild Things Are*, 10

Setterington, Ken, 185

Seuss, Dr., *The Lorax*, 10

Sewell, Anna, *Black Beauty*, 10

sexuality: auto-eroticism, and desire, 181; Pan, and gay sexuality, 176, 182–83

Shadow in Hawthorn Bay (Lunn), 16–17, 139, 141, 144–46, 147

Shakespeare, William, 46

Sherwood, Mary, *The Infant's Progress*, 99n7

Shippey, Tom, 226–27

Showalter, Elaine, 99n9

Sigler, Carolyn, "Wonderland to Wasteland," 157

Silent Spring (Carson), 9

Silkenat, David, 99n3

Singh, Rashna B., *Goodly Is Our Heritage*, 30

Slipperjack, Ruby, *Little Voice*, 12, 131

Smart, Christopher, 48

Smith, David, 99n7

Smith, G.B., 227

Smith, Lane, *Grandpa Green*, 154, 158–61, 168n5

Smollett, Tobias, 48

The Snow Queen (Andersen), 14, 18, 197

The Snow Queen (Kernaghan), 14; death symbolism, 203–4; as female *Bildungsroman*, 198, 212n4; female heroism, 210–12, 213n8; gender roles, subversion of, 198, 199–200, 201, 212n5, 213n6; hut symbolism, 202; and the *Kalevala*, 18–19, 198, 201, 202, 207, 213n7; liminality as "in-betweenness, 199; quest symbols, 205–7, 213n6; retelling of Andersen's story, 198, 212n3; shamanic magic, and female cooperation, 207–8, 213n7; threshold between Natue and Culture, 201; women, power of, 200–201

Sorfleet, John R., 103, 119–20

Space and Place in Children's Literature (Cecire, Field, Roy), 21

speculative fiction: *Lightfinder* (Paquette), 16, 122–24; Tribe series (Kwaymullina), 123

spirituality: existential questions of, 243–44; and space-time relativity, 247

Spitz, Ellen Handler, 283, 284, 293, 298n3

Stafford, Margot: imperialism in children's literature, 6; *Journeys Through Bookland*, 6

steampunk fiction, 19, 229–30; aesthetics, and science of, 241; and age of sail, 19, 230, 233; *Airborn* (Oppel), 13, 229; juxtaposition of historical periods, 236; *Larklight* (Reeve), 229; *Leviathan* (Westerfeld), 229; "Maker Movement," 239; mobility, pleasure and danger of, 236, 241; re-enchantment with flight, 229–31; and wonder, 239–41

Stevenson, Robert Louis: *A Child's Garden of Verse*, 51–52; *Strange Case of Dr. Jekyll and Mr. Hyde*, 170; "The Northwest Passage," 52–57, 61

The St. Lawrence and the Saguenay (Sangster), 103–4

The Story of Canada (Lunn, Moore), 140

The Strange and Dangerous Voyage of Captaine Thomas James (Franklin), 46

the sublime and the marvelous, 47

Sullivan, T.D., "Song from the Backwoods," 79

survival tales, of Indigenous identity, 109, 122–24

Tales of Ireland for Irish Children (Bayne), 69, 83n2

Taylor, Drew Hayden: *The Night Wanderer*, 124, 131–32, 138n10; *Raven Stole the Sun*, 21, 294–96

Taylor, Ken, identity and landscape, 68, 82

Tehanu (Le Guin), 265; approach to place, 277–79

theatre: *Beneath the Banyan Tree*, 290; Bharatanatyam dance, 290, 291, 294; *Binti's Journey*, 290, 291–93; coevalness, 296; dance spectatorship research, 282, 299nn9–11; as embodied space, 20, 281, 287–90, 296–97; empathy, and imaginary "Otherness," 20, 283–87, 296–97, 298nn3–4; fictional reality of theatre, 282–83, 286–87; impact of dance training on muscle memory, 286; jingju (Peking/Beijing Opera), 294; kinaesthetic empathy, 284–86, 287, 290, 293–94, 298n5, 298n8; "Monster Under the Bed," 289–90; movement development processes, 293–94; performance as earth-writing, 284, 296–97; *Raven Stole the Sun* (Taylor), 294–96; *Sanctuary Song*, 290, 291; "soul speak," 293–94; spatial relationship of actor/audience, 287–90, 292–97; "Wanda T. Grimsby, Detective Extraordinaire," 288, 290

Theatre Direct: *Binti's Journey*, 290, 291–93; *Sanctuary Song*, 290, 291

Theatre for Young Audiences (TYA), audience statistics, 282, 297–98n2

Thomas, Joyce, 205

Thompson, Dawn, *Knowing Their Place?*, 21

Thomson, James, *Winter*, 50

Thwaite, Ann, 89

time, as signified by space, 216

Time Quintet (L'Engle), 3, 243–44; "inscape" of place, 19–20

time travel, 139–40, 143–44; portal, 139, 143

Tolkien (Rosebury), 11–12

Tolkien, J.R.R., 266; comparison to *Deathly Hallows*, 216, 217–19; contrasts to *Deathly Hallows*, 220–21; on fantasy, 232, 240, 241; First World War experience, influence of, 11–12, 226–28; on landscape of the imagination, 264; place, and myth, 3, 11–12, 21; storyteller as sub-creator, 103

Tolkien, J.R.R. – WORKS: *The Hobbit*, 11, 216, 218; *The Lord of the Rings*, 11, 12, 110, 216, 218–19; "On Fairy-Stories," 11, 108, 119, 232, 240, 241, 245; *Two Towers*, 12

The Tombs of Atuan (Le Guin), 264; centrality of location, 272; visualization, reader's powers of, 273–74; and woman's space, 272–75

Toohey, Peter, 220

topiary, 22, 158–63, 168n7

topoi: gardens, and green places, 17–18; of the green world, 87; of imaginative geography, 4

Traill, Catharine Parr, 145; *Canadian Crusoes*, 102, 105, 110, 111, 114, 120n2; on lack of superstition, 104

transgeneric fiction, 125–26

travel and exploration narratives, 48

Trevor, William, 74

Trites, Roberta Seelinger, 208–9

tropes: of noble savage vs. colonized native, 34–36; North as entrance to hell, 46, 57, 59, 65n9, 65n10; nostalgic patriotism of emigrants, 80; of self-conquest, 15, 51, 52; of the "sublime quest," 47–48, 51, 52, 64n3; theoretical construct of "empire," 26–27, 43n1; use of in manipulation of readers, 68; of wildness, 172

Two Little Pilgrim's Progress (Burnett),
15; as celebratory book, 98;
children's reaction to City
Beautiful, 91–93, 99n4; *Dial*
review, 88; emotional/intellectual
neglect, 86–87; natural
landscape, moral rejection of,
85–86; oppressive space of natural
landscape, 87–89; pilgrimage
journey, 93–94; as pilgrimage/
journey narrative, 86; World's
Fair, association with Celestial
City, 95–97
Two Towers (Tolkien), 12
Tynan, Kathleen, *Peeps at Many
Lands—Ireland*, 71–72

Ungava (Ballantyne), 61
universalism: incarnational view
of reality, 248–49, 259–60n5; of
L'Engle, 244, 259–60n5, 260n7
Upton, Bertha and Florence, *The
Golliwogg's Polar Adventure*, 61–63
urban landscapes: greening of, 17, 21,
161–64, 168n7

vampirism, 132, 138n10
von Slatt, Jake, 239

Wadsworth, Sarah, 98n1
Watson, Nancy, *The Politics and Poetics
of Irish Children's Literature*, 68
Weaver, Roslyn, 123
Wells, H.G.: *First Men in the Moon*, 233;
The Island of Doctor Moreau, 175
Welty, Eudora, 2, 21
West, Alan, 169–83
Westerfeld, Scott: female characters,
235; *Leviathan*, 19, 229, 230, 231, 238
West/non-West dualism of
"ours"/"theirs," 4–5
Where the Wild Things Are (Sendak), 10

White, Andrea, 48
Wiegand, Wayne A., 98n1
Wilde, Lady, oral tradition in Ireland, 74
Wilde, Oscar, 73
wilderness: Canadianism of, 105–6;
as Perilous Realm, 106
wildness: and de-wilding, 170–71,
172, 173–74; trope of, 172;
unpredictability of, 170, 183n1
Wild Things (Dobrin, Kidd), 10, 21,
157–58
Willoughby, Sir Hugh, 50
Will's Garden (Maracle), 12
The Wind in the Willows (Grahame), 10,
13; apocalyptic theme, 172, 183n2;
blending of human and animal,
170–71, 173–75, 177; comparison
to Nancy Drew series, 169–70;
"our land-barbarian land"
duality, 17–18, 169; sexuality,
177, 181, 182–83; stability, and
impulsiveness, 178–82; violence,
178; wildness, and de-wilding,
170–71, 172, 173–74
Winnie the Pooh (Milne), 63
Winter Evenings (Hack), 48–51, 60–61
Winters, Sarah, 19, 215–28
Wiseman, Christopher, 227
The Wizard of Oz (Baum), 102
Wolf, Casey, 127
Wolff, Doris, 12–13
women's spaces, academic literature
on, 265–66
wonder, senses of, 239–41
worlding, 7–8; double-worlding, 8
World's Columbian Exposition.
See Chicago World's Fair
World War I, and *The Lord of the Rings*,
226–28
World War II, and Partisan
Resistance, 216, 224–26, 228
Wynne-Jones, Tim, *The Maestro*, 2

Yates, David, 215, 216, 228
Yeats, W.B., 70, 74, 80
Yerxa, Leo, *Last Leaf First Snowflakes to Fall*, 13
Young Adult novels: *All Night* (Cumyn), 302; Australian Indigenous dystopias, 123; home-away-home narrative, 81; *Hot Pterodactyl Boyfriend* (Cumyn), 302; migrant tales, 81; *The Snow Queen* (Kernaghan), 198; *Tilt* (Cumyn), 302; transmission of cultural Indigenous knowledge, 131–32
Young People's Theatre, "Monster Under the Bed," 289–90